Closed-Loop/Ground-Source Heat Pump Systems

Installation Guide

NRECA Research Project 86-1

Supported by

NATIONAL RURAL ELECTRIC COOPERATIVE ASSOCIATION
1800 MASSACHUSETTS AVE., N.W.
WASHINGTON D.C. 20036

Prepared by

OKLAHOMA STATE UNIVERSITY
DIVISION OF ENGINEERING TECHNOLOGY
STILLWATER, OKLAHOMA 74078

Distributed by

INTERNATIONAL GROUND SOURCE HEAT PUMP ASSOCIATION
P.O. BOX 1688
STILLWATER, OKLAHOMA 74076-1688

The National Rural Electric Cooperative Association (NRECA) is the national service organization of nearly 1000 rural electric systems. These cooperatively owned utilities own and operate about 44% of the miles of distribution lines in the nation to provide power to less than 10% of the nation's people, primarily in the sparsely populated, rural areas of 46 states.

NRECA was founded in 1942 to unite rural electric systems in a way that would permit them to develop the services and support they needed to properly serve rural America. NRECA is one of the largest, rural-oriented cooperative organizations in the United States.

The NRECA Research Fund which was used to support this project was created to conduct studies and carry out research of special interest to rural electric systems and their consumers.

Oklahoma State University was Founded December 25, 1890 as the Oklahoma Agricultural and Mechanical College. It became Oklahoma State University on July 1, 1957. Oklahoma's land-grant institution has three goals: to instruct, to do research, and to offer educational assistance to the public through extension.

The Division of Engineering Technology at Oklahoma State University is part of the College of Engineering, Architecture, and Technology.

The International Ground Source Heat Pump Association (IGSHPA) formed in 1987 was established as an integral part of Oklahoma State University. The purpose of IGSHPA is to promote the efficient use of ground-source heat pumps, to develop and promote sound industry standards, to support the effective marketing of ground-source heat pumps, to identify and to support industry-related research, to develop and distribute internationally recognized training materials, to enable its members to have direct input into any published materials, to provide a forum for information interchange, to enable association members to benefit from the advantage of large numbers, and to represent the association's members in matters of local, state, national, and international interest.

Library of Congress Catalog Card Number: 88-28980
ISBN: 0-929974-01-8

Closed-Loop/Ground-Source Heat Pump Systems: Installation Guide

NOTICE

This publication was prepared by the organization named below. Neither NRECA, the organization named below, nor any person acting on behalf of either of them: (a) makes any warranty, expressed or implied, with respect to the use of any information, apparatus, method, or process disclosed in the publication or that such use may not infringe privately owned rights; or (b) assumes any liabilities with respect to the use of, or for damages resulting from the use of, any information, apparatus, method, or process disclosed in this publication.

Prepared by:
Oklahoma State University
Division of Engineering Technology
Stillwater, Oklahoma 74078

CONTENTS

CONTENTS

ILLUSTRATIONS

ILLUSTRATIONS

ILLUSTRATIONS

ILLUSTRATIONS

TABLES

TABLES

FOREWORD

The manual is intended as a source of field-proven installation practices for heating and air conditioning contractors, architects, engineers, utility representatives or anyone who desires to correctly install ground-coupled heat pumps.

This manual was developed with funds provided by the NRECA Research Program to which approximately one-quarter of U.S. rural electric cooperatives belong. These cooperatives recognize the importance of research to the future of electrification and, consistent with cooperative philosophy, have elected to pool their research funds at NRECA so that more significant projects can be funded than those which they could undertake individually.

Concern for their consumer members has always driven electric cooperatives to assume the lead in implementing innovative energy conservation programs and load-control technologies. Because of the favorable operating characteristics for both consumers and electric utilities of ground-coupled heat pumps (GCHP), promoting them is a logical outgrowth of this philosophy. Electric cooperatives will continue to pioneer this and other technologies beneficial to consumers on an individual basis and collectively through the NRECA Research Program.

For nearly three decades, heat pumps have been the electric utilities' strategic weapon in the space heating war. Nearly one-fourth of all homes built since 1977, and one-third of the homes built in the South, use heat pumps, primarily air source. However, the highest demands from air-source heat pumps coincide with utility peak periods. Advanced, more efficient air-source heat pump systems are being developed, but even these advanced heat pumps can only moderate demand driven by the wide temperature variations of the outside air.

Research to find better, more efficient heat pumps has shown that ground-coupled heat pumps offer many advantages over their air-source cousins. Consumers save money because GCHPs consume about one-third less energy; utilities benefit because GCHP demand can be 50% lower during critical utility peak periods. This occurs because ground-coupled heat pumps tap a more stable energy source than the air, the earth. Moderate underground temperatures improve operating efficiency, help maintain capacity even during extreme weather and assure an increase in the total number of heat pump applications, especially in northern climates.

The benefits that GCHPs provide consumers and utilities often justify the extra effort and expense of burying heat exchangers in the earth. There is also a local economic benefit: 50% of the consumers' expenditure for the GCHP system is for trenching or drilling, money that remains in the local economy.

There are many factors that will assure the increased use of GCHPs. As new, lower cost drilling and trenching methods and techniques are found, the cost of GCHPs will decrease and the technology will become even more competitive. Full-condensing water heating, multispeed compressors and other advanced technologies that are currently used in air-source units will eventually be applied to GCHPs, where the technology will produce more dramatic benefits.

One of the goals of the rural electric cooperatives and their Research Program is to encourage and influence the wider use of ground-coupled heat pumps. This goal will become easier to achieve as the factors noted above come into play. Consumers' desire for the safest and lowest cost means of conditioning homes and heating domestic water assures increased use of ground-coupled heat pumps.

Peyton Collie
Research Engineer
Energy Research Division
Energy and Environmental Policy Department
National Rural Electric Cooperative Association

ACKNOWLEDGMENTS

The development of this Installation Guide for Closed-Loop/Ground-Source Heat Pump Systems is the result of many years of effort primarily by the members of the Division of Engineering Technology at Oklahoma State University. The author was James E. Bose, Professor and Director of the Division of Engineering Technology. Don Adams, Eric Price, and Bob Weaver prepared Appendix A.

Other contributors include:

Phil Albertson, Charles Machine Works (Ditch Witch)
Phil Amsterdam, Climate Master
Al Bierbaum, Iowa Association of Electric Cooperatives
Peyton Collie, National Rural Electric Cooperative Association
Frank Eckhart, Oklahoma State University
Larry Eitelman, McElroy Manufacturing
Dan Ellis, WaterFurnace
Michael G. Glasgow, Phillips Driscopipe
Griff Hawkinson, Vanguard Plastics
Howard Newton, Florida Heat Pump
Fred Paepke, Bard Manufacturing
Ted Striplin, McElroy Manufacturing
Phil Rawlings, Command-Aire
Bernard Woller, Buckeye Power
Tom Warner, Mammoth Heat Pumps

The editor was Irene Fischer. Her assistance, comments, and contributions are greatly appreciated.

Lastly, so many others contributed helpful suggestions, comments, and criticisms that is impossible to show a complete list. However, their assistance is acknowledged and appreciated.

Errors in the manuscript may have occurred in spite of our efforts. These errors are attributable to the authors. Any suggested corrections and/or comments should be made in writing to:

Oklahoma State University
Ground Source Heat Pump Publications
Stillwater, Oklahoma 74078
J.E. Bose, Project Director

1 Introduction and Overview

In This Section: Overall concept; energy source; basic heat pump system; heat pump configurations; heat pump operation; domestic hot water options; system materials and components; who's involved

This manual is an **INSTALLATION GUIDE** for **Closed-Loop Ground-Source (cl/gs) Heat Pump Systems**. It provides current technical information on installing closed-loop/ground-source heat pump systems. It summarizes correct practices of many successfully operating systems in the United States and Europe. Generally, the information is applicable to residential and light commercial buildings.

This manual provides specific recommendations for:

1. selecting system components including plastic pipe, antifreeze solutions, circulating pumps, and water-source heat pumps
2. design configurations, pipe sizing, and do's and don'ts of ground heat exchanger fabrication
3. design procedures for the buried piping system to achieve long, trouble-free life
4. procedures for burying a closed loop of plastic pipe below the earth's surface

The manual is in furtherance of the NRECA member network's continuing effort to provide state-of-the-art techniques in conserving energy and reducing the risks associated with introducing a new technology. The manual also is provided in support of an evolving manufacturer-distributor-dealer network that favors the development of this technology.

Overall Concept

The closed-loop portion of a ground-source heat pump system consists of a long plastic pipe buried below the earth's surface. This plastic pipe is buried in the ground, or ground-coupled, to allow heat transfer between fluid in the pipe and the earth. The heat pump transfers thermal energy to and from the closed, buried pipe and the building's thermal load. The system consists of the closed-loop buried pipe, a water-source (water-to-air) heat pump, and an air distribution system for directing heated or cooled (conditioned) air to specific locations in the building. In some units, an extra heat exchanger called a desuperheater is added to provide a portion of the domestic hot water (DHW).

1

Energy Source

The heat pump system uses solar energy stored in the earth's crust. Energy is transferred daily to and from the earth's surface by solar radiation, rainfall, wind, etc. As a consequence of this solar energy, the earth's temperature at depths greater than 30 feet approaches the annual average air temperature. Between the surface and a depth of six feet (where most horizontal buried pipe systems are placed), the ground temperature will swing above and below the annual average air temperature depending on geographic location, soil type, and moisture.

Because of its own insulation, the earth's soil temperature is more moderate year-round than the outside air. Extreme air temperatures, above and below the annual average temperature at a location, may range ± 50 degrees Fahrenheit. However, for cl/gs heat pumps the entering liquid temperatures for buried pipe systems can be half this at depths of three to five feet. Figure 1.1 illustrates air temperature data for Stillwater, Oklahoma. The average annual air temperature is 59.2°F with extreme winter and summer values of -5° and 105°F, respectively. Fig-

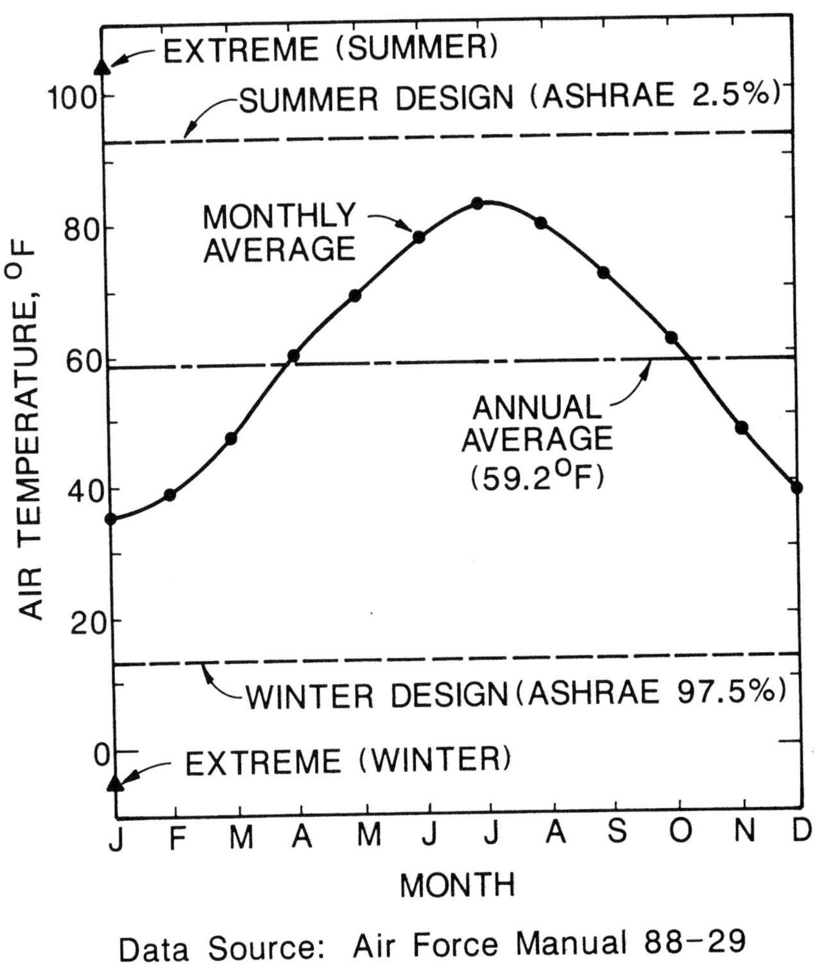

FIGURE 1.1: Annual Air Temperature Variation, Stillwater, Oklahoma

ure 1.2 shows ground temperatures for the same location. In this case, normal undisturbed (no heat removal or addition) ground temperatures at a depth of five feet range from of 49°F (February 18) to 74°F (August 20). In the summer and winter, the maximum and minimum soil temperatures are delayed four weeks compared to soil surface temperatures, giving the earth an additional advantage. For this reason, the ground-coupled heat pump will have its greatest advantage in areas where the annual air temperature swing is high.

Only a small part (less than 2 to 3%) of the stored energy in the earth's crust comes from its hot core. Oil and gas drilling logs show the normal increase in earth's temperature to be from one to three degrees Fahrenheit for each 100 feet in depth.

Solar energy stored in the earth's crust supplies the energy for space conditioning and heating of domestic hot water. Since the earth's underground temperature in the summer is cooler than outside air, loads from summer cooling can be more efficiently rejected underground.

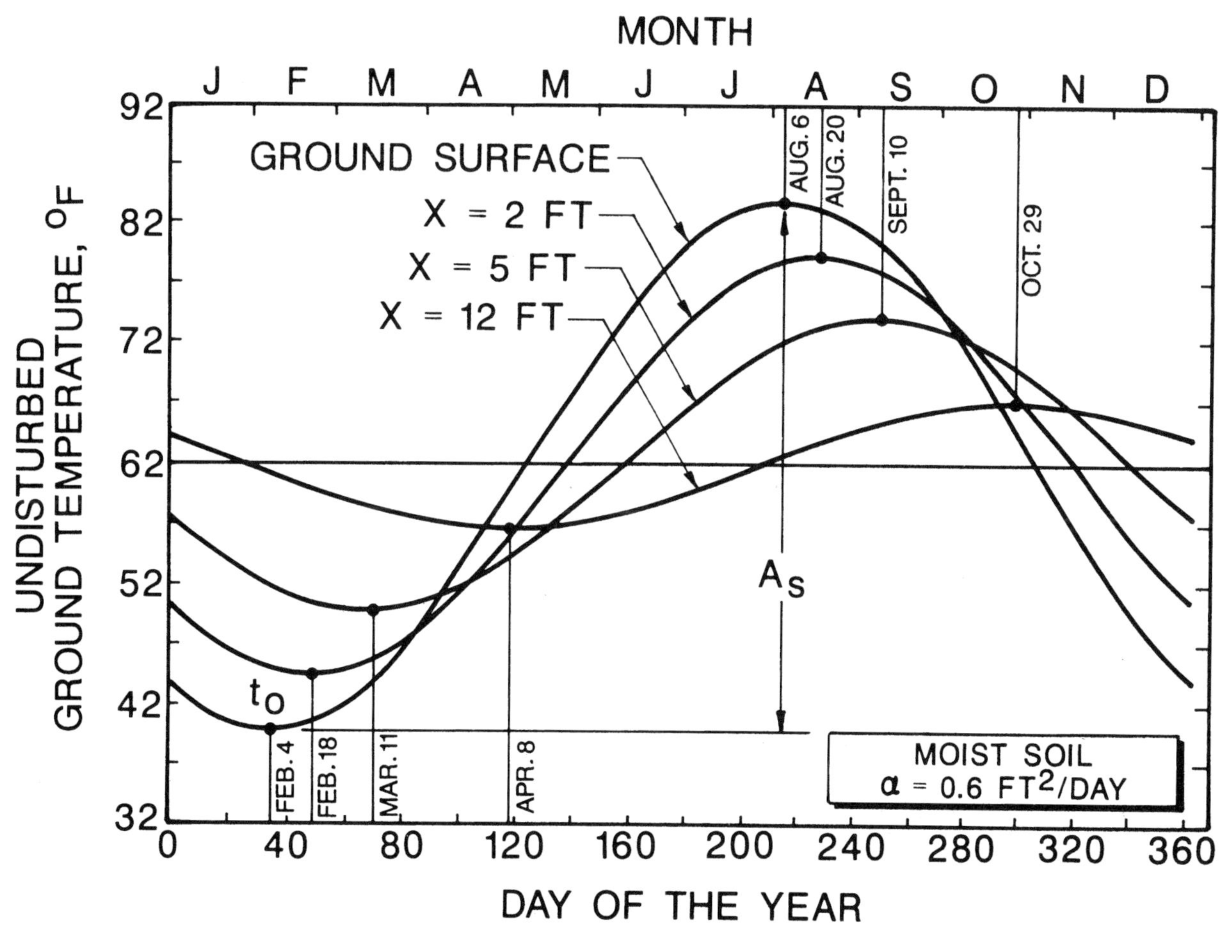

FIGURE 1.2: Annual Soil Temperature Variation, Stillwater, Oklahoma

1

Basic Heat Pump System

In the cl/gs heat pump system, a long-life high-strength plastic pipe is buried beneath the earth's surface. Depending on available land area, the pipe will be placed in either a horizontal or vertical configuration (Figure 1.3). Water or a water and antifreeze solution circulates through the closed-loop pipe, transporting heat to the heat pump in the winter and away from the heat pump during the summer.

The heat energy exchange between the circulating fluid in the ground heat exchanger and the heat pump occurs in the heat pump's water-to-refrigerant heat exchanger. Hence, the name water-source heat pump since water or another fluid is used as the heat exchange medium. Other than this water heat exchanger, which replaces the outdoor air unit, the basic components in a water-source heat pump are the same as an air-source unit. Because there is no requirement for an outdoor fan unit for heat exchange, all outdoor aboveground components are eliminated. Also, there are no defrost components since defrost cycles are eliminated.

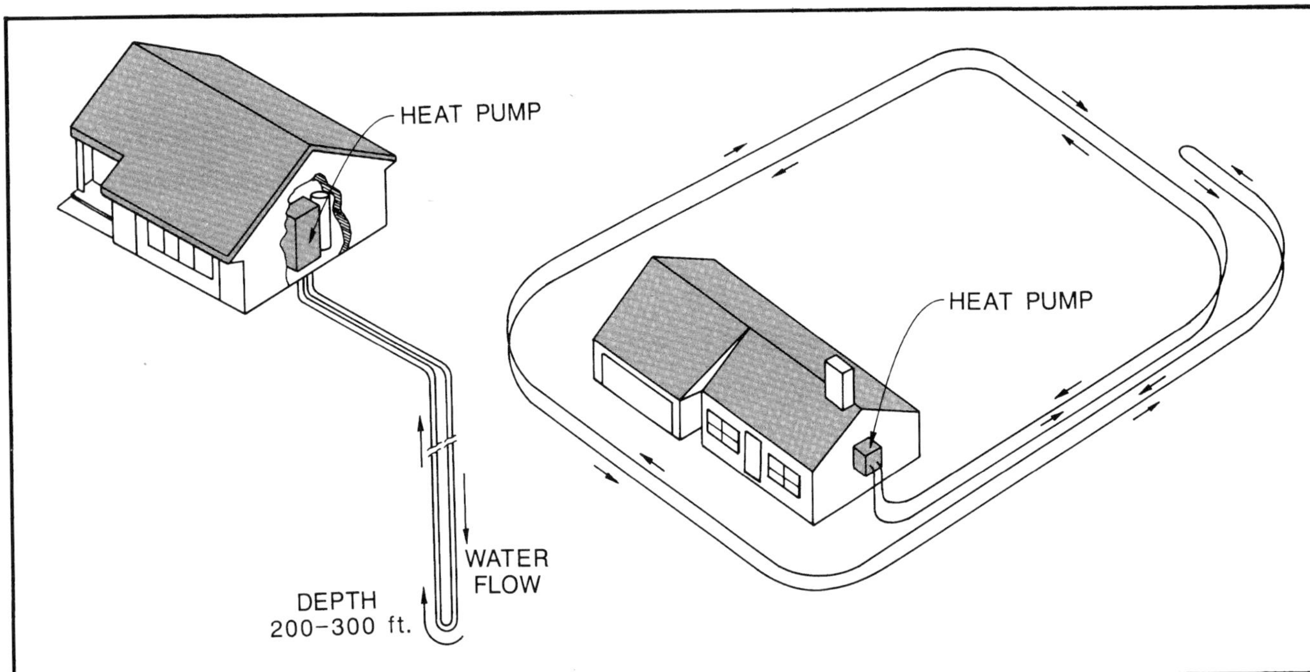

FIGURE 1.3: System Configurations

Heat Pump Configurations

Heat pumps are described in terms of their inputs and outputs. In this manual the inputs or sources are air and water (liquid) and the outputs are air and/or water. Consequently, there are four types of heat pumps:

1. air-to-air
2. air-to-water
3. water-to-air and
4. water-to-water

The most common type of heat pump to date is the air-to-air unit. Its greatest market penetration has been in areas where outside air temperatures are moderate, such as southern climates. Air-to-water units are common in domestic hot water systems and in areas with extremely cold weather where a secondary heating system is available to back up the air-source system.

Water-to-air is the most prevalent water-source heat pump system in the United States where heating and cooling are required. The water source can be an open system (water well, direct use of pond or lake water, etc.) or a closed-loop ground heat exchanger.

The water-to-air heat pump unit can be operated with a boiler and cooling tower. It is used frequently in commercial buildings to "move" heat from one area to another. A closed loop inside the building is connected to individual units. If the loop temperature becomes too hot, the cooling tower rejects the heat to the atmosphere. If the loop temperature becomes too cold, the boiler is used.

The water-to-water heat pump is best applied as a heating only system in northern climates. Europe, where hydronic systems are used in under-the-floor configurations, is where most of these heat pumps have been installed. These systems have high efficiency since the required output temperature is lower than forced air systems.

Heat Pump Operation

The cl/gs heat pump system consists of three loops that operate during all heat pump cycles and an optional fourth loop that preheats domestic hot water. The four loops are shown in Figures 1.4 and 1.5 and are defined below:

1. *AIR LOOP*: A loop used to distribute conditioned air to the building. A squirrel cage blower is used to move air through a duct distribution system. Duct sizes (diameter and length) are designed to distribute the conditioned air to specific locations depending on their heat loss or gain.
2. *REFRIGERANT LOOP*: A sealed and pressurized loop that transports thermal energy from point to point in the circuit. Refrigerant flow is forced through the circuit by a compressor in the vapor portion of the loop.
3. *GROUND LOOP*: A sealed and pressurized loop of water or antifreeze solution circulated below the earth's surface. It absorbs heat from the surrounding earth in the winter and rejects heat in the summer. The fluid is circulated by a low-power circulating pump.
4. *DOMESTIC HOT WATER LOOP*: A sealed and pressurized loop that circulates water from the domestic hot water tank to a heat pump's desuperheater for preheating domestic hot water. The water is circulated in this loop by a low-power circulating pump. Newer heat pumps may have full condensing loops that provide 100% of the domestic hot water.

During heat pump operation, thermal energy is transferred from one loop to another providing heating, cooling, dehumidification, and domestic hot water preheating.

HEATING CYCLE

During heating cycles (see Figure 1.6) the heat inputs and outputs are as follows:

Inputs:

1. thermal energy from the earth
2. heat pump compressor energy
3. circulating pump and blower energy

Outputs:

1. space heating
2. domestic hot water

The major heat inputs are the earth's thermal energy and compressor energy. For a three-ton (36,000 Btu/hr) heat pump system operating with a COP of 3, two-

1

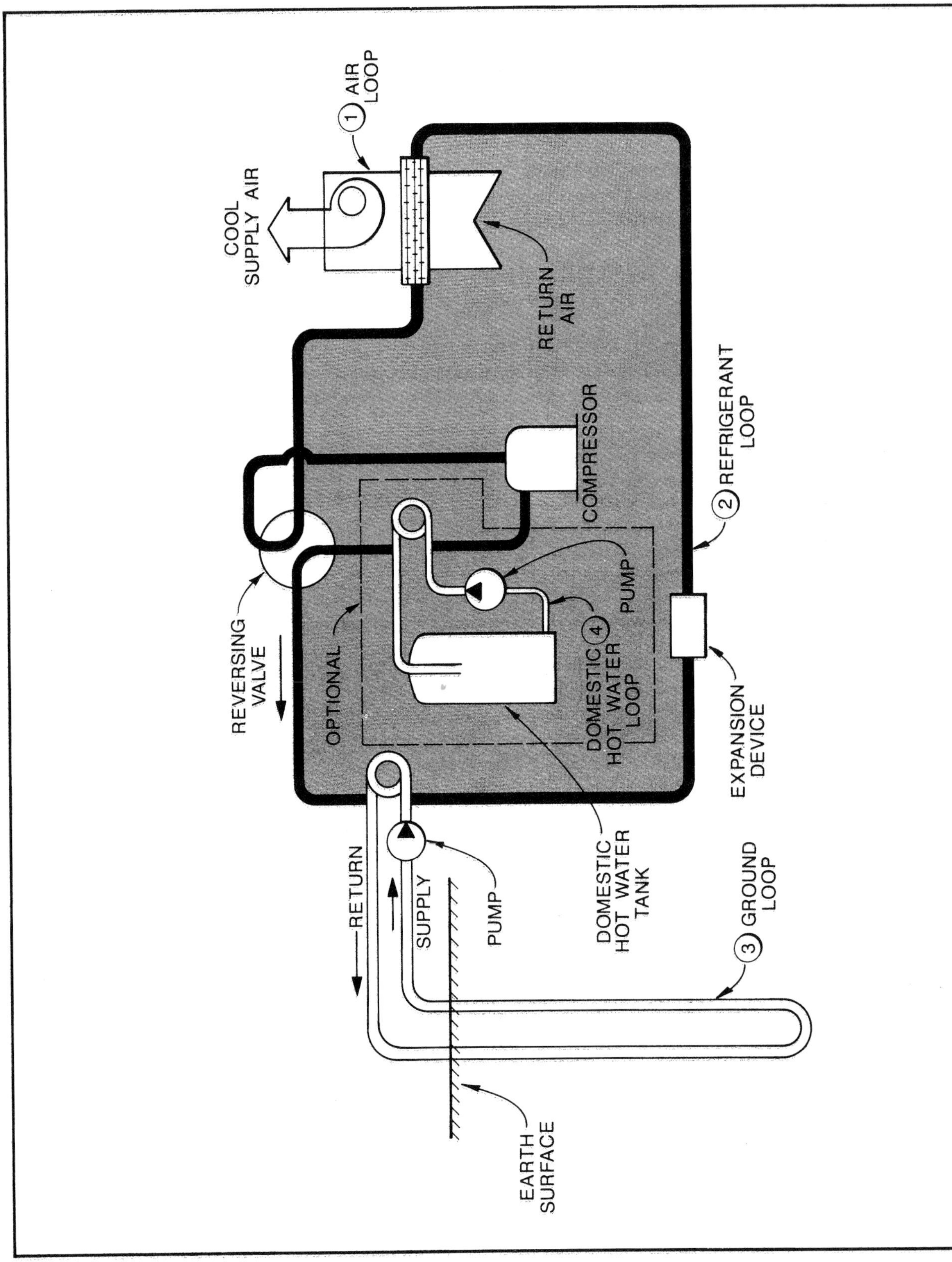

FIGURE 1.4: Heat Pump Operation — Cooling Mode

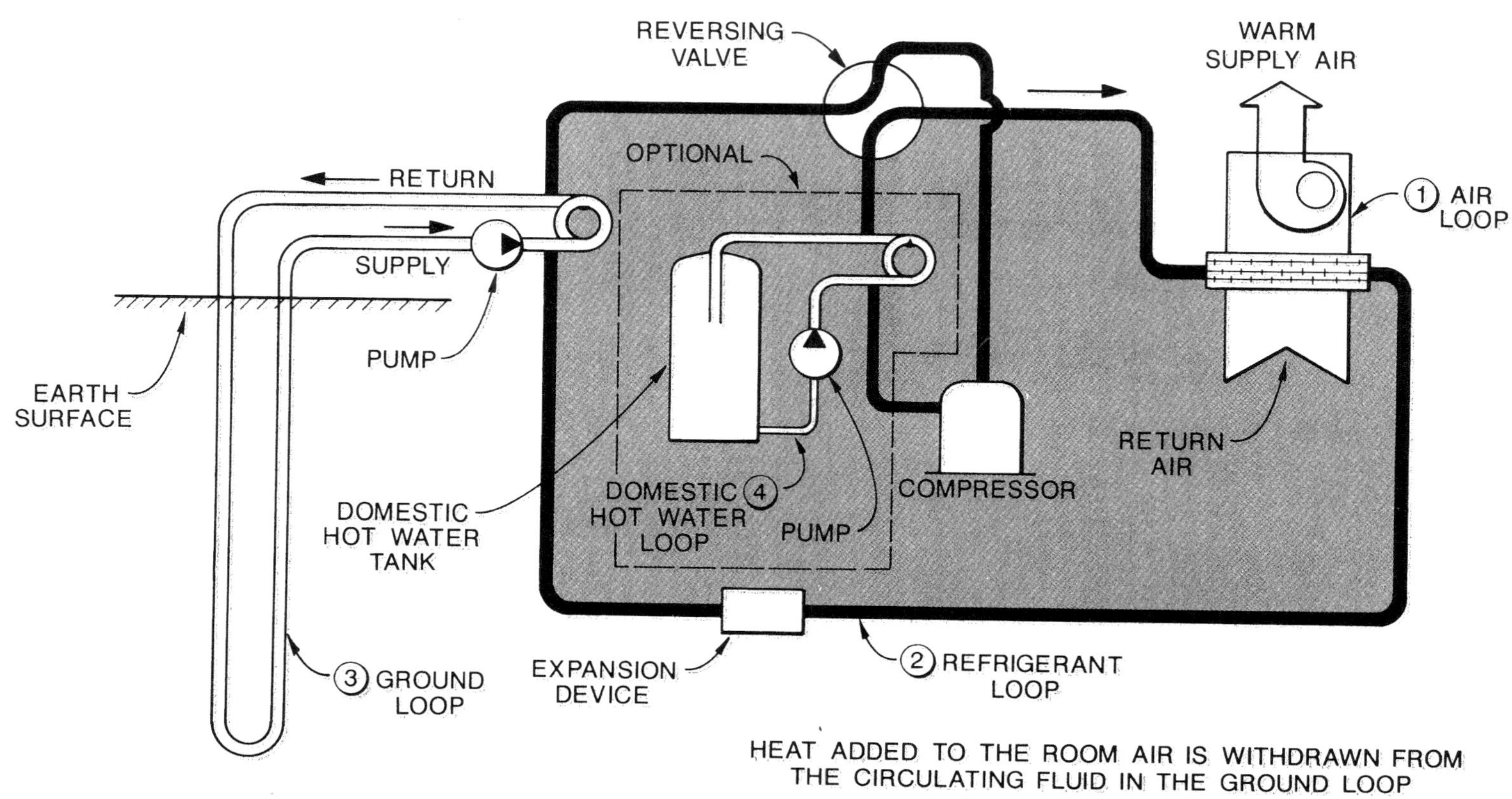

FIGURE 1.5: Heat Pump Operation — Heating Mode

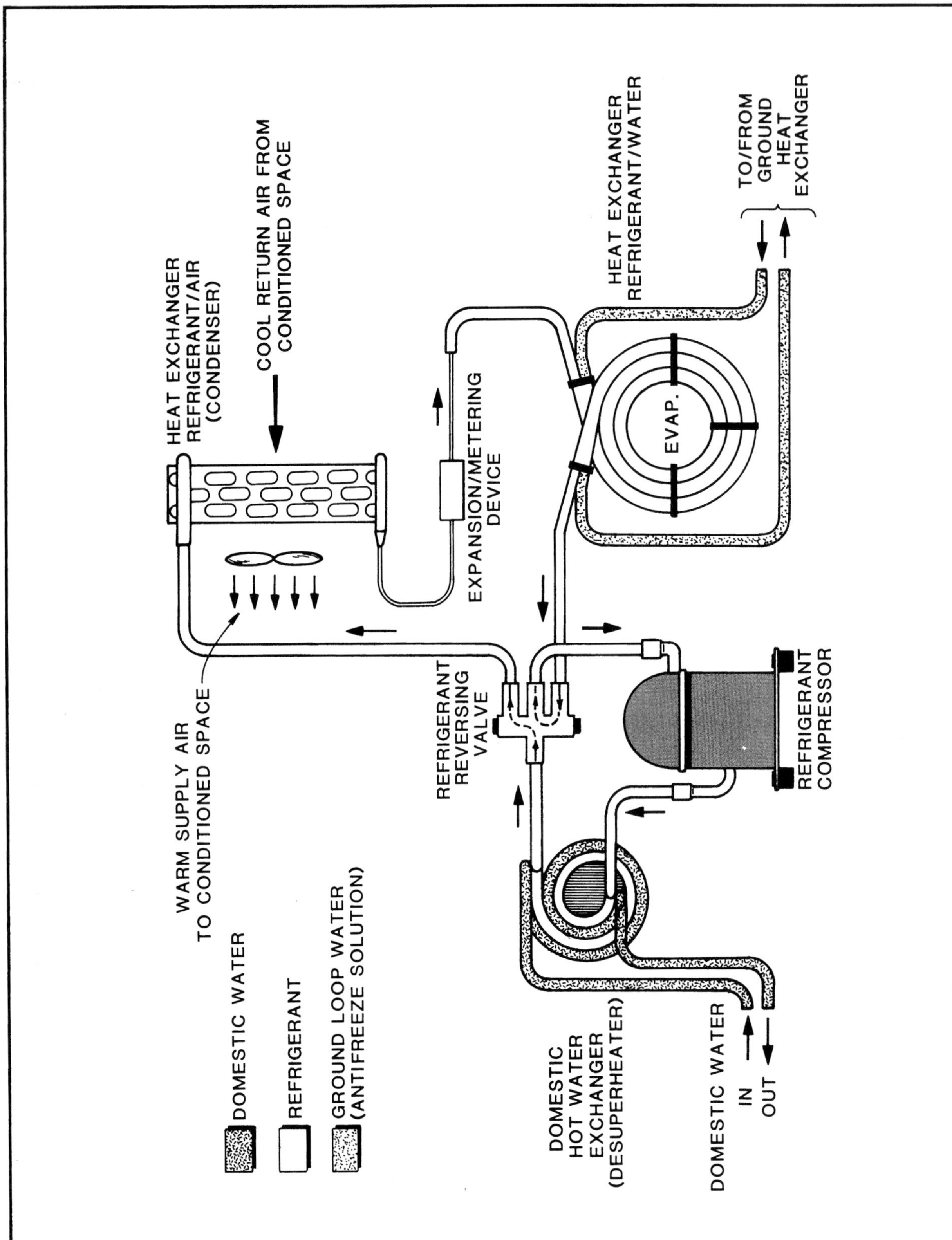

FIGURE 1.6: Heating Cycle

thirds (24,000 Btu/hr) of the delivered energy will come from the earth and one-third (12,000 Btu/hr) from the electrical energy inputs. If the total energy required is greater than this, some form of supplemental heat is required. Normally the amount of heat input to the system from blowers, pumps, etc., will be small compared with the earth and compressor energy. The blower fan motor energy will, however, be taken into consideration in the rating of the heat pump's capacity and performance.

The refrigerant vapor in the circuit is first compressed, which raises its temperature and pressure. This increased pressure forces the vapor through the refrigerant system.

The hot vapor continues to a refrigerant-to-air heat exchanger which raises the temperature in the building because the cooler air absorbs heat from the hot refrigerant vapor. As heat is removed from the refrigerant in the refrigerant-to-air heat exchanger, the vapor condenses to a liquid. Thus, during the heating cycle, the indoor air coil serves as the system condenser.

The warm liquid returning from the condenser passes through a metering device. This reduces its pressure and causes a corresponding reduction in temperature. The low-pressure, low-temperature liquid then flows to the evaporator where thermal energy from the ground loop vaporizes the refrigerant and the cycle continues.

During the heating cycle a circulating pump moves water or antifreeze solution through the buried ground heat exchanger. As this water circulates through the ground loop, the water is warmed by the earth's higher temperature. The heat is transferred from the water to the refrigerant heat exchanger in the water-source heat pump. During this part of the cycle the water-to-refrigerant heat exchanger serves as an evaporator, changing the liquid refrigerant to vapor.

For units equipped with desuperheaters, the hot gas from the discharge of the compressor passes through a second water-to-refrigerant heat exchanger that preheats domestic hot water. The domestic hot water, which is to be preheated, is moved by a circulating pump. At this point, only a small portion of the total heat energy available (the thermal energy of the vapor in a superheated condition) is removed.

COOLING CYCLE

During cooling cycles, the heat inputs and outputs are as follows:

Inputs:

1. building heat gain
2. compressor energy
3. pump and blower energy

Outputs:

1. domestic hot water
2. waste heat to the earth

The heat output during cooling generally will be much greater than that required for domestic hot water. For a nominal three-ton (36,000 Btu/hr) cooling unit, the net amount of rejected heat is approximately 50,000 Btu/hr. Of this, only about 12,000 to 15,000 Btu/hr is available or needed for domestic hot water if the unit has the water heating option. The remaining heat must be rejected through the ground heat exchanger to the earth. Compared to the heating cycle, which absorbs only 24,000 Btu/hr for every hourly run, cooling cycles generally reject almost twice the heat for each hour of operation.

During cooling operations (Figure 1.7), the hot gas leaving the compressor preheats the domestic hot water, if the unit has the water heating option, and then is routed to the water heat exchanger where excess heat is rejected to the earth.

Space cooling is accomplished by the warm indoor air passing over the cold evaporator coil. The high pressure hot liquid in the condenser was forced through a metering device which results in a low-pressure and low-temperature liquid. Evaporation of this liquid to a vapor is the cooling mechanism.

1

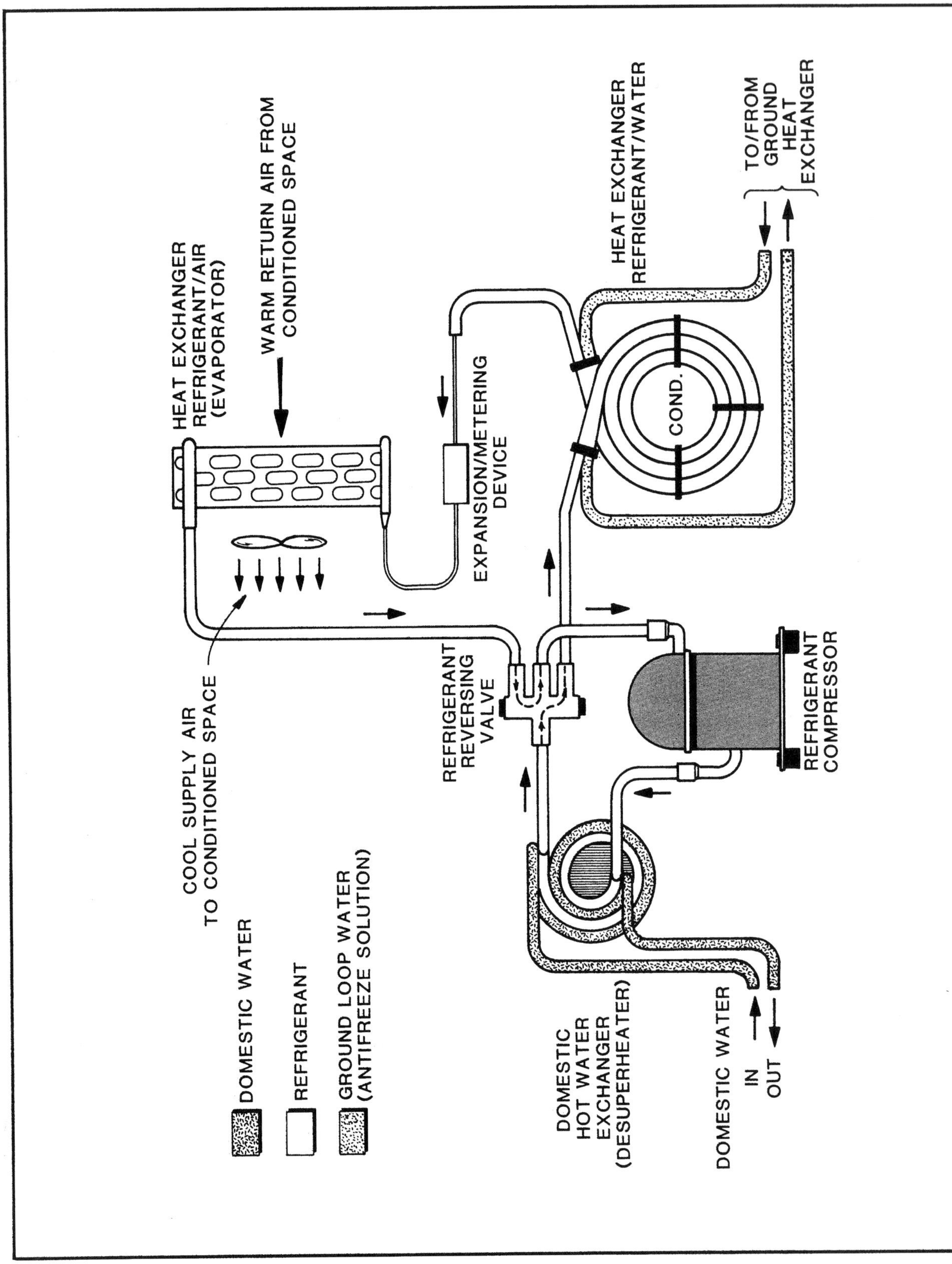

FIGURE 1.7: Cooling Cycle

Figures 1.6 and 1.7 show the components and their arrangement in a commercial water-source heat pump. Included in these drawings is the reversing valve that changes the direction of refrigerant flow to the indoor air coil and the water coil heat exchanger between the heating and cooling cycles. As noted in these figures, the domestic hot water preheating coil is between the compressor and the reversing valve. This allows for heating domestic hot water in both the heating and cooling cycles.

Domestic Hot Water Options

In addition to space heating and cooling, water-source heat pump units are available with three specific domestic hot water options:

1. *NONE*: No hot water preheating capability.
2. *SUPPLEMENTAL*: Partial preheating (desuperheater) during heat pump operation in the heating or cooling cycle.
3. *DEDICATED OR DEMAND*: Year-round total (full condensing water heating) hot water heating on a first priority basis.

OPTION 1 — A conventional water-source heat pump without provision for domestic hot water preheating. This configuration represents the lowest cost water-source heat pump. It is used in buildings with more than one heat pump and where there is no requirement for domestic hot water.

OPTION 2 — Uses a refrigerant-to-water heat exchanger (desuperheater) installed at the discharge of the heat pump compressor. The hot gas at this point is in a "superheated" condition giving rise to the common name desuperheater. In southern climates, "free hot water" is available during the cooling season and typically can supply 100% of the domestic hot water. In the winter, the desuperheater generating domestic hot water competes with the heating load. Since there is generally excess heat pump capacity for many winter hours, domestic hot water efficiency equals heat pump efficiency. If the hot water demand has been met, the circulating pump is shut off.

OPTION 3 — Total or 100% domestic hot water on a first priority basis simply means that, before any space heating or cooling is accomplished, all domestic hot water requirements are met by the heat pump system. These units have a completely different design. Heat pumps with this feature will have the longest-run cycles. This usually improves heat pump performance and generally reduces the total annual demand and energy costs.

System Materials and Components

Plastic Pipe: The most common materials used in underground heat exchangers are polyethylene and polybutylene. These materials are considered flexible, allowing easy installation. Both materials can be joined by heat fusion, which gives high reliability. The grade of pipe used should be carefully chosen. See Section 4 for grade selection and the appropriate standards for pipe selection.

Antifreeze Solutions: Except in the southernmost climates, an antifreeze solution will be required to prevent freezing in the heat pump heat exchanger. The choices are:

1. salts — calcium chloride and sodium chloride
2. glycols — ethylene and propylene
3. alcohols — methyl, isopropyl, and ethyl

1

These solutions can and have been used successfully. For any fluid to be used, it must be:

1. safe
2. non-toxic
3. non-corrosive
4. a good heat-transfer medium
5. low cost
6. long lasting

Generally speaking, the salts are safe and non-toxic and have good heat-transfer characteristics, low cost, and long life. However, they are corrosive in the presence of air and most metals. The major drawbacks of salts are their corrosive nature and their difficulty to clean up. One strong advantage is they are considered non-toxic and environmentally safe. With proper metal component selection and air purging of the system, salts can and have been used successfully. In northern climates where heat pump loop operating temperatures go as low as 25°F, calcium chloride and methanol have been used successfully.

Glycols are relatively safe and generally non-corrosive, but are considered toxic, have fair heat transfer characteristics, medium cost, and a finite life. For low-temperature operation, the glycols can become viscous, requiring greater pumping power and, therefore, reducing the heat pump system efficiency. In southern climates where heat pump closed-loop operating temperatures remain above 35°F, the glycols have been used successfully.

Improperly handled, alcohols are unsafe (burn and explode when mixed with air), toxic, and relatively non-corrosive, and have fair heat transfer characteristics, medium cost, and a long life. Their major drawbacks are explosiveness and toxicity. Diluting (50% water and alcohol) the alcohol solution before taking it on site somewhat reduces the risk of explosion. Since alcohols are non-corrosive, their use is very popular. Alcohols are used in northern and southern climates.

CAUTION: Many anti-corrosion inhibitors are toxic and care must be taken to prevent a non-toxic antifreeze solution from becoming toxic after the solution is inhibited.

Metal Components: The system's metal components must be compatible with the cl/gs circulating fluid. The components that must be carefully selected are:

1. circulating pump and pump flanges
2. all metal piping
3. sensing ports
4. any metal component in contact with the circulating fluid

Depending on the antifreeze used, the metals chosen for the system must be carefully selected. The manufacturer's and dealer's recommendations must be followed carefully. See Appendix A for a discussion of materials that are compatible with the various antifreeze solutions.

Heat Pumps: There are a number of excellent water-source heat pumps on the market designed for closed-loop/ground-source operation. For closed-loop operation the following characteristics should be considered:

1. manufacturers' entering liquid temperature limits
2. features (domestic hot water, insulated water lines, etc.)
3. safety listings (UL, ETL, or ARL)
4. type (split, package, water-to-water, etc.)
5. warranty
6. performance rating (ARI 320 & 325, etc.)

Water-source heat pumps can be classified by their application and consequently their inlet water operating temperature range. For example, in commercial buildings the water temperature may range from 60° to 95°F. If the inlet water temperature drops below 60°F, a boiler is used to raise the temperature. If the temperature goes above 95°F, a cooling tower is used to reject heat. Heat pumps used for this type of application generally cannot be used for ground-coupled systems where the inlet temperature may range from 25°F to 105°F.

A second type of heat pump is a ground-water heat pump that operates on well water temperatures ranging from 42° to 74°F. The ground-coupled system will operate from 25°F to 95°F in northern climates and from 40°F to 105°F in southern climates. Care should be taken to pick a unit suitable for the expected or design inlet liquid temperature. The installing contractor should be aware of the differences and make recommendations accordingly.

Who's Involved

Simply stated, the building owner, the utility company, and all the trades in between are involved. The decision to choose a particular energy source or one system over another ultimately rests with the building's owner. The decision will be based on information from a number of sources. It should probably be based on economics, which requires an understanding by all involved of the possible benefits. Included in the economic evaluation should be expected life-cycle costs, first costs, predicted maintenance, expected life, projected fuel increases, etc.

Given that the system reduces peak demand and lowers energy consumption, what roles do the following groups play?

Utility Company: The utility's involvement with the ground-coupled heat pump system represents new opportunity in the following ways:

1. reduced peak demand in heating, cooling, and domestic hot water
2. increased sales to new consumers
3. leadership role in a new technology
4. strengthening role as the consumers' "trusted advisor"

The utility company generally has served as a source of information on new products and concepts for its consumers. The utility becomes involved when a new product is mutually beneficial to it and its consumers. By promoting systems that reduce peak generating demand, a utility is able to control consumer costs by adding new accounts with existing generating capacity. If peak generating demand is not a problem, then the system can be marketed to consumers using other fuels or to retain existing consumers.

In some instances other fuels are displaced because of the higher efficiency of the electric system and the versatility of the water-source heat pump. Examples where increased sales have occurred include dairies switching from gas water heating to electric heat pumps that generate hot water by cooling milk. In many cases, electric sales increase.

To remain competitive with other fuel producers, many utilities offer training programs, seminars, mail inserts with billings, etc., to keep consumers aware of available energy-saving systems. By taking a leadership position, the utility maintains its position as a "trusted advisor." Leadership roles have been taken in setting minimum standards on equipment and installation to assure delivery of the high efficiency promised.

In areas where dealers and contractors have not seen the need or share the utility's concern to reduce peak demand costs and consumer energy consumption, the utility can and has assumed the leadership role. This can be accomplished by purchasing materials and equipment in bulk and passing savings to the local dealers. If necessary, the utility can install the ground heat exchangers as part of its incentive program.

Installing Contractor or Dealer: Building owners rely on dealers to provide service and information on new concepts. The

1

dealer, given sufficient information about a new concept, can develop new markets. The dealer, by working cooperatively with the utility, can deliver energy systems that meet consumer expectations. By agreeing to a minimum standard, dealers are able to compete among themselves and still deliver what the consumer expects.

Heat Pump Manufacturers and Distributors: Manufacturers, aware of the interest by the utilities, can convince investors to produce water-source heat pump equipment for this market. At this time, water-source heat pumps are on the market which will operate at the colder loop temperatures in the North and the higher loop temperatures in the South. Many manufacturers have design criteria and recommended practices for installation of their equipment.

Plastic Pipe Manufacturers: Plastic pipe manufacturers have products and appropriate standards to provide the industry the necessary assurance of a long life and trouble-free installation. Product warranties of 50 years are available.

The market for plastic pipe in ground heat exchangers could easily be many times larger than the market for distributing natural gas. For a typical residence, a thousand feet of plastic pipe is common whereas hookups for natural gas are a tenth of that.

Fusion Joining Manufacturers: Butt and socket fusion joining procedures have been developed for the natural gas distribution market which have shown the joints to be stronger than the pipe itself. Training programs are available from the pipe manufacturers and distributors. The natural gas industry has provided the development of a product for joining of plastic pipes for the fabrication of ground heat exchangers.

Trenching and Drilling Manufacturers: These industries are well developed. What is not clear in many locations is the installation procedures for ground heat exchangers. Training programs are available from some manufacturers. Trenching and drilling contractors are available at most locations and have excellent experience in local underground conditions. They are a good source of information.

Universities and Research Organizations: The role of universities is to provide research that ultimately reduces the system's cost, increases the system's efficiency, and safeguards the public's health. Important roles include development of design criteria, methods of identifying soil thermal properties and stability, and training.

The local utility's member services representative can be contacted for a list of qualifying heat pump manufacturers for its service area. The National Rural Electric Cooperative Association, Research Division, 1800 Massachusetts Avenue, N.W., Washington D.C. 20036, is interested in innovative installation methods or products. The International Ground Source Heat Pump Association (101 Industrial Building, Stillwater, Oklahoma 74078) can be contacted for training and appropriate standards.

2 Economics, Marketing, and Demand Reduction

In This Section: Benefits to homeowner; benefits to HVAC contractor; benefits to utility; additional benefits from demand hot water heat pump

Before any new heating, ventilation, and air conditioning (HVAC) system will have significant market penetration, it must satisfy a number of economic criteria for those involved in the energy distribution and utilization business.

The homeowner must be convinced any additional system cost will be returned in a reasonable time and, in the long run, that any new system will save a significant amount of money.

The contractor must be convinced the new business opportunity will justify any additional cost in new equipment and/or training.

The electric utility must be convinced the system will be consistent with its goal of serving its consumers with the lowest electric power cost available.

Benefits to Homeowner

The ground-coupled system reduces demand and lowers energy consumption. For the consumer this means lower utility bills. In cases where utilities are providing low-interest loans or incentive programs, the consumer often sees an immediate positive cash flow.

In many applications, the ground temperature during mid-winter can be 40 degrees warmer than the coldest air temperature. This significantly increases both the capacity and efficiency of the heat pump system. Depending on geographic location, cl/gs heat pump systems have reduced heating costs 50% more than electric furnaces and about 25% more than air-source units. In extreme northern climates, the advantage over air-source heat pumps for heating has been even greater. Also, improvements in cooling efficiency have been around 25%.

To demonstrate consumer benefits of switching to the cl/gs heat pump system, Figure 2.1 and Table 2.1 give the total monthly electric power consumption of a residence converted from an electric furnace and conventional air conditioning to a cl/gs heat pump system. Prior to conversion of the existing HVAC system, the rural electric cooperative performed an energy audit and made recommendations about weather stripping, caulking, and extra insulation. The homeowner followed these suggestions as part of the retrofit package.

Electric power usage for the first year

2

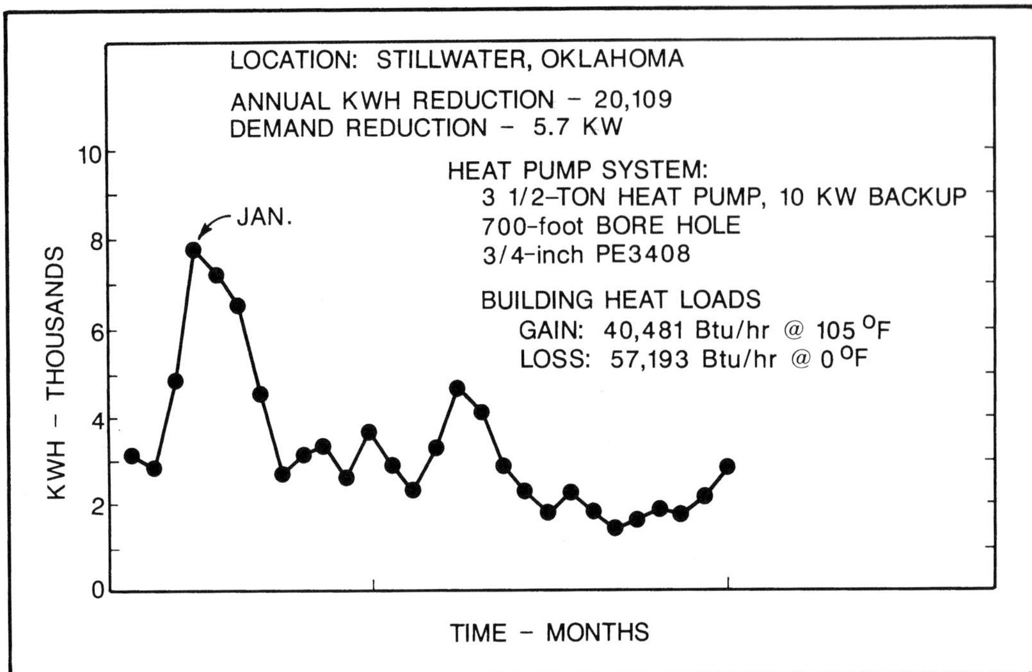

FIGURE 2.1: Electric Furnace/AC to CL/GS Heat Pump

of operation was reduced by 20,109 kWh from the previous year's usage of 50,100 kWh. The particular heat pump's capacity at 40°F was 30,000 Btu/hr with a COP of 2.8. Using these values, a demand reduction of 19,286 Btu/hr or 5.7 kW was achieved. The homeowner sees a $336.72 positive cash flow the first year with a simple payback period of seven years.

Expected power reductions for this location have shown:

1. a 40% reduction in total electric consumption for this residence after converting from an electric furnace and air conditioning to a ground-source heat pump system
2. 100% of the domestic hot water requirement met with the desuperheater during summer cooling periods; domestic water heating costs reduced up to 60% each year with a properly operating system
3. up to a 25% improvement in summer cooling when these systems are compared with conventional air-source equipment

Northern locations should have greater energy use reductions in the winter since the number of operating hours on the heat pump system would be higher. Performance tests at a location in Brookings, South Dakota, indicated a 50% improvement over an air-source unit.

An additional advantage of the cl/gs system is reduced maintenance. The unit is indoors, which should give many of the electrical components a greater life expectancy. There are fewer components. For example, the defrost cycle required in an air-source system has been eliminated. Finally, the compressor operates under less severe conditions than air-source heat pumps which have a proven life expectancy that approaches 20 years.

Water requirements of the closed-loop system consist simply of a single filling of the ground heat exchanger system. For a residence, this is approximately 50 to 100

TABLE 2.1: Case Study of a Residential Retrofit

LOCATION: STILLWATER, OKLAHOMA
ANNUAL KH REDUCTION: 20,109 @ $0.08 = $1,608.72/YR DEMAND REDUCTION: 5.7 KW DEMAND CHARGE: (100% Ratchet @ $8.23/KW/MO) = $563/YR FINANCING: $6,648.92 @ 5% FOR 7 YEARS LOAN PAYMENT: $95.39/MO OR $114.68/YR
HEAT PUMP SYSTEM: 3½-TON HEAT PUMP, 10 KW BACKUP DOMESTIC HOT WATER DESUPERHEATER
GROUND HEX SYSTEM: 700-FOOT BORE HOLE ¾-INCH POLYETHYLENE, PE355434C PARALLEL FLOW, U-BEND HEX
BUILDING LOADS: HEAT GAIN: 40,481 BTU/HR HEAT LOSS: 57,193 BTU/HR

gallons. For an open system, where water is discharged to a pond or drainage ditch, a typical heat pump system in the Midwest using nine gallons per minute would use about 1.5 million gallons each year for both heating and cooling. In locations with limited underground water, the closed-loop system may be the only choice.

Benefits to HVAC Contractor

The contractor benefits from having available to him a total electric system that:

1. competes in areas with low-cost fossil fuels for those consumers who desire electric heating
2. is a higher priced system with excellent paybacks, thus allowing higher profit margins when compared to conventional equipment
3. is simple in operation with an expected lifetime exceeding that of conventional gas and electric systems

All contractors and businessmen who have excellent products with demonstrated desirable performance will benefit from having newly developed products for sale.

Benefits to Utility

For the electric utility, this system will provide sufficient improvements in efficiency that other fuel users can be convinced to use electric heating and cooling systems. In many cases this can be done without increasing the utility's service size. Gas customers with electric air conditioning can generally be provided with heating without an increase in electric service size.

Domestic hot water preheating occurs only during heat pump usage and leaves

2

the demand valleys open for conventional hot water heating. Properly marketed, this system should increase kilowatt-hour sales by attracting other fuel users.

Table 2.2 gives data for a one-year cl/gs heat pump program for a rural electric cooperative in Oklahoma.

The following comments are pertinent to the data presented in Table 2.2.

1. Seventy-two percent of these cl/gs heat pumps represent a new load. A $300 rebate is granted on each cl/gs heat pump system.
2. The heat pumps on which rebates were given (70 total units) will use 630,000 kWh annually, producing revenues of $35,000 ($0.06/kWh in summer and $0.055/kWh in winter).
3. The cooperative will save 106 kW in peak load demand at a cost savings of $10,513 per year (based on an average three-ton unit at $9/kW).

TABLE 2.2: CL/GS Heat Pump Program Summary (Verdigris Valley Electric Cooperative)

Number of Units Installed: 70

40	replaced natural gas or propane (heat)
14	replaced electric furnaces
5	replaced air-source heat pumps
11	new homes

In all cases except new homes, the water-source heat pump replaced less efficient air conditioning (SEER of 6.5 was the average for the replaced equipment).

Additional Benefits from Demand Hot Water Heat Pump

Figure 2.2[1*] details the demand reduction capability of a recently developed **Demand Hot Water Heat Pump**. This heat pump has the capacity for space heating, space cooling, and producing 100% of the domestic hot water on demand. This allows the unit to produce domestic hot water with the unit's full capacity at any time.

Four modes of operation are integrated in the design:

1. space heating
2. space cooling
3. combination space cooling and domestic hot water heating
4. dedicated hot water heating on demand

The demand reduction potential is excellent. During summer peak demand periods, 100% of the domestic hot water is generated by the heat pump system as waste heat. This eliminates the water heating load during the summer peak demand period.

*See numbered references at end of volume.

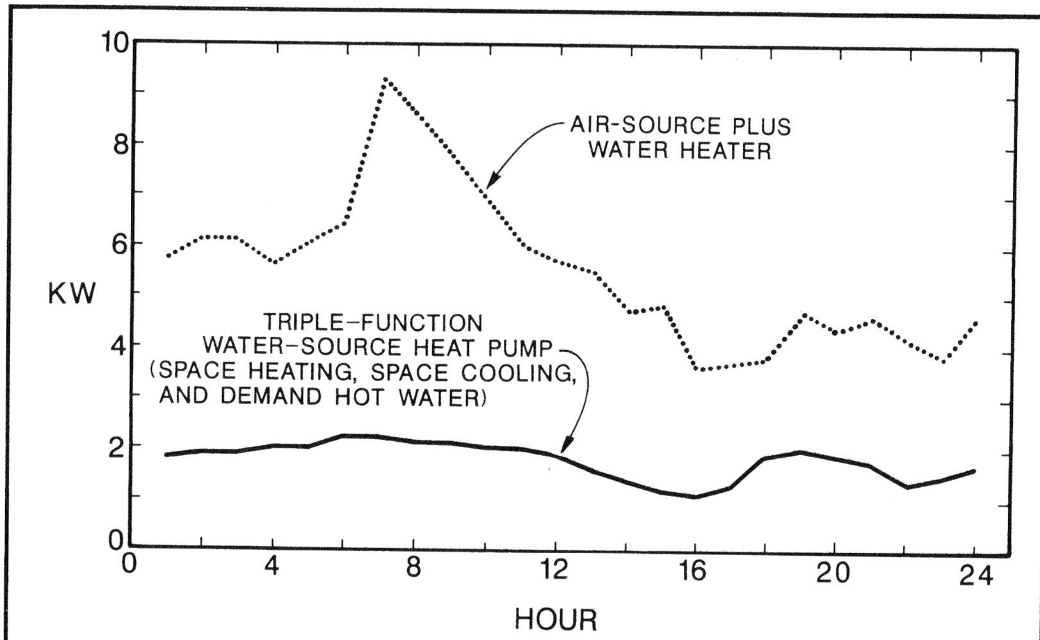

FIGURE 2.2: Heat Pump Load Shape for Winter Peak: Waste Heat Recovery Project, 1985 Source: *Waste Heat Recovery Project.*[1]

3 Selecting, Sizing, and Designing the Heat Pump System

In This Section: Design procedure for the heat pump system; determining building heating and cooling loads; load calculation procedures; energy calculations; performance of an air-source heat pump; performance of a ground-source heat pump; ground load calculations; equipment selection; air filtering; air distribution system; air diffusers or supply registers; duct layout

Figure 3.1 shows a typical water-source heat pump installation. The system consists of:

1. a water-source heat pump
2. a conventional domestic hot water tank and heat pump desuperheater
3. a loop pump kit (circulating pump(s), connection hoses, service valves, etc.) to circulate water or an antifreeze solution through the heat pump and ground heat exchanger
4. supply and return air duct system
5. optional plenum heater for winter supplemental heating or emergency heat in the event of heat pump failure
6. air filter assembly

The selection and sizing of the heat pump system requires:

1. the building's heating and cooling loads and domestic hot water requirements to be provided by the heat pump system
2. an estimate of the ground heat exchanger loads given the heat pump operating hours

With specific ground heat exchanger loads, the methods presented in Section 4 can then be used to design the ground heat exchanger system.

Design Procedure for the Heat Pump System

The design procedure makes possible the easy study of alternative designs. The steps are as follows:

1. Determine the building's heating and cooling design loads.
2. Select a properly sized heat pump system.
3. Select a type of indoor air distribution system.
4. Select the proper air supply diffusers and registers and return grilles for the air distribution system.
5. Size the indoor air distribution system.
6. Estimate the building's energy requirements given:
   a. the building's heating and cooling loads
   b. the type and size of heat pump equipment selected

3

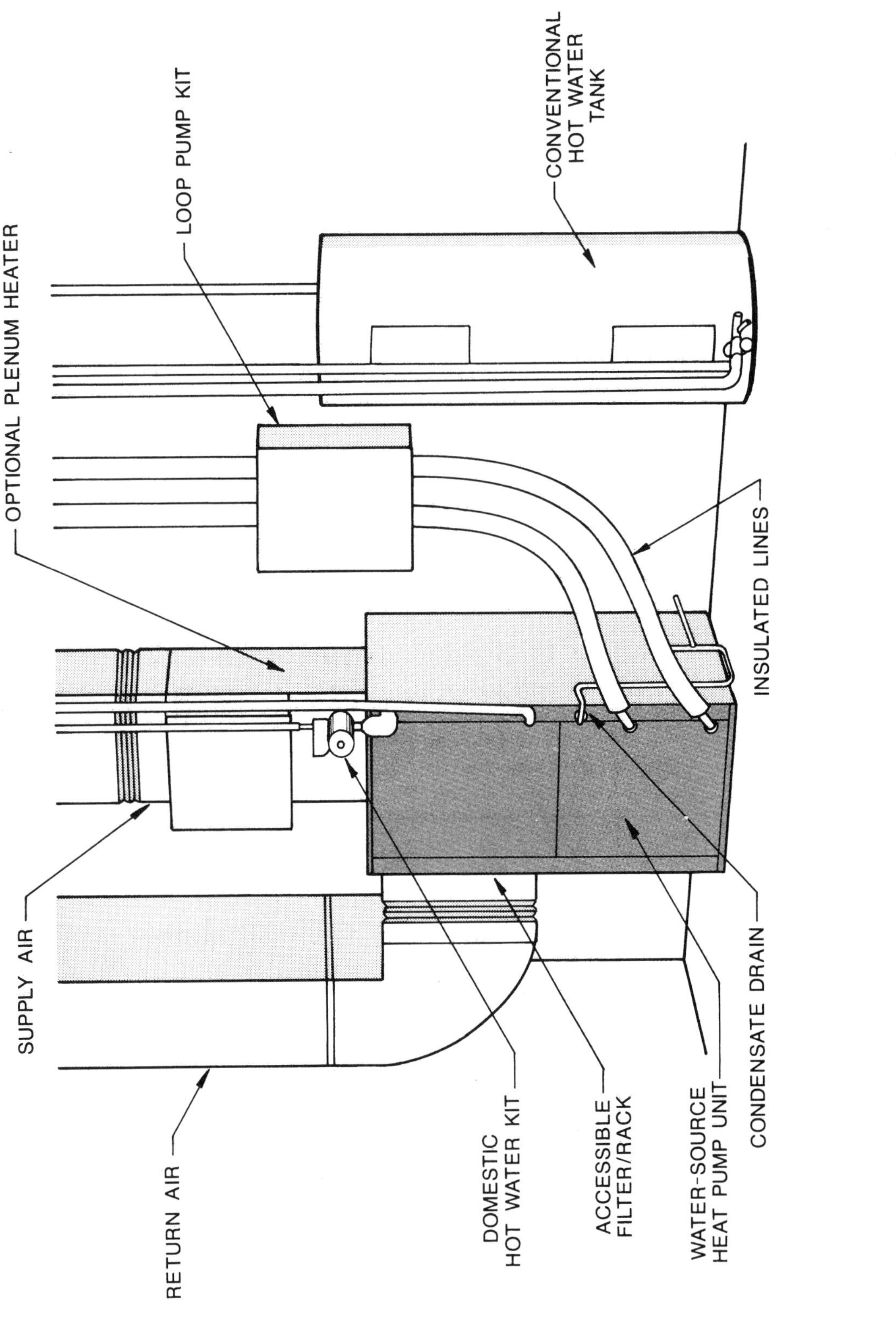

FIGURE 3.1: Typical Water-Source Heat Pump Installation

3

c. the climatic and soil thermal characteristics
7. Estimate the ground heat exchanger loads:
   a. annual load
   b. design month's load

This Section 3 details the procedures used to accomplish the above requirements. The better the estimates, the more likely the consumer will be satisifed with the system.

Determining Building Heating and Cooling Loads

As just noted, the complete design of a building's heating and cooling system involves several steps. Each is important to satisfactory system performance. However, the cooling and heating load estimates are the most crucial because everything else in the system is dependent on them. The most basic load calculations are referred to as design loads. Two other types of loads will be discussed: energy loads and ground loads. The purposes of the different load types are as follows:

Design Loads are used to size and select the equipment for a system (such as a heat pump) and to design the air distribution system (supply air diffusers, return air grilles, and the duct system). Design loads are based on standard or accepted conditions for a given locality (a design day).

Energy Loads are used in predicting the energy necessary to operate the system for some prescribed time such as a month, a year, or a season. The basic calculation methodology may be the same as for the design load; however, actual operating and weather data are used instead of design conditions.

Ground Loads are associated with ground-source systems and relate to the design of the ground heat exchanger coupling device. In principle these calculations are similar to the energy loads except the ground load is the heat rejected to the ground (cooling mode) or removed from it (heating mode).

BASIC DESIGN LOAD DETERMINATION BY BUILDING TYPE

The basic calculations to estimate cooling loads are the same for all types of buildings. However, the final load estimate depends on how the basic results are handled for different building types. Heating load procedures (winter) are the same for all buildings because only the building envelope is considered under nighttime conditions. Therefore, the remainder of this discussion pertains to the cooling load.

The general building categories are:

1. single-family detached, one- or two-story
2. multifamily:
   a. apartments
   b. duplexes
   c. town houses
3. commercial, light and heavy.

Various building types require different methods for load estimation because the heating and cooling systems generally used and the geometrical features of each vary considerably.

A **single-family detached residence** will have walls facing in at least four directions and a roof. It may be more than one story. The heating and cooling system will be a single zone type with a constant amount of air distributed to each room and a single thermostat for temperature control. The rooms are reasonably open to each other with a central air return. Each floor may have a separate system.

This configuration results in mixing the air from all rooms and a leveling effect that requires a distribution of air to each room different from other building types. Since the amount of air supplied to each room is based on the room's load, the calculation procedure used must account for this fact. Figure 3.2 illustrates this type of structure.

A **commercial building** may take on almost any geometric configuration but may be best characterized by large, variable internal loads requiring temperature control in each space with similar load characteristics. This space could be each room. These spaces usually are separated so air mixing from space to space is limited. Therefore, very little load leveling among spaces is achieved, and the peak load for each space is needed for the system design. Figure 3.3 shows a commercial-type building.

A **multifamily residential building** is made up of many living units. These have most of the same features as a single-family detached house except each apartment will have only one or two exposed walls and possibly have a roof. In the case of two exposed walls, they are at right angles to each other. Each living unit has a heating/cooling unit with central air distribution, rooms open to each other, and one thermostat.

This configuration does not have the load–leveling effect of the single-family detached house, but it is not a pure commercial type where each room may have temperature control. Therefore, a procedure that gives loads somewhere between the single-family detached and pure commercial is required. Figure 3.4 shows a typical apartment-type building.

There are many buildings that do not fall exactly into one of the three categories. Critical to the designation "single-family detached" is the presence of both East and West exposed walls. Therefore, some multifamily-type structures should be treated as single-family detached when the exposed surfaces are oriented in certain ways. For example, a duplex or apartment with exposed East, West, and South or East, West, and North walls with or without a roof should be treated as single-family detached (Figure 3.5). Another example (which should be treated as single-family detached) would be an apartment, town house, or condominium with only East and West or North and South exposed walls (Figure 3.6).

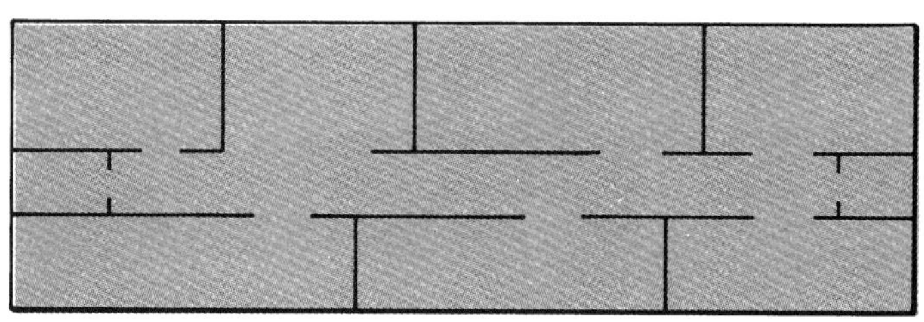

- HEAT AND AIR CAN MOVE EASILY FROM ROOM TO ROOM
- ALL FOUR WALLS ARE EXPOSED
- ANY ORIENTATION
- SINGLE ZONE TEMPERATURE CONTROL

TREAT AS SINGLE-FAMILY DETACHED FOR LOAD CALCULATIONS

FIGURE 3.2: Single-Family Detached House

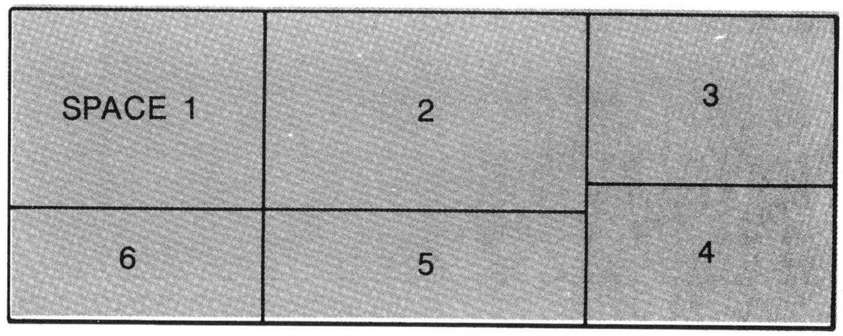

- HEAT AND AIR CANNOT MOVE EASILY FROM SPACE TO SPACE
- ISOLATED ZONES OR SPACES
- TEMPERATURE CONTROL IN EACH SPACE
- ANY ORIENTATION

TREAT AS COMMERCIAL BUILDING FOR LOAD CALCULATIONS

FIGURE 3.3: Commercial Building

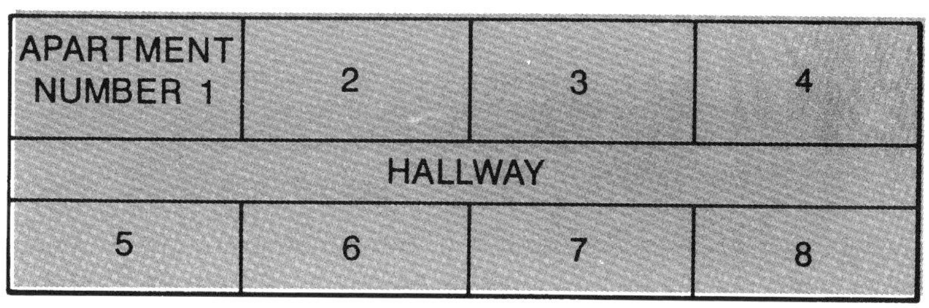

- EACH APARTMENT HAS ONLY 1 OR 2 EXPOSED WALLS
- ANY ORIENTATION
- AIR MOVES EASILY WITHIN EACH APARTMENT UNIT

TREAT AS MULTIFAMILY BUILDING FOR LOAD CALCULATIONS

FIGURE 3.4: Multifamily Building — Apartment

3

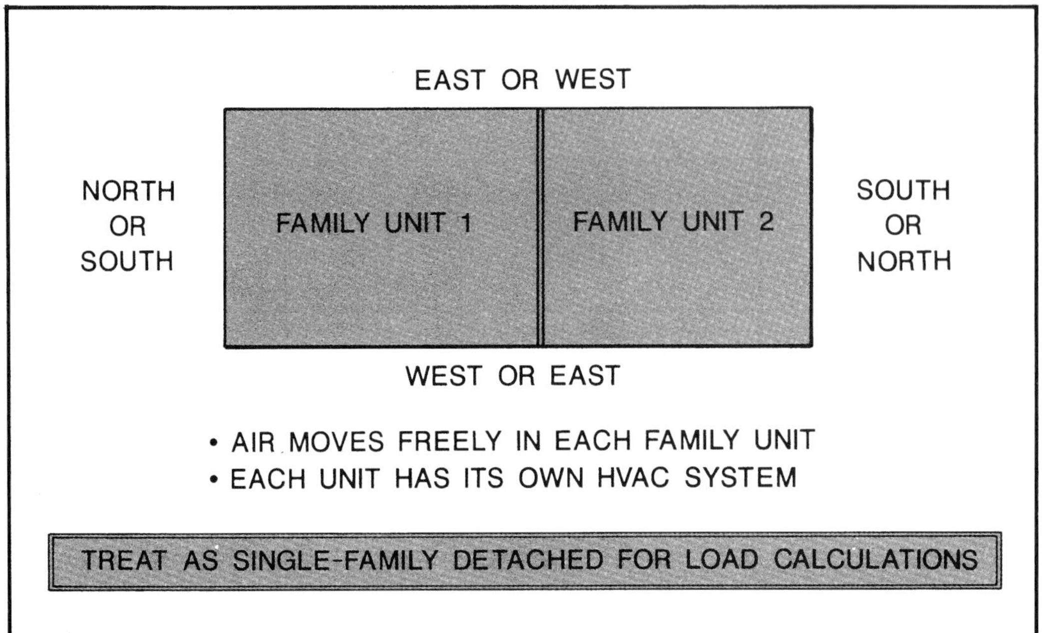

FIGURE 3.5: Multifamily Unit — Duplex

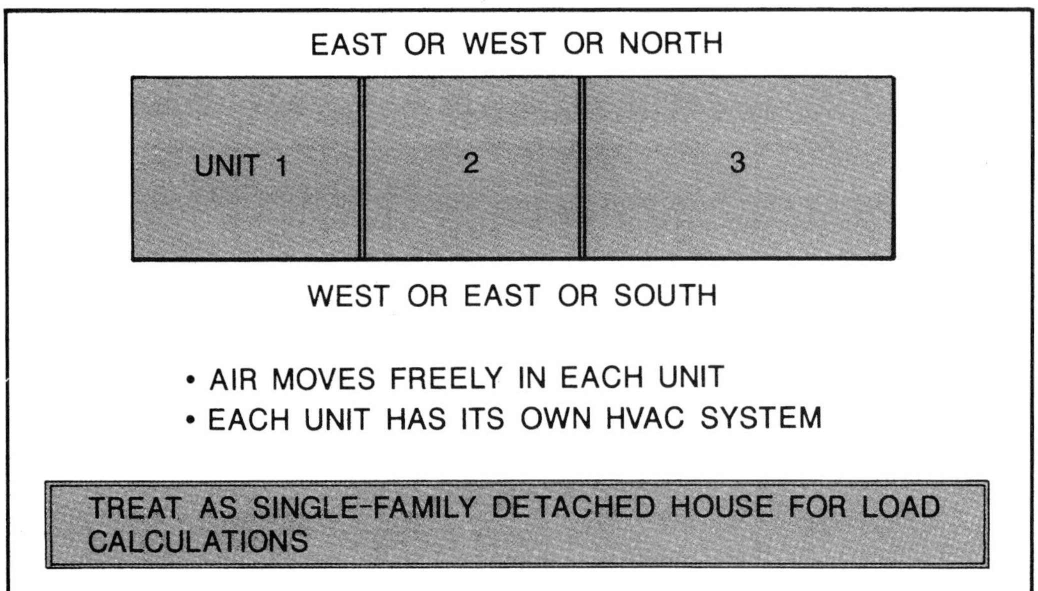

FIGURE 3.6: Multifamily Unit — Apt./Condominium/Town House

Many light commercial buildings with single zone heating/cooling equipment and three or more exposed walls should be treated as single-family detached except for internal loads. **Any residential structure without an East and a West**

3

exposed surface should be treated as a multifamily structure. Examples are: duplexes and apartments with North, South, and East or North, South, and West walls (Figure 3.7); single-family detached houses zoned so that East and West exposures are not in the same zone (Figure 3.8); and light commercial buildings with a limited number of zones which separate East and West rooms.

Regardless of the building type, the design load must be completed on a room-by-

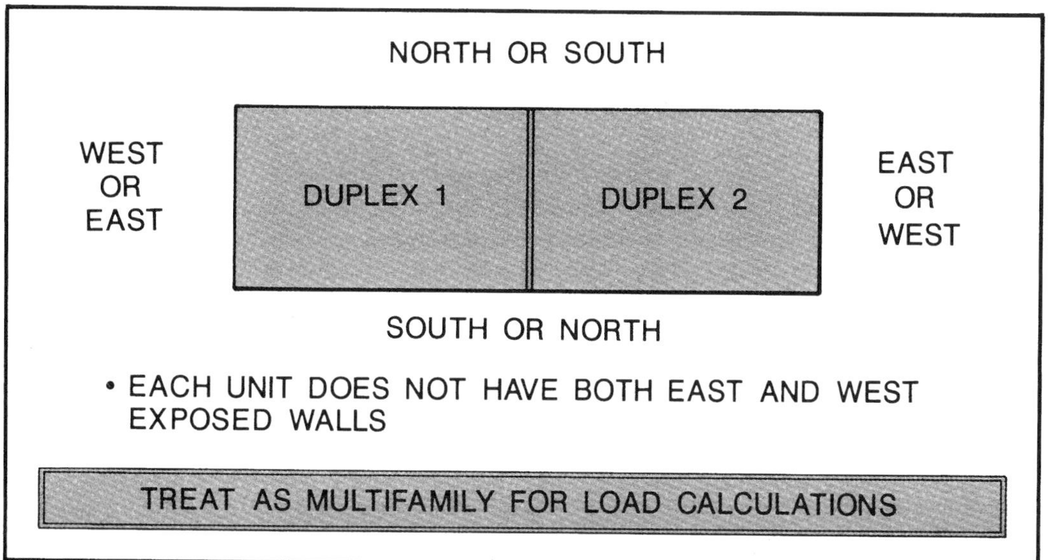

FIGURE 3.7: Multifamily Unit

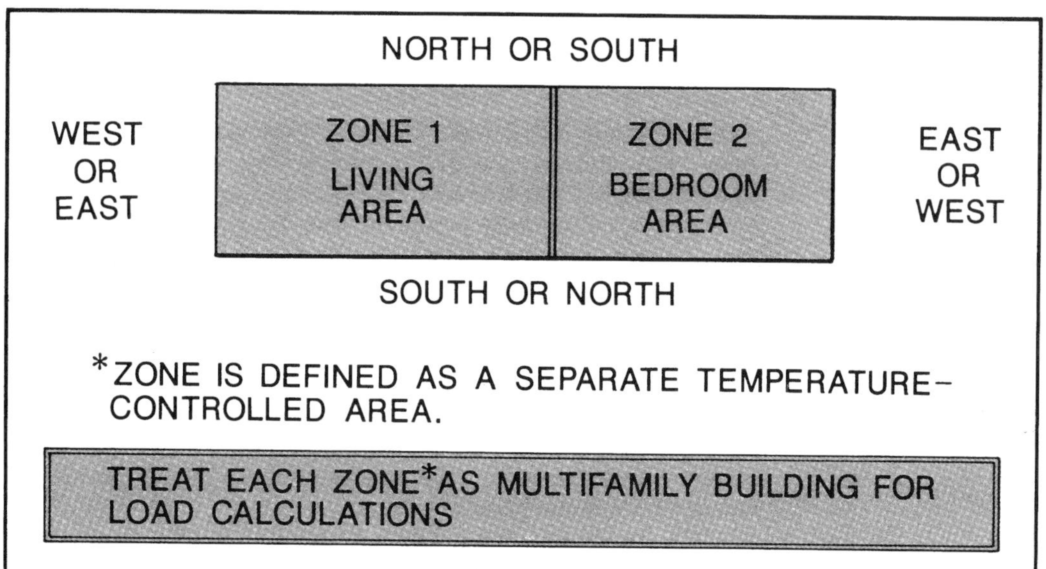

FIGURE 3.8: Single-Family Detached

3

room basis so the proper amount of air for each room can be determined for design of the air distribution system. To select a properly sized heating/cooling unit, the peak or maximum load (block load) for each zone must be computed. This procedure may vary considerably for different types of buildings.

BLOCK LOADS

The block load for a single-family detached house with one central system is simply the sum of all the room loads. If the house is zoned with a completely separate system for each zone, each zone block load is required which is the sum of the loads for all rooms in each zone. If the house is zoned with one central heating/cooling system, the block load for the house must be computed as if it were one zone.

In the case of multifamily structures, each living unit has a zone load equal to the sum of the room loads.

For apartments with their own separate systems, the load for each unit establishes the system size. Apartment buildings with a central heating/cooling system for the whole building (such as a hydronic system with fan coils in each apartment) require a block load calculation for the complete structure to size the central system. Each unit load establishes the size of the fan coil and air distribution system for individual apartments.

Commercial buildings usually require hour-by-hour calculations which, when totaled, give all the information necessary for each room, zone, and building.

Load Calculation Procedures

There are many acceptable calculation procedures for single-family detached residences. ASHRAE (*Handbook*),[2] ACCA (*Manual J*),[3] and many manufacturers publish these methodologies.

Commercial procedures are published by ASHRAE in its *Handbook of Fundamentals, 1985*[2] and a special publication entitled *Cooling and Heating Load Calculation Manual, GRP 158*[4] that contains hand calculation methods for commercial and residential buildings. ASHRAE published updated methodologies in 1984 for single-family and multifamily residential structures.[5] The *GRP 158* residential section should be updated with this new data, which is the only known methodology for multifamily structures.

Excellent microcomputer software is available to carry out all types of design load calculations; however, care should be taken that a given piece of software is not based on obsolete procedures and limited in scope. A recommended software program is *Cooling and Heating Load Calculations (LDCAL5)*.[6] Developed at Oklahoma State University, it includes methodologies for all types of buildings. The procedure is based on the ASHRAE Transfer Function Methodology and the latest research and development.

Other excellent programs are available and have found wide use in the industry. Recommended load calculation procedures are:

Single-Family Detached:
1. ASHRAE *GRP 158* (updated)
2. ASHRAE *GRP 158*
3. ACCA *Manual J*
4. manufacturers' programs

Multifamily:

Hand Method
1. ASHRAE *GRP 158* (updated)
2. manufacturers' programs

Computer Software
1. LDCAL5 (Oklahoma State University)
2. manufacturers and software companies

Commercial:

Hand Method
1. ASHRAE *GRP 158*
2. manufacturers and software companies

3

Computer Software
1. LDCAL5 (Oklahoma State University)
2. manufacturers and software companies

Computer methods are strongly recommended because the results generally are easier to obtain, more reliable, and more consistent from one job to another.

Energy Calculations

The primary reason for buying and installing a ground-source heating and cooling system is to save energy and reduce costs. Therefore, it is important to estimate the energy requirements of various alternative systems. There are three basic ways to do this: degree day, bin, and hour-by-hour.

The **degree day** procedure is the simplest, but generally produces the poorest results. It cannot be used with any reliability on systems where the efficiency of operation depends on the outdoor environment. A heat pump is a good example.

The **bin method** is relatively simple and takes into account the effect of outdoor temperature and partial load conditions. The bin method can be refined as much as necessary to meet a particular system's needs.

Hour-by-hour methods generally are used for large commercial systems where considerable detail is required.

Chapter 28 of the *ASHRAE Handbook of Fundamentals, 1985* describes in detail the above methods and includes examples. A special ASHRAE publication, *Simplified Energy Analysis Using the Modified Bin Method,* is especially useful when refined bin calculations are needed.

The bin method is recommended for all energy calculations except large commercial structures because it is easy to use with hand or computer calculations.

The bin method is based on the concept that all the hours during the year when a particular temperature band occurs can be grouped together. Then an energy calculation is made for those hours with the equipment operating under those particular conditions. Each temperature band is called a bin and the band is 5 degrees F. Depending on the geographical location, 20 or more bins may exist. Bin weather data are available from two main sources:

1. *Air Force Manual 88-29,*[7] which is widely used but is now about eight years old
2. ASHRAE's 1986 *Bin Weather Data,* which is based on current studies and development of weather data for energy calculations; this newer data is recommended

The bin method requires a load profile for the building, that is, the amount of heating or cooling required, as a function of outdoor temperature, to maintain the conditioned space at the desired level. Figure 3.9 is a plot of such a profile. In some cases more than one profile may be required to accommodate different building uses such as occupied and unoccupied periods. The bin weather data is given in three shifts of eight hours each so it can be modified to the actual situation.

The load profile for a building may be determined in a number of ways. A comprehensive approach is described in the modified bin method. It is recommended for commercial-type buildings where internal heat gains are fairly large and variable. For most residential and light commercial buildings using unitary equipment, a simplified method described as follows is quite satisfactory:

Two points on the heating or cooling profile are known or can be calculated. It is well documented that neither heating or cooling is generally required for a residential building when the outdoor temperature is around 60° to 65°F. Bin

3

temperatures in this range are 62° and 67°F; therefore, it may be may assumed that cooling is required at 67°F and above while heating is required at 62°F and below. These points are shown on Figure 3.9 as Oc and Oh.

The peak cooling load will occur during the daylight hours and corresponds closely to the design cooling load at the outdoor design temperature. This is point dc in Figure 3.9. The peak heating load generally occurs during early morning in the absence of any significant internal loads. This corresponds to the design heating load at the outdoor design temperature. This is point dh in Figure 3.9. The points are then joined by a straight line to obtain the desired load profile as shown.

Table 3.1 is an example of some bin data given in three eight-hour periods. Column 1 is the bin temperature and columns 2, 3, and 4 are the annual hours that each temperature occurs during each period. Suppose a building is occupied from 7 A.M. to 7 P.M. on weekdays and unoccupied during other hours and weekends. Figure 3.10 shows diagramatically how the hours are distributed during the week and how the hours are assigned to each building shift. Then, referring to Table 3.1, columns 5 and 6 are computed using the occupied and unoccupied fraction derived from Figure 3.10. The occupied area (hours x days) for bin ranges I, II, and III are:

I 1 hr. x 5 days = 5
II 8 hrs. x 5 days = 40
III 3 hrs. x 5 days = 15

The total area (hours x days) is equal to (7 x 8 = 56) for bin ranges I, II, and III, respectively. Therefore, the occupied and unoccupied fractions for each bin range are as follows:

I	0.089	0.911
II	0.714	0.287
III	0.268	0.732

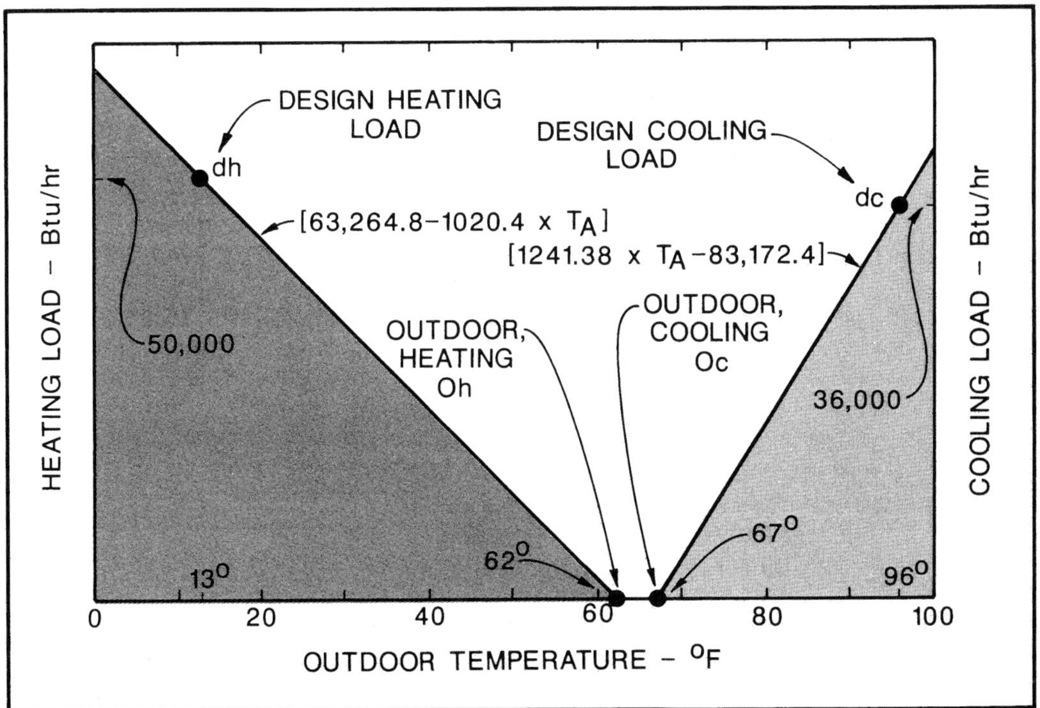

FIGURE 3.9: Building Heating and Cooling Load Profile

3

TABLE 3.1: Annual Bin Hours of Occurrence of Outdoor Dry Bulb Temperature

(1) BIN	(2) 1-8	(3) 9-16	(4) 17-24	(5) OCCUPIED	(6) UNOCCUPIED
102°F					
97		5	1	3.8	2.2
92		64	8	47.9	24.2
87	0	189	54	149.5	93.6
82	10	282	136	238.8	189.3
77	88	296	247	285.4	345.6
72	318	270	337	311.5	613.6
67	337	229	292	271.9	586.2
62	292	218	245	247.4	507.7
57	255	207	226	231.2	456.9
52	233	213	225	233.2	437.8
47	236	201	228	225.7	439.3
42	257	229	248	252.9	481.1
37	268	200	240	231.1	476.9
32	264	155	202	188.4	432.6
27	160	86	116	106.8	255.2
22	101	48	63	60.2	151.8
17	55	19	27	25.7	75.3
12	31	8	12	11.7	39.3
7	10	1	2	2.1	10.9
2	1	0	0	.1	0.9

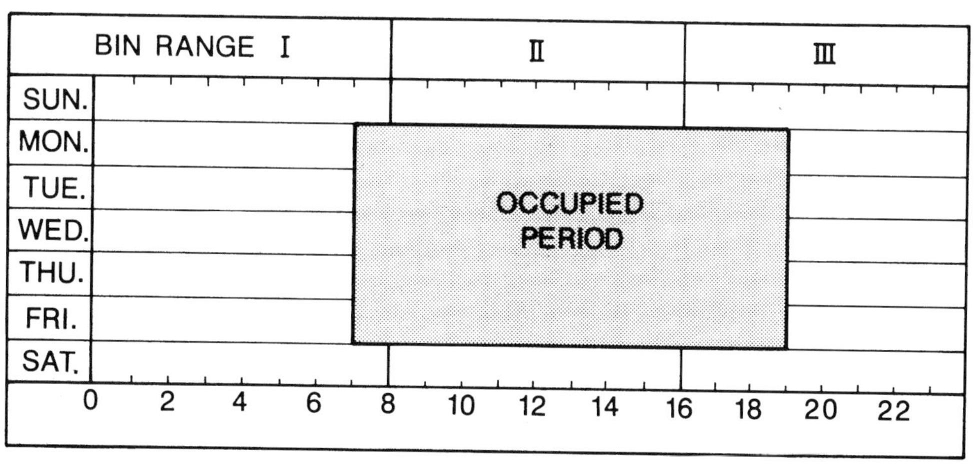

FIGURE 3.10: Graphical Interpretation of Bin Hours in a Commercial Building

For the 47°F bin:

Occupied Hrs. = 0.089(236) + 0.714(201) + 0.268(228) = 225.7

Unoccupied Hrs. = 0.911(236) + 0.286(201) + 0.732(228) = 439.3

Table 3.1 is completed using the same

3

procedure for each bin. The use of a spreadsheet and a microcomputer makes this a simple calculation. Then the occupied and unoccupied hours for each bin are used in conjunction with corresponding load profiles for the occupied and unoccupied periods. It is usually adequate to use one shift for a residential structure. In that case, only one load profile is needed and the bin hours in columns 2, 3, and 4 of Table 3.1 are simply summed.

The operating characteristics of the heating/cooling equipment as a function of the outdoor temperature are required. Indoor conditions are sufficiently constant so they have no effect on system performance. Figure 3.11 shows the performance characteristics of a typical air-to-air heat pump as an example. The loss in efficiency due to partial load operation when the heat pump (or other type equipment) cycles on and off must also be considered. This cycling increases run time for the equipment. A partial load factor (PLF) is defined as follows:

Equation 3.1

$$PLF = \frac{\text{Theoretical Equipment Run Time}}{\text{Actual Equipment Run Time}}$$

Studies at the National Bureau of Standards have resulted in an analytical expression for PLF which uses a "degradation" factor, C_D.

Equation 3.2

$$PLF = 1 - C_D \left[1 - \frac{\text{Structure Load}}{\text{Unit Capacity}} \right]$$

C_D has an assumed value of about 0.25 unless the equipment manufacturer has tested his unit and determined a lower value. When the unit capacity is less than the load, PLF will be greater than one. In this case PLF is set equal to 1.0 since the unit will be running continually. Note that:

Actual Run Time = Theoretical Run Time / PLF

The bin calculation method can now be outlined for a typical bin:

1. Determine the building load from the profile shown in Figure 3.9.
2. Determine the unit capacity from Figure 3.11.
3. Compute the theoretical run time fraction, Xr, as the ratio of building load to unit capacity.

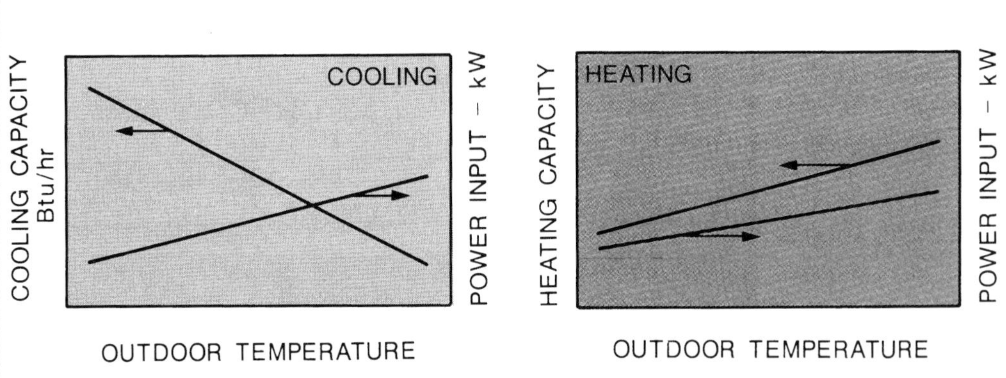

FIGURE 3.11: Heat Pump Performance Characteristics (Air-Source)

4. Compute the partial load factor, PLF, from Equation 3.2.
5. Compute the actual run time fraction, Xa, as Xr/PLF.
6. Compute actual run time, r, as bin hours x Xa.
7. Determine the rate of unit input, Rp, from Figure 3.11.
8. Compute the energy use, P, for this bin.
9. Determine energy cost per unit of energy, $/kWh, from the local utility rate schedule.
10. Compute energy cost for this bin as $/kWh x P.
11. Repeat steps 1 through 10 for all bins.

Steps 1 through 11 constitute the basic bin method. Certain refinements may be required. For example, the heat pump may not be able to supply enough heat when the outdoor temperature is very low and electric resistance heat may be required. Often the fan energy associated with the heat pump may not be accounted for in the performance data and must be added to the calculation. Also, when the building load exceeds the unit capacity, the PLF is assumed to be 1.0 because the unit will run continually.

Performance of an Air-Source Heat Pump

The small microcomputers now so readily available lend themselves to the bin method. An example for an air-to-air heat pump is presented below. Figure 3.9 shows the load profile for a residence in Stillwater, Oklahoma. In this case it was assumed that 62°F was the temperature where no heating was required. It was assumed that cooling is required above 67°F. Points dh (design heating load) and dc (design cooling load) were determined from the load calculations.

Figure 3.12 shows the performance characteristics of an air-to-air heat pump

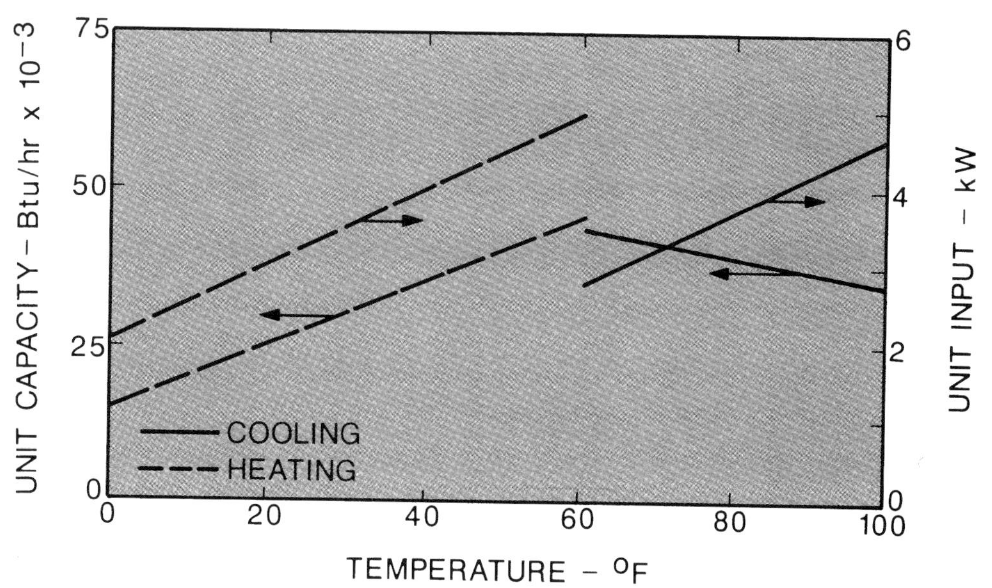

FIGURE 3.12: Heat Pump Operating Characteristics (Air-Source)

3

for the residence. This data was taken from the manufacturer's catalog and includes all fan energy.

Table 3.2 shows the results of the bin calculations for a house in Stillwater, Oklahoma. The calculations were made with a personal computer using a spreadsheet program. A column-by-column explanation of Table 3.2 follows:

Col. 1 Bin temperatures at 5°F.

Col. 2 Annual occurrence of the bin temperatures in Col. 1 from *Air Force Manual 88-29*.

Col. 3 Building load from Figure 3.9. The loads are calculated from an equation of the straight lines in Figure 3.9. The equation is stored in each cell of Col. 3 (Load = A + B x T).

Heating: $[63{,}264.8 - 1020.4 \times T_A]$

Cooling: $[1241.38 \times T_A - 83{,}172.4]$

Col. 4 Heating or cooling capacity of the heat pump from Figure 3.12. Again equations of the straight lines for the heating or cooling capacity are stored in the cells of Col. 4.

Heating: $[12.05 + 0.445 \times T_W] \times 1000$

Cooling: $[48.6 - 0.18 \times T_W] \times 1000$

Col. 5 The fraction of the total bin hours that the heat pump must run is the ratio of the building load to the unit capacity. The equation for this ratio is stored in the cells of Col. 5. If the ratio is greater than 1.0, it is set equal to 1.0.

Col. 6 The partial load factor is computed from Equation 3.2 with C_D = 0.25 with Equation 3.2 stored in the cells of Col. 6.

TABLE 3.2: Energy Calculation by the Bin Method: Air-to-Air Heat Pump

RESIDENCE IN STILLWATER, OKLAHOMA

BIN TEMP-°F 1	BIN HOURS 2	BLDG. LOAD BTU/HR 3	UNIT CAP. BTU/HR 4	THEO. RUN TIME-% 5 3/4	PARTIAL LOAD FAC. 6	ACTUAL RUN TIME-% 7 5/6	UNIT INPUT KW 8	EL. RES. KW 8A	UNIT INPUT KWH 9 2x7(8+8A)	ENER. COST $/KWH 10	ENER. COST $ 11 9x10
112	1	55,862.16	32,171.4	1.00	1.00	1.00	4.72		4.72	0.075	0.35
107	13	49,655.26	33,367.9	1.00	1.00	1.00	4.55		59.13	0.075	4.43
102	69	43,448.36	34,564.4	1.00	1.00	1.00	4.37		301.78	0.075	22.63
97	185	37,241.46	35,760.9	1.00	1.00	1.00	4.20		776.83	0.075	58.26
92	309	31,034.56	36,957.4	0.84	0.96	0.87	4.02		1,087.86	0.075	81.59
87	420	24,827.66	38,153.9	0.65	0.91	0.71	3.85		1,152.87	0.075	86.47
82	565	18,620.76	39,350.4	0.47	0.87	0.54	3.68		1,131.69	0.075	84.88
77	731	12,413.86	40,546.9	0.31	0.83	0.37	3.50		947.92	0.075	71.09
72	800	6,206.96	41,743.4	0.15	0.79	0.19	3.33		502.65	0.075	37.70
67	712	0.00	42,939.9	0.00	0.75	0.00	3.15		0.00	0.075	0.00
62	678	0.00	45,100.0	0.00	0.75	0.00	5.0258	0.000	0.00	0.050	0.00
57	654	5,102.00	42,725.0	0.12	0.78	0.15	4.8113	0.000	481.82	0.050	24.09
52	625	10,204.00	40,350.0	0.25	0.81	0.31	4.5968	0.000	893.42	0.050	44.67
47	631	15,306.00	37,975.0	0.40	0.85	0.47	4.3823	0.000	1,310.05	0.050	65.50
42	622	20,408.00	35,600.0	0.57	0.89	0.64	4.1678	0.000	1,663.58	0.050	83.18
37	568	25,510.00	33,225.0	0.77	0.94	0.82	3.9533	0.000	1,830.32	0.050	91.52
32	495	30,612.00	30,850.0	0.99	1.00	0.99	3.7388	0.000	1,839.98	0.050	92.00
27	311	35,714.00	28,475.0	1.00	1.00	1.00	3.5243	2.122	1,755.88	0.050	87.79
22	184	40,816.00	26,100.0	1.00	1.00	1.00	3.3098	4.313	1,402.60	0.050	70.13
17	101	45,918.00	23,725.0	1.00	1.00	1.00	3.0953	6.504	969.57	0.050	48.48
12	49	51,020.00	21,350.0	1.00	1.00	1.00	2.8808	8.696	567.25	0.050	28.36
7	14	56,122.00	18,975.0	1.00	1.00	1.00	2.6663	10.887	189.75	0.050	9.49
2	5	61,224.00	16,600.0	1.00	1.00	1.00	2.4518	13.079	77.65	0.050	3.88
-3	1	66,326.00	14,225.0	1.00	1.00	1.00	2.2373	15.270	17.51	0.050	0.88

TOTAL COOLING COST: 447.41

TOTAL HEATING COST: 649.97

Col. 7 The actual run time is the ratio of the theoretical run time to the partial load factor with the function stored in the cells of Col. 7.

Col. 8 The unit power input (kW) is obtained from Figure 3.13 in the form of a straight line equation and stored in each cell.

Heating: $I = 1.755 + 0.0235 \times T_W$

Cooling: $I = 1.47 + 0.018 \times T_W$

Col. 8A The supplemental electric resistance heat required is equal to the load (Col. 3) minus the unit capacity (Col. 4) divided by 3412 Btu/hr/kW and stored in the cells of Col. 8A for heating only. If the actual run time is less than or equal to 1.0, this quantity is 0.

Col. 9 The total unit input in kWh is the product of the bin hours, the actual run time fraction and the sum of the heat pump and electric resistance power input (Col. 2 x Col. 7) x (Col. 8 + Col. 8A). Again this function is stored in each cell of Col. 9.

Col. 10 The electric utility rates are assumed constant and stored in this column.

Col. 11 The energy cost is the cost per kWh times the total kWh for each bin (Col. 9 x Col. 10). The sums for heating and cooling are shown at the bottom of Col. 11.

NOTE: The above analysis for the air-to-air heat pump does not include defrosting losses which may be significant.

Performance of a Ground-Source Heat Pump

Once a spreadsheet like Table 3.2 has been developed it can easily be modified to accommodate a different type of system, such as a ground-source heat pump. Consider the previous example using a ground-source system.

The bin data and the building load profile will be unchanged. The performance of a ground-source heat pump depends on the entering water temperature, assuming other variables such as air volume flow rate, air temperature, etc., are constant. To accommodate the bin method there needs to be a relation between the outdoor temperature and the entering water temperature.

The entering water temperature depends on many factors such as the time of year, type of soil, size of the ground heat exchanger, etc. It is not easy to determine at a particular time. However, a properly designed system is sized so the entering water temperature stays within designer specified limits for the complete season. Briefly:

1. The water temperature will vary from the minimum entering water temperature (EWT_{MIN}) during the coldest month to maximum entering water temperature (EWT_{MAX}) during the warmest month. A first estimate of EWT_{MIN} is 40°F above the coldest outdoor temperature or 25°F, whichever is the greater. EWT_{MAX} is usually taken as 10° to 20°F below the highest outdoor temperature at a specific location.
2. When the seasons change, the ground temperature is near its mean value and the entering water will be about the same because the system is operating infrequently.

With these facts in mind, assume a relation between the entering water temperature and the outdoor temperature (TA) as shown in Figure 3.13. This relation is rational and adaptable to local conditions. Two straight-line segments represent the

3

Consider the following two straight-line segments which need to be represented in equation form:

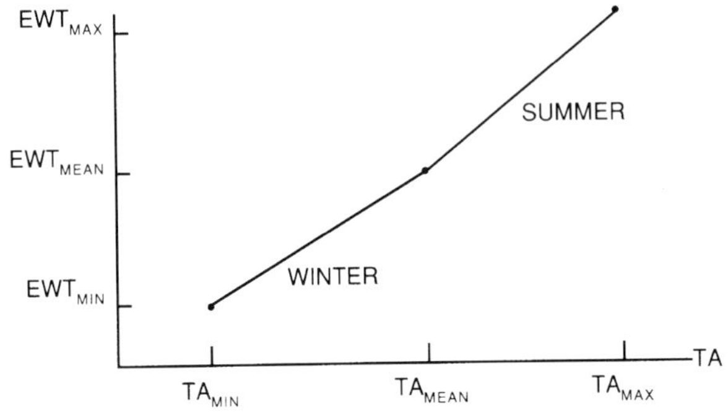

WINTER (HEATING)

$$EWT_H = EWT_{MIN} + \left[\frac{EWT_{MEAN} - EWT_{MIN}}{TA_{MEAN} - TA_{MIN}}\right]\left[TA - TA_{MIN}\right]$$

SUMMER (COOLING)

$$EWT_C = EWT_{MEAN} + \left[\frac{EWT_{MAX} - EWT_{MEAN}}{TA_{MAX} - TA_{MEAN}}\right]\left[TA - TA_{MEAN}\right]$$

For Stillwater, Oklahoma:

$EWT_{MAX} = 100°F$	$TA_{MAX} = 112°F$
$EWT_{MEAN} = 62°F$	$TA_{MEAN} = 60°F$
$EWT_{MIN} = 37°F$	$TA_{MIN} = -3°F$

$$EWT_H = 37 + \{ (62 - 37) / (60 - (-3)) \} \{TA - (-3)\}$$
$$= 38.19 + 0.397\ TA$$

$$EWT_C = 62 + \{ (100 - 62) / (112 - 60) \} \{TA - 60\}$$
$$= 18.15 + 0.731\ TA$$

FIGURE 3.13: Heat Pump Entering Water Temperature Approximation

performance for the winter and summer seasons. The heat pump performance can now be represented as a function of the entering water temperature, and Figure 3.13 can be used to predict the heat pump entering water temperature for heating (EWT_H) and cooling (EWT_C) given the outdoor air temperature (TA).

Figure 3.14 shows water-source heat pump data from a manufacturer's catalog suitable for the previous Stillwater, Oklahoma, residence example. Equations for the various lines are shown. Suitable data to express EWT versus outdoor temperature are as follows:

Entering Water Temperature Data:
$EWT_{MAX} = 100°F$
$EWT_{MEAN} = 62°F$
$EWT_{MIN} = 37°F$

Air Temperature Data:
$TA_{MAX} = 112°F$
$TA_{MEAN} = 60°F$
$TA_{MIN} = -3°F$

Table 3.3 shows the results for this type of system.

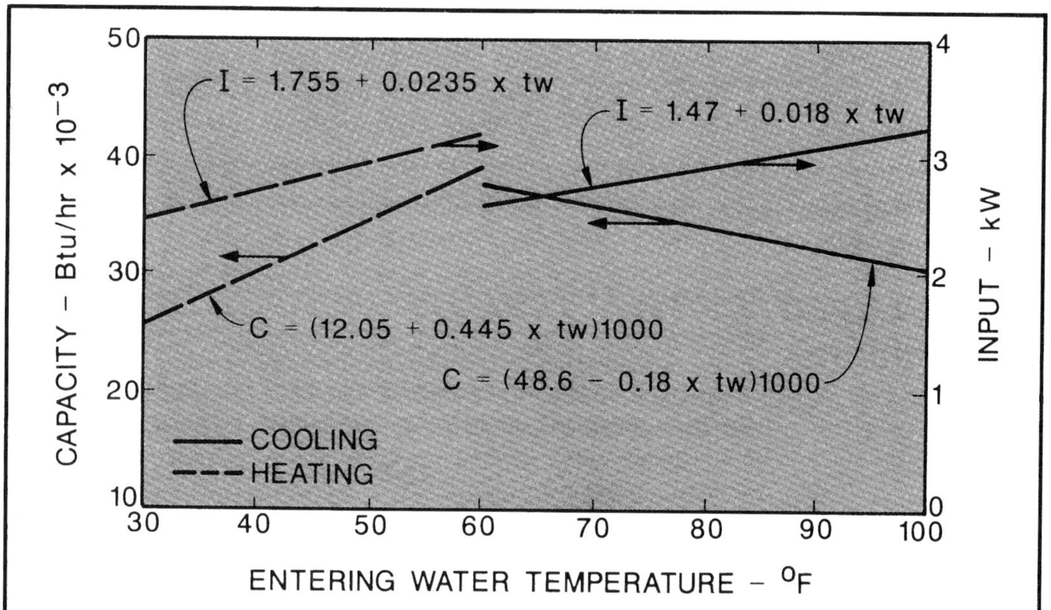

FIGURE 3.14: Water-Source Heat Pump Performance

TABLE 3.3: Energy Calculation by the Bin Method: Ground-Source Heat Pump

RESIDENCE IN STILLWATER, OKLAHOMA

BIN TEMP-°F 1	BIN HOURS 2	BLDG. LOAD BTU/HR 3	EWT F 3A	UNIT CAP. BTU/HR 4	THEO. RUN TIME-% 5 3/4	PARTIAL LOAD FAC. 6	ACTUAL RUN TIME-% 7 5/6	UNIT INPUT KW 8	ELEC. RES.-KW 8A	UNIT INPUT KWH 9 2x7(8+8A)	ENER. COST $/KWH 10	ENER. COST $ 11 9x10
112	1	55,862.16	100.0	30,596.04	1.00	1.00	1.00	3.27		3:27	0.075	0.25
107	13	49,655.26	96.4	31,253.94	1.00	1.00	1.00	3.20		41.66	0.075	3.12
102	69	43,448.36	92.7	31,911.84	1.00	1.00	1.00	3.14		216.58	0.075	16.24
97	185	37,241.46	89.1	32,569.74	1.00	1.00	1.00	3.07		568.51	0.075	42.64
92	309	31,034.56	85.4	33,227.64	0.93	0.98	0.95	3.01		882.47	0.075	66.18
87	420	24,827.66	81.7	33,885.54	0.73	0.93	0.79	2.94		969.99	0.075	72.75
82	565	18,620.76	78.1	34,543.44	0.54	0.88	0.61	2.88		989.90	0.075	74.24
77	731	12,413.86	74.4	35,201.34	0.35	0.84	0.42	2.81		864.22	0.075	64.82
72	800	6,206.96	70.8	35,859.24	0.17	0.79	0.22	2.74		479.01	0.075	35.93
67	712	0.00	67.1	36,517.14	0.00	0.75	0.00	2.68		0.00	0.075	0.00
62	678	0.00	62.8	39,997.8	0.00	0.75	0.00	3.23	0.00	0.00	0.050	0.00
57	654	5,102.00	60.8	39,114.5	0.13	0.78	0.17	3.18	0.00	347.09	0.050	17.35
52	625	10,204.00	58.8	38,231.1	0.27	0.82	0.33	3.14	0.00	640.85	0.050	32.04
47	631	15,306.00	56.8	37,347.8	0.41	0.85	0.48	3.09	0.00	937.66	0.050	46.88
42	622	20,408.00	54.9	36,464.5	0.56	0.89	0.63	3.04	0.00	1,190.86	0.050	59.54
37	568	25,510.00	52.9	35,581.2	0.72	0.93	0.77	3.00	0.00	1,313.69	0.050	65.68
32	495	30,612.00	50.9	34,697.8	0.88	0.97	0.91	2.95	0.00	1,327.83	0.050	66.39
27	311	35,714.00	48.9	33,814.5	1.00	1.00	1.00	2.90	0.56	1,076.39	0.050	53.82
22	184	40,816.00	46.9	32,931.2	1.00	1.00	1.00	2.86	2.31	951.03	0.050	47.55
17	101	45,918.00	44.9	32,047.9	1.00	1.00	1.00	2.81	4.07	694.49	0.050	34.72
12	49	51,020.00	43.0	31,164.5	1.00	1.00	1.00	2.76	5.82	420.60	0.050	21.03
7	14	56,122.00	41.0	30,281.2	1.00	1.00	1.00	2.72	7.57	144.08	0.050	7.20
2	5	61,224.00	39.0	29,397.9	1.00	1.00	1.00	2.67	9.33	59.99	0.050	3.00
-3	1	66,326.00	37.0	28,514.6	1.00	1.00	1.00	2.62	11.08	13.71	0.050	0.69

TOTAL COOLING COST: 376.17
TOTAL HEATING COST: 455.91

3

Many computer programs are available on the open market to perform this type of energy study. Also available are programs to perform hour-by-hour calculations. It is recommended that software be purchased to do the hour-by-hour calculations because those procedures are quite complex.

Ground Load Calculations

The design of the ground-coupled heat exchanger for a ground-source system requires information regarding the amount of heat transferred to or from the ground during a specified period of time. Usually the warmest and coldest months of the year are adequate. The energy calculations discussed above are close to what is required, and the bin method lends itself nicely to this need.

During the cooling season all the energy inputs to the system must be rejected to the ground. These are the unit cooling capacity (Col. 4, Table 3.3), the unit power input (Col. 8, Table 3.3) and the ground heat exchanger circulating pump power. The circulating pump power has not been included in this calculation table but can be assumed to be equal to the pump power multiplied by the heat pump operating hours.

During the winter months heat is removed from the ground in an amount equal to the unit capacity minus the electrical power input. The power input includes the blower motor and the ground heat exchanger circulating pump power. The calculation procedure sums this for all bin temperatures and bin hours.

It is, therefore, a simple matter to modify the procedure in Table 3.3 to obtain these quantities for the complete year. To obtain the ground loads for just January and July as an example, the bin hours and temperatures must be changed to reflect just those months. The yearly ground loads will first be computed for the example shown in Table 3.3 with the results given in Table 3.4. Fluid pumping power has been neglected. The same type of calculation is shown in Table 3.5 for July and January.

Equipment Selection

Sizing the heating/cooling equipment for a comfort conditioning system is fundamentally the most important step in the design process. This has already been stressed in the section on load calculations since equipment size is based on the load. For residential and light commercial systems, the equipment is usually the unit type (a matched set of components designed to serve as a whole); once a unit is selected, a number of parameters are fixed or have little room for adjustment. For example, the unit is designed for one nominal airflow rate, and only a small adjustment is possible. A water-source heat pump is no exception; it is designed for a limited range of water flow rates. Therefore, other parts of the system such as the duct system or ground heat exchanger and brine circulating pump must match the capacity requirements of the heat pump.

In the case of a large commercial system, a secondary delivery system (such as hot and chilled water from a central plant) probably will be used. In this case, the central plant would be sized to meet the peak building block load but the air handlers in the secondary system would be sized for the local zone loads. Some residential-sized water-to-water heat pumps allow the same type of zone control and sizing considerations. Emphasis in this guide is on unit-type equipment.

TABLE 3.4: Annual Ground Load Using the Bin Method — Ground-Source Heat Pump

RESIDENCE IN STILLWATER, OKLAHOMA

BIN TEMP-°F 1	BIN HOURS 2	BLDG. LOAD BTU/HR 3	EWT F 3A	UNIT CAP. BTU/HR 4	THEO. RUN TIME-% 5 3/4	PARTIAL LOAD FAC. 6	ACTUAL RUN TIME-% 7 5/6	UNIT INPUT KW 8	ELEC. RES.-KW 8A	UNIT INPUT KWH 9 2x7(8+8A)	GROUND LOAD-BTU/HR 10 Summer	GROUND LOAD-BTU/HR 11 Winter
112	1	55,862.16	100.0	30,596.04	1.00	1.00	1.00	3.27		3.27	41,755	
107	13	49,655.26	96.4	31,253.94	1.00	1.00	1.00	3.20		41.66	548,445	
102	69	43,448.36	92.7	31,911.84	1.00	1.00	1.00	3.14		216.58	2,940,882	
97	185	37,241.46	89.1	32,569.74	1.00	1.00	1.00	3.07		568.51	7,965,157	
92	309	31,034.56	85.4	33,227.64	0.93	0.98	0.95	3.01		882.47	12,761,541	
87	420	24,827.66	81.7	33,885.54	0.73	0.93	0.79	2.94		969.99	14,483,989	
82	565	18,620.76	78.1	34,543.44	0.54	0.88	0.61	2.88		989.90	15,268,538	
77	731	12,413.86	74.4	35,201.34	0.35	0.84	0.42	2.81		864.22	13,775,390	
72	800	6,206.96	70.8	35,859.24	0.17	0.79	0.22	2.74		479.01	7,893,964	
67	712	0.00	67.1	36,517.14	0.00	0.75	0.00	2.68		0.00	0	
62	678	0.00	62.8	39,997.8	0.00	0.75	0.00	3.23	0.00	0.00		0
57	654	5,102.00	60.8	39,114.5	0.13	0.78	0.17	3.18	0.00	347.09		3,079,295
52	625	10,204.00	58.8	38,231.1	0.27	0.82	0.33	3.14	0.00	640.85		5,622,048
47	631	15,306.00	56.8	37,347.8	0.41	0.85	0.48	3.09	0.00	937.66		8,130,418
42	622	20,408.00	54.9	36,464.5	0.56	0.89	0.63	3.04	0.00	1,190.86		10,200,799
37	568	25,510.00	52.9	35,581.2	0.72	0.93	0.77	3.00	0.00	1,313.69		11,110,758
32	495	30,612.00	50.9	34,697.8	0.88	0.97	0.91	2.95	0.00	1,327.83		11,082,000
27	311	35,714.00	48.9	33,814.5	1.00	1.00	1.00	2.90	0.56	1,076.39		7,434,400
22	184	40,816.00	46.9	32,931.2	1.00	1.00	1.00	2.86	2.31	951.03		4,265,241
17	101	45,918.00	44.9	32,047.9	1.00	1.00	1.00	2.81	4.07	694.49		2,268,106
12	49	51,020.00	43.0	31,164.5	1.00	1.00	1.00	2.76	5.82	420.60		1,064,884
7	14	56,122.00	41.0	30,281.2	1.00	1.00	1.00	2.72	7.57	144.08		294,114
2	5	61,224.00	39.0	29,397.9	1.00	1.00	1.00	2.67	9.33	59.99		101,420
-3	1	66,326.00	37.0	28,514.6	1.00	1.00	1.00	2.62	11.08	13.71		19,560
										TOTALS:	75,679,662	64,673,044

TABLE 3.5: Ground Load for July and January Using the Bin Method — Ground-Source Heat Pump

RESIDENCE IN STILLWATER, OKLAHOMA

BIN TEMP-°F 1	BIN HOURS 2	BLDG. LOAD BTU/HR 3	EWT F 3A	UNIT CAP. BTU/HR 4	THEO. RUN TIME-% 5 3/4	PARTIAL LOAD FAC. 6	ACTUAL RUN TIME-% 7 5/6	UNIT INPUT KW 8	ELEC. RES.-KW 8A	UNIT INPUT KWH 9 2x7(8+8A)	GROUND LOAD-BTU/HR 10 JULY	GROUND LOAD-BTU/HR 11 JANUARY
112	1	55,862.16	100.0	30,596.04	1.00	1.00	1.00	3.27		3.27	41,755	
107	6	49,655.26	96.4	31,253.94	1.00	1.00	1.00	3.20		19.23	253,128	
102	27	43,448.36	92.7	31,911.84	1.00	1.00	1.00	3.14		84.75	1,150,780	
97	62	37,241.46	89.1	32,569.74	1.00	1.00	1.00	3.07		190.53	2,669,404	
92	90	31,034.56	85.4	33,227.64	0.93	0.98	0.95	3.01		257.03	3,716,954	
87	104	24,827.66	81.7	33,885.54	0.73	0.93	0.79	2.94		240.19	3,586,512	
82	128	18,620.76	78.1	34,543.44	0.54	0.88	0.61	2.88		224.26	3,459,067	
77	153	12,413.86	74.4	35,201.34	0.35	0.84	0.42	2.81		180.88	2,883,221	
72	125	6,206.96	70.8	35,859.24	0.17	0.79	0.22	2.74		74.84	1,233,432	
67	40	0.00	67.1	36,517.14	0.00	0.75	0.00	2.68		0.00	0	
62	20	0.00	62.8	39,997.8	0.00	0.75	0.00	3.23	0.00	0.00		0
57	31	5,102.00	60.8	39,114.5	0.13	0.78	0.17	3.18	0.00	16.45		145,960
52	48	10,204.00	58.8	38,231.1	0.27	0.82	0.33	3.14	0.00	49.22		431,773
47	61	15,306.00	56.8	37,347.8	0.41	0.85	0.48	3.09	0.00	90.65		785,983
42	94	20,408.00	54.9	36,464.5	0.56	0.89	0.63	3.04	0.00	179.97		1,541,600
37	105	25,510.00	52.9	35,581.2	0.72	0.93	0.77	3.00	0.00	242.85		2,053,925
32	116	30,612.00	50.9	34,697.8	0.88	0.97	0.91	2.95	0.00	311.17		2,596,994
27	89	35,714.00	48.9	33,814.5	1.00	1.00	1.00	2.90	0.56	308.04		2,127,529
22	72	40,816.00	46.9	32,931.2	1.00	1.00	1.00	2.86	2.31	372.14		1,669,008
17	48	45,918.00	44.9	32,047.9	1.00	1.00	1.00	2.81	4.07	330.06		1,077,912
12	31	51,020.00	43.0	31,164.5	1.00	1.00	1.00	2.76	5.82	266.10		673,702
7	11	56,122.00	41.0	30,281.2	1.00	1.00	1.00	2.72	7.57	113.20		231,090
2	4	61,224.00	39.0	29,397.9	1.00	1.00	1.00	2.67	9.33	48.00		81,136
-3	1	66,326.00	37.0	28,514.6	1.00	1.00	1.00	2.62	11.08	13.71		19,560
										TOTALS:	18,994,253	13,436,173

3

SIZING THE HEAT PUMP

There is often a question as to whether heat pumps should be sized for the heating or the cooling load. The nature of the thermodynamic cycle used will not usually accommodate both loads in most climates. A heat pump sized for cooling generally will be too small for heating and, when sized for the heating load, the unit will be grossly oversized for cooling.

A primary consideration in resolving this question is comfort. An improperly sized unit will not provide adequate comfort if either over- or undersized. When undersized for cooling, the conditioned space will obviously be uncomfortable.

When oversized for cooling, the unit will cycle frequently, leading to a high average coil temperature and inadequate removal of moisture from the air, the cause of mildew problems and consumer complaints about stuffiness. Frequent cycling reduces equipment life and operating efficiency and increases summer energy use and operating cost during cooling. Oversizing also increases initial equipment cost.

Nationally recognized HVAC organizations recommend that heat pump oversizing to accommodate the heating load be limited to 10–15% of the cooling load. This sizing recommendation assures sufficient dehumidification during summer and keeps the equipment from short-cycling. In fact, slight undersizing is widely encouraged to improve humidity control, particularly in new "super-insulated" homes. The majority of the HVAC industry's installation experience, and therefore its sizing recommendations, revolves around air-source heat pumps.

However, a growing body of experience with ground-source heat pumps encourages sizing based either on the calculated heating load or substantial oversizing based on the cooling load. These newer sizing recommendations vary widely but cluster around equipment sizing of no less than 75% of the design heat loss or 150–200% of the cooling load.

Utilities, especially the northern, winter-peaking ones, most frequently encourage or require HVAC contractors to install oversized units in an effort to reduce the winter peak impact. There is no disagreement that oversizing can substantially enhance the economics of wintertime operating cost and reduce electric heating demand. Furthermore, sources in the field claim that no dehumidification problems have been reported in hundreds of "oversized" ground-source heat pumps.

No nationally recognized HVAC organization has yet endorsed the practice of grossly oversizing ground-source heat pumps. This guide is similarly unable to offer support for the practice. Only time, additional installations, and more experience can lead to its endorsement.

The most accepted practice is to size heat pumps on the basis of cooling load. This will ensure that the system will provide a comfortable environment year-round, but may require some kind of supplemental heating to handle any excess heating requirement.

The experience of Oklahoma State University has shown that, in northern regions where the heating load is relatively high and summer latent loads may be relatively low, moderate oversizing may be justified. However, it is Oklahoma State's belief that a ground-source heat pump should not be oversized by more than 25% of the design cooling load.

A number of manufacturers have begun to produce and market HVAC equipment with two-speed compressors which would accommodate oversizing on the heating side while allowing equipment to operate in a desirable manner for cooling.

This equipment will allow utilities and consumers to best meet both conditions: long-run cycles during cooling and sufficient capacity for heating.

SENSIBLE AND LATENT LOAD CONSIDERATIONS

As a general rule, unit-type equipment is designed to operate with a space **sensible**

heat factor (SHF) of about 0.75 to 0.78. The SHF is defined as:

Equation 3.3

$$SHF = \frac{\text{Sensible Cooling Load}}{\text{Total Cooling Load}}$$

This may be varied somewhat by adjusting the airflow over the evaporator coil or changing condenser conditions in some way. However, there is little control after a unit is selected. A unit SHF of about 0.75 is adequate to handle the latent loads in most parts of the United States. The conditions in Gulf Coast cities are the most severe from this standpoint. Most other regions of the country do not have such severe latent load conditions, and the space SHF may be in the 0.85 to 0.95 range.

Caution must be exercised in sizing equipment when the SHF for the unit does not match the SHF for the conditioned space. For example, a 10,000 Btu/hr unit with an SHF of 0.75 is designed to remove 7,500 Btu/hr of sensible heat and 2,500 Btu/hr of latent heat. Assuming the space has a cooling load of 10,000 Btu/hr with an SHF of 0.9, to maintain the space at the design temperature the unit needs to remove 9,000 Btu/hr of sensible heat and 1,000 Btu/hr of latent heat. The unit generally cannot do this because of factors which control the heat and moisture transfer processes of the cooling coil. The conclusions are:

1. The SHF for the unit must be less than or equal to the SHF for the space.
2. The cooling unit should be sized based on the sensible cooling load for the space and the sensible cooling capacity of the unit.

This will ensure that both the sensible and latent loads can be handled.

Consider a house which has a total cooling load of 26,600 Btu/hr, of which 22,300 Btu/hr is sensible load. The space sensible heat factor (SHF) is 22,300/26,600 = 0.84. A water-source-type unit as represented in Tables 3.6 and 3.7 is to be selected for the house. The following data (Table 3.8) are taken from these tables:

Assume the ground-source heat exchanger is designed to return 90°F water to the heat pump at the rate of 6 to 7 gpm at the peak cooling load. Air returns to the heat pump at 80°F dry bulb and 67°F wet bulb. Note in Table 3.6 the 850 cfm unit has a total cooling capacity of 26,700 Btu/hr and a sensible cooling capacity of 18,500 Btu/hr. On the basis of total capacity this unit is a good match but its sensible capacity is 3,800 Btu/hr (1/3 ton) too low. The unit will probably be too small.

Referring to Table 3.7 which lists data for a 1000 cfm unit, it is noted that the sensible capacity is 22,200 Btu/hr and the total capacity is 31,800 Btu/hr. This is a

TABLE 3.8: Heat Pump Summary Table

MODEL NUMBER	25	33
Entering Air Conditions (°F)	80/67	80/67
Heat Pump Airflow (CFM)	850	1,000
Sensible Capacity (Btu/hr)	18,500	22,200
Total Capacity (Btu/hr)	26,700	31,800
Flow (GPM)	6	7
Water Pressure Drop (inches)	7.2	6.2
Entering Water Temp (°F)	90	90

3

TABLE 3.6: Typical Water-Source Heat Pump Data, Model 25, 850 CFM

			COOLING						HEATING				
EWT	GPM	WPD	EA	TC	SC	KW	HR	EER	EA	HC	KW	HE	COP
30	3.0	2.7	75/63	27.9	18.7	1.83	34.2	15.2	60	20.0	1.82	13.8	3.23
			80/67	30.3	19.5	1.87	36.7	16.2	70	19.3	1.91	12.8	2.96
			85/71	32.8	20.3	1.91	39.3	17.2	80	18.6	2.01	11.7	2.72
	4.5	5.7	75/63	28.5	18.9	1.75	34.5	16.3	60	20.9	1.87	14.5	3.27
			80/67	31.0	19.7	1.78	37.0	17.4	70	20.2	1.97	13.4	3.00
			85/71	33.4	20.5	1.82	39.6	18.4	80	19.4	2.07	12.4	2.75
	6.0	9.7	75/63	29.1	19.1	1.70	34.9	17.1	60	21.6	1.91	15.1	3.32
			80/67	31.6	19.9	1.73	37.5	18.3	70	20.9	2.01	14.0	3.04
			85/71	34.1	20.7	1.76	40.2	19.3	80	20.1	2.11	12.9	2.79
50	3.0	2.3	75/63	27.3	18.3	2.09	34.4	13.1	60	26.6	2.15	19.2	3.62
			80/67	29.7	19.1	2.13	36.9	13.9	70	25.7	2.26	18.0	3.33
			85/71	32.1	19.9	2.17	39.5	14.7	80	24.8	2.38	16.7	3.06
	4.5	4.8	75/63	27.9	18.5	1.99	34.6	14.0	60	27.6	2.22	20.0	3.65
			80/67	30.3	19.3	2.03	37.2	14.9	70	26.7	2.33	18.7	3.35
			85/71	32.7	20.1	2.07	39.8	15.8	80	25.8	2.45	17.4	3.08
	6.0	8.2	75/63	28.4	18.7	1.93	35.0	14.7	60	28.4	2.26	20.7	3.68
			80/67	30.9	19.5	1.97	37.6	15.7	70	27.4	2.38	19.3	3.38
			85/71	33.4	20.3	2.01	40.2	16.6	80	26.5	2.50	17.9	3.10
70	3.0	2.1	76/63	25.4	17.9	2.38	33.6	10.7	60	32.2	2.44	23.9	3.87
			80/67	27.7	18.6	2.43	36.0	11.4	70	31.3	2.57	22.5	3.57
			85/71	29.9	19.4	2.48	38.3	12.0	80	30.4	2.69	21.2	3.30
	4.5	4.5	75/63	26.0	18.1	2.27	33.7	11.4	60	33.5	2.51	24.9	3.90
			80/67	28.2	18.8	2.32	36.1	12.2	70	32.5	2.65	23.5	3.60
			85/71	30.5	19.6	2.36	38.5	12.9	80	31.5	2.78	22.1	3.33
	6.0	7.6	75/63	26.5	18.2	2.21	34.0	12.0	60	34.4	2.57	25.7	3.93
			80/67	28.8	19.0	2.25	36.5	12.8	70	33.4	2.70	24.2	3.63
			85/71	31.1	19.8	2.30	38.9	13.6	80	32.4	2.84	22.7	3.35
90	3.0	2.0	75/63	23.6	17.4	2.68	32.7	8.8	60	35.9	2.59	27.1	4.06
			80/67	25.6	18.1	2.74	35.0	9.4	70	35.1	2.73	25.8	3.77
			85/71	27.7	18.9	2.79	37.2	9.9	80	34.3	2.86	24.5	3.51
	4.5	4.2	75/63	24.1	17.6	2.55	32.8	9.4	60	37.3	2.67	28.2	4.09
			80/67	26.2	18.3	2.61	35.1	10.0	70	36.5	2.81	26.9	3.80
			85/71	28.3	19.0	2.66	37.3	10.6	80	35.6	2.95	25.5	3.53
	6.0	7.2	75/63	24.6	17.8	2.48	33.0	9.9	60	38.4	2.73	29.1	4.13
			80/67	26.7	18.5	2.53	35.3	10.6	70	37.5	2.87	27.7	3.83
			85/71	28.8	19.2	2.58	37.6	11.2	80	36.6	3.01	26.3	3.56
110	3.0	1.9	75/63	20.6	16.2	3.02	30.9	6.8					
			80/67	22.4	16.9	3.08	32.9	7.3					
			85/71	24.2	17.5	3.14	34.9	7.7					
	4.5	4.0	75/63	21.0	16.3	2.88	30.8	7.3					
			80/67	22.8	17.0	2.94	32.9	7.8					
			85/71	24.7	17.7	2.99	34.9	8.2					
	6.0	6.8	75/63	21.4	16.5	2.79	31.0	7.7					
			80/67	23.3	17.2	2.85	33.0	8.2					
			85/71	25.2	17.9	2.91	35.1	8.7					

CFM = airflow, cubic feet per minute
EWT = entering water temperature, °F
GPM = gallons per minute water flow
WPD = water pressure drop, feet of water
EA = entering air temperature, °F (dry bulb/wet bulb)
HC = total air heating capacity, Btu/hr (1000)
TC = total cooling capacity, Btu/hr (1000)
SC = sensible cooling capacity, Btu/hr (1000)
KW = total unit power input, kilowatts
HR = total heat rejection, Btu/hr (1000)
HE = total heat absorption, Btu/hr (1000)
EER = Energy Efficiency Ratio = Btu output/Watt input
COP = Coefficient of Performance = Btu output/Btu input

1. Performance data at EWT below 35°F are based upon 15% methanol antifreeze solution.

Source: WaterFurnace International.

3

TABLE 3.7 Typical Water-Source Heat Pump Data, Model 33, 1,000 CFM

			COOLING						HEATING				
EWT	GPM	WPD	EA	TC	SC	KW	HR	EER	EA	HC	KW	HE	COP
30	4.5	3.5	75/63	35.5	24.2	2.17	42.9	16.3	60	24.1	2.20	16.6	3.21
			80/67	38.6	25.2	2.22	46.2	17.4	70	23.2	2.31	15.3	2.94
			85/71	41.7	26.2	2.26	49.4	18.4	80	22.4	2.43	14.1	2.70
	7.0	8.4	75/63	36.2	24.4	2.11	43.4	17.2	60	25.4	2.27	17.6	3.28
			80/67	39.4	25.4	2.15	46.7	18.3	70	24.5	2.39	16.3	3.00
			85/71	42.5	26.5	2.20	50.0	19.4	80	23.6	2.51	15.0	2.76
	9.0	13.5	75/63	37.0	24.7	2.05	44.0	18.1	60	26.3	2.34	18.4	3.30
			80/67	40.2	25.7	2.09	47.3	19.2	70	25.4	2.46	17.0	3.02
			85/71	43.4	26.7	2.13	50.7	20.4	80	24.5	2.58	15.6	2.77
50	4.5	3.0	75/63	34.0	23.5	2.52	42.6	13.5	60	32.7	2.62	23.7	3.65
			80/67	37.0	24.5	2.57	45.7	14.4	70	31.6	2.76	22.2	3.36
			85/71	39.9	25.5	2.62	48.9	15.2	80	30.5	2.89	20.6	3.09
	7.0	7.1	75/63	34.7	23.8	2.44	43.0	14.2	60	34.2	2.70	25.0	3.71
			80/67	37.7	24.8	2.49	46.2	15.1	70	33.0	2.84	23.3	3.40
			85/71	40.7	25.7	2.54	49.4	16.0	80	31.9	2.98	21.7	3.13
	9.0	11.4	75/63	35.4	24.0	2.37	43.5	14.9	60	35.5	2.78	26.0	3.74
			80/67	38.5	25.0	2.42	46.8	15.9	70	34.3	2.93	24.3	3.43
			85/71	41.6	26.0	2.47	50.0	16.8	80	33.1	3.08	22.6	3.15
70	4.5	2.8	75/63	31.8	22.7	2.84	41.5	11.2	60	41.9	3.12	31.2	3.93
			80/67	34.6	23.6	2.90	44.5	11.9	70	40.7	3.28	29.5	3.63
			85/71	37.3	24.6	2.95	47.4	12.6	80	39.5	3.45	27.7	3.35
	7.0	6.6	75/63	32.5	22.9	2.76	41.9	11.8	60	43.8	3.22	32.9	3.99
			80/67	35.3	23.9	2.81	44.9	12.5	70	42.6	3.39	31.0	3.68
			85/71	38.1	24.8	2.87	47.9	13.3	80	41.3	3.55	29.1	3.40
	9.0	10.6	75/63	33.1	23.1	2.68	42.3	12.4	60	45.6	3.32	34.2	4.03
			80/67	36.0	24.1	2.73	45.3	13.2	70	44.2	3.49	32.3	3.71
			85/71	38.9	25.1	2.78	48.4	14.0	80	42.9	3.66	30.4	3.43
90	4.5	2.6	75/63	28.6	21.1	3.21	39.6	8.9	60	48.5	3.48	36.7	4.09
			80/67	31.1	22.0	3.28	42.3	9.5	70	47.4	3.66	34.9	3.80
			85/71	33.6	22.8	3.34	45.0	10.1	80	46.3	3.84	33.2	3.53
	7.0	6.2	75/63	29.2	21.3	3.12	39.9	9.4	60	50.8	3.58	38.6	4.16
			80/67	31.8	22.2	3.18	42.6	10.0	70	49.6	3.77	36.8	3.85
			85/71	34.3	23.1	3.25	45.4	10.6	80	48.4	3.96	34.9	3.58
	9.0	10.0	75/63	29.8	21.5	3.03	40.1	9.8	60	52.4	3.70	39.8	4.16
			80/67	32.4	22.4	3.09	42.9	10.5	70	51.2	3.89	37.9	3.85
			85/71	35.0	23.3	3.15	45.7	11.5	80	49.9	4.08	36.0	3.58
110	4.5	2.5	75/63	25.4	19.8	3.62	37.8	7.0					
			80/67	27.7	20.6	3.69	40.3	7.5					
			85/71	29.9	21.4	3.77	42.7	7.9					
	7.0	5.9	75/63	26.0	20.0	3.51	38.0	7.4					
			80/67	28.2	20.8	3.58	40.5	7.9					
			85/71	30.5	21.6	3.66	43.0	8.3					
	9.0	9.5	75/63	26.5	20.2	3.41	38.1	7.8					
			80/67	28.8	21.0	3.48	40.7	8.3					
			85/71	31.1	21.8	3.55	43.2	8.8					

CFM = airflow, cubic feet per minute
EWT = entering water temperature, °F
GPM = gallons per minute water flow
WPD = water pressure drop, feet of water
EA = entering air temperature, °F (dry bulb/wet bulb)
HC = total air heating capacity, Btu/hr (1000)
TC = total cooling capacity, Btu/hr (1000)
SC = sensible cooling capacity, Btu/hr (1000)
KW = total unit power input, kilowatts
HR = total heat rejection, Btu/hr (1000)
HE = total heat absorption, Btu/hr (1000)
EER = Energy Efficiency Ratio = Btu output/Watt input
COP = Coefficient of Performance = Btu output/Btu input

1. Performance data at EWT below 35°F are based upon 15% methanol antifreeze solution.

Source: WaterFurnace International.

much better choice because of the close match between the sensible cooling load and the unit's sensible capacity. Further, the larger airflow rate will lead to a better air distribution system.

EFFICIENCY CONSIDERATIONS

In general, efficiency is defined as follows:

Equation 3.4

$$EFF = \frac{\text{Useful Energy Delivered}}{\text{Energy Supplied}}$$

In the case of a heat pump, this parameter is presented in slightly different ways for the heating and cooling modes. For cooling, the efficiency parameter is the Energy Efficiency Ratio (EER):

Equation 3.5

$$EER = \frac{\text{Total Cooling Capacity in Btu/hr}}{\text{Power Input in Watts}}$$

This parameter is given in Table 3.7. For the selection example above, EER = 10 and the power input is then:

$$\text{Power} = \frac{31{,}800 \text{ Btu/hr}}{10 \times 100} = 3.18 \text{ kW}$$

For heating, the efficiency parameter is the Coefficient of Performance (COP):

Equation 3.6

$$COP = \frac{\text{Heating Capacity in Btu/hr}}{\text{Power Input in Btu/hr}}$$

From Table 3.7 for the unit selected in the above example with EWT = 50°F, the COP is 3.4 and the power is:

$$\text{Power} = \frac{33{,}000 \text{ Btu/hr}}{3.4 \times 3412} = 2.85 \text{ kW}$$

It should be noted that the EER and COP are dependent on the EWT, water flow rate, airflow rate, and entering air temperatures. Therefore, there is a need to be specific when comparing EER and COP of different units. Obviously, equipment is wanted that is as efficient as possible; however, high efficiency usually has a high price tag.

It is difficult to state a general rule about selecting equipment on the basis of efficiency because of economic considerations and the merits of each system.

INDOOR AIR QUALITY (IAQ)

To maintain proper indoor air quality the designer must consider:

1. the air filtering system
2. the use of outside air to dilute pollutants

Indoor air quality has received too little attention in the past. Most unit-type equipment has a very nominal (inefficient) filter system, and most duct systems are not designed to provide outdoor air.

CAUTION: Bringing in too much outdoor air over a direct expansion coil can quickly overload the unit.

Modern construction practices have drastically reduced the infiltration for most buildings which aggravates the indoor air quality problem because the most effective way of removing pollutants and odors is to simply vent them to the outdoors. Infiltration and exfiltration accomplish this quite well. However, when infiltration is reduced below about 0.5 air changes per hour (ACH), this method becomes inadequate and positive ventilation methods are required.

Air Filtering

To generate good indoor air quality the air distribution system should have a high-efficiency filter and have provisions for bringing in outdoor air. The typical filter supplied with a unit will remove less than 20% of the small atmospheric dust particles that enter the filter.

Filters which will remove 80 to 90% of pollutants in the size range of smoke particles are available at a reasonable cost. These filters are usually pleated to have a large surface area and must be built into the system. Figure 3.15 is a schematic of a heat pump showing how a high-efficiency filter and outdoor air intake may be designed. This arrangement assures a clean, healthy environment in the space, provided the outdoor air is clean. When the weather is mild this system also permits circulation of large quantities of outdoor air under controlled conditions.

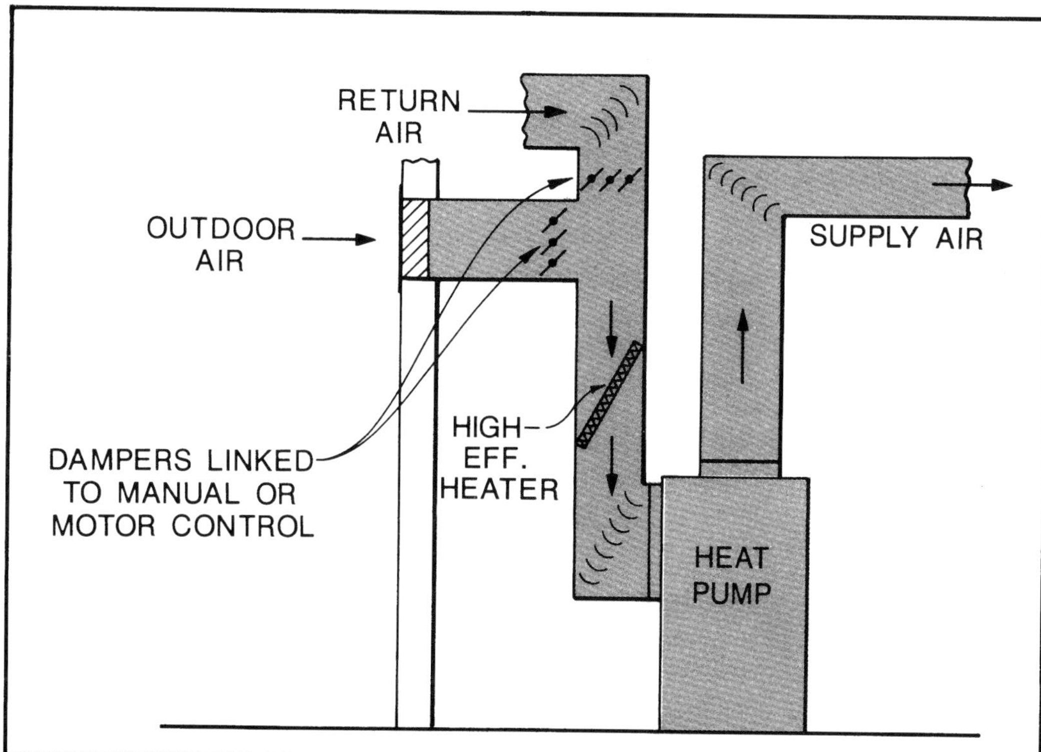

FIGURE 3.15: Outdoor Air Supply

Air Distribution System

After the loads have been correctly computed, the unit properly sized, and provisions for IAQ made, the air distribution system must be designed. Heat pumps do not deliver air for heating at as high a temperature as fossil-fueled furnaces; therefore, the duct system design becomes even more important. There are three basic steps in designing an air distribution system:

1. Select the best location for the duct system: crawl space, basement, or attic.
2. Select and locate the supply air diffusers and return air grilles based

3

on individual room heat gain/loss calculations.

3. Lay out and size the ducts to match the heat pump airflow rate and static pressure capabilities.

The air distribution system provides both the heating and cooling for the system. As a result, attention must be given to both. Further, a system that does not provide a comfortable environment cannot be operated efficiently. Buoyancy of the air in the space and the manner in which air is introduced to the space play important roles with respect to comfort. The object is to introduce and mix the air in the space so the temperature will be uniform throughout and occupants will be unaware of air movement.

AIR DIFFUSERS

Discussions below will focus on the following two types of air supply systems used with CL/GS heat pumps:

1. through-the-floor perimeter
2. ceiling

Low sidewall will not be covered since it is not recommended for heat pump systems.

Through-the-Floor Perimeter

In a single-story structure, the ideal location for the supply registers for a heat pump system is through the floor along the outside wall or perimeter. This placement provides good air distribution and comfort in the winter and summer by projecting conditioned air in a vertical spreading pattern toward the ceiling.

Through-the-floor systems normally will have ceiling or high wall returns. The high return system will assure that during the cooling operation the warm air is siphoned off from the ceiling; this will fill the room with cool air. When the system starts operating, the cool air will fill the room from the floor up. By returning the warmest air back to the system, the heat pump system will operate at a higher efficiency during the cooling operation.

Type: Floor baseboard
Location: Perimeter
Diffuser discharge: Vertical spreading
Throw: Toward the ceiling
Return location: Ceiling for heating and cooling
Note: Vertical nonspreading diffusers are acceptable but not recommended.

Ceiling

A ceiling supply system is a second choice to a through-the-floor perimeter system. In some cases, such as existing slab or two-story house construction, this duct location is unavoidable. Ceiling supply systems can be acceptable with proper system design. Diffuser placement includes:

1. perimeter
2. centered
3. inside walls

There are geographic climates and building thermal protection methods for which all of the above are satisfactory and recommended. Before arbitrary placement, the designer should consider local climate and successful practice in the area.

The use of step-down or round-type ceiling diffusers, although widely used throughout the industry, is not recommended for heat pump applications and is especially to be avoided in northern climates. Step-down diffusers tend to disperse the air near the ceiling, and the warm air will stay there unless some other means (such as a fan) is used to move, mix, and force the warm air downward.

Ceiling Perimeter

Ceiling perimeter duct systems are similar to floor duct systems and are placed along the perimeter of the exterior wall at the ceiling. This type of diffuser placement places the warm air against the cold outside wall.

Type: Ceiling
Location: Perimeter at points of greatest load

Diffuser discharge: Downward and parallel to the wall

Throw: Toward the floor and in a fan-shaped pattern

Return location: Floor or low wall

Centered

Centered systems (or slightly offset toward the outer wall) are used in moderate heating climates. With this type of centered placement, a diffuser should be selected that will throw a portion of the air toward the outside wall where the greatest load occurs.

Type: Ceiling

Location: Ceiling pattern depending on number of outlets required. May be offset toward the outside wall with a portion of the air thrown down the wall

Diffuser discharge: Horizontal and downward

Throw: To walls or to adjacent air pattern

Return location: Floor or low wall

Inside Wall

Newer home designs with highly insulated walls and duct work in the conditioned space have used inside wall diffusers which throw the air towards the outside wall. The advantages of the inside wall duct system are that shorter, less costly duct runs are required and duct energy loss is minimal.

Type: High sidewall

Location: Interior

Diffuser discharge:

Horizontal Throw: To outside wall

Return location: Low central

Note: Ideally, ceiling-type systems should only be used where floors are not exposed to cold temperatures, such as second-story floors, or where highly insulated outside walls exist.

RETURN GRILLES

There is considerable debate, and almost an equal division, among HVAC contractors as to the ideal location for return grilles. Some argue that no system will operate properly unless a return register is located in each room along with a supply diffuser. While this may improve a system's design, it requires more attention to detail since a diffuser, a return, and the location of each must be determined. If a contractor is convinced to use this method, here again it is good practice to always locate the return grille opposite the supply diffuser in terms of vertical orientation.

However, many contractors have been extremely successful in providing comfortable systems using a central return. There are a number of advantages to the central return system. First and most important is that the return grille can also contain the filter. Anyone who has replaced a heating system that was 20 to 30 years old is familiar with the thick layer of dust that collects in an unfiltered, small return-type duct. This situation can be avoided by having the filter grille at the entrance of the return duct system.

Another advantage is that high/low combination returns can be employed. A schematic high/low (ceiling and floor) return system as shown in Figure 3.16 has the advantage of returning stagnant air from the ceiling in the summer and from the floor in the winter, which improves heat pump performance.

Referring to Tables 3.6 and 3.7, the improvements in heat pump performance are shown in Table 3.9. Improvements in both heat pump capacity and efficiency can be realized by returning the coolest air in the winter and the hottest air in the summer. From Table 3.9 the efficiency improvement from 75° to 80°F return air is an increase of the EER from 9.8 to 10.5 and the capacity from 29,800 to 32,400 Btu/hr.

Some means of blocking one or the other of the return grilles must be provided. Use of the double return air system has little effect on system performance as long as the air return path is not controlled. One method is to use motorized dampers on the high and low grilles.

A second method, where filter grilles are used, requires the homeowner to simply

3

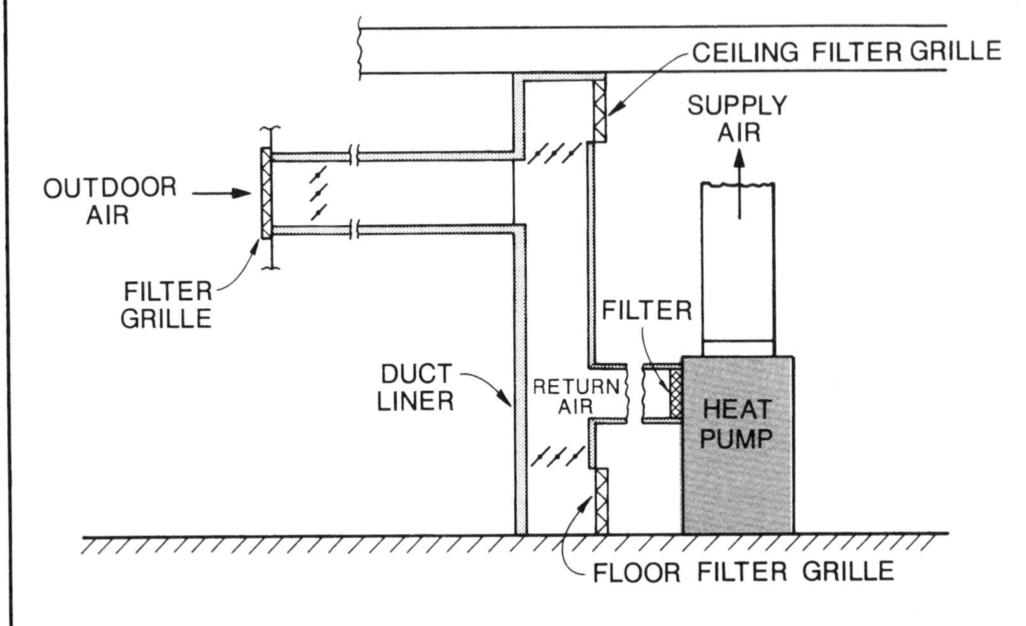

FIGURE 3.16: Stacked Return

TABLE 3.9: Performance Improvements Using a Stacked Return Air System (Model 33)

COOLING							
EWT °F	FLOW GPM	EA °F	TC	SC	KW	HR	EER
90	9	75/63 80/67 85/71	29.8 32.4 35.0	21.5 22.4 23.3	3.03 3.09 3.15	40.1 42.9 45.7	9.8 10.5 11.1
HEATING							
EW °F	FLOW GPM	EA °F	HC	KW	HE	COP	
30	9	60 70 80	26.3 25.4 24.5	2.34 2.46 2.58	18.4 17.0 15.6	3.03 3.02 2.77	

change the lower filter at the start of the heating season so that cool air is drawn from the floor and change the higher return filter so that warm air is drawn from the ceiling during the cooling season.

However, the designer must size the return grilles so that either grille can independently handle the total airflow requirements for that particular heat pump system. This translates into the following return grille schedule:

Heat pump size (tons)	Grille Sizes (inches)
2	20 x 20
2½	25 x 20
3	30 x 20 or
3	Two filters each 16 x 20

Generally, it is good practice to use standard-size filters to encourage the homeowner to purchase filters and change them regularly. If the filters are an unusual size and difficult to obtain, the homeowner may attempt to use filters that "almost" fit, and a good deal of unfiltered air will pass back to the unit, which will eventually block the coil. This can result in decreased efficiency and probably cause a compressor unit failure.

Where the central filter grille is used, the designer must be sure there is an air path from each individual room back to the filter grille. Typically, a one-inch undercut on doors will provide sufficient air return. However, if the homeowner carpets an uncarpeted floor, or replaces existing carpet with very thick pile, the air passage must be maintained by cutting off enough of the door to maintain the one-inch gap.

Also, the airflow path from a supply diffuser to the return grilles should not be through kitchens or bathrooms. State codes prohibit locating a return grille in either a kitchen or a bathroom area.

If an outside fresh air system is used, as shown in Figure 3.16, the homeowner must be made aware of the necessity of replacing this filter periodically. Even if special filters such as electronic air cleaners are used, it is still good practice to use filter return grilles. The return grille filter will remove larger airborne particles and allow the air cleaner to do the job it was installed for, that is, the removal of very small particles. This practice reduces the frequency of cleaning the electronic air cleaner while allowing the electronic air cleaner to operate at its highest possible efficiency.

With properly selected air supply diffusers that adequately mix the supply and room air, the need for dual returns is minimized. The additional cost of a dual return system must also be considered. A dual return should not be used to overcome poor diffuser selection.

Return grilles should be selected on the basis of the velocity in the grille (minimum area). For small grilles (12 inches by 6 inches approximate) a maximum velocity of 500 ft/min is recommended. As the grille increases in size to about 14 inches by 36 inches, the maximum velocity should be reduced to 300 ft/min.

The biggest problem with return grilles is placing them too close to the unit. This may result in a noisy, unsatisfactory system. The return duct should have at least two changes in direction between the grille and the unit. Also, duct liner or a section of flexible pipe should be used to reduce noise transfer from equipment to the space.

RETROFIT SYSTEMS

In cases where heat pumps are replacing a fossil fuel furnace (northern heating-only systems), it may be necessary to add additional duct work and diffusers to supply sufficient air to provide a comfortable environment. The air coming from a properly designed water-source heat pump system will be warmer than an air-source unit but cooler than a fossil furnace. With proper duct design, both the air-source and water-source systems can provide the necessary comfort.

The reader is referred to the following handbooks for more complete treatment of selection of the air distribution system and duct design:

ASHRAE Handbooks (American Society of Heating, Refrigerating, and Air-

3

Conditioning Engineers, Inc., 1791 Tullie Circle N.E., Atlanta, GA 30329)

1. *ASHRAE Handbook of Fundamentals, 1985.* Chapter 23, "Space Air Diffusion"
2. *ASHRAE Handbook of Fundamentals, 1983 Equipment,* Chapter 2, "Air-Diffusing Equipment"

ACCA Manuals (Air Conditioning Contractors of America, 1513 16th Street, N.W., Washington D.C. 20036)

1. ACCA Manual G, *Selection of Distribution System*
2. ACCA Manual D, *Duct Design for Residential Winter and Summer Air Conditioning and Equipment Selection*
3. ACCA Manual H, *Heat Pump Systems: Principles and Applications*
4. ACCA Manual E, *Room Air Distribution Considerations*

SMACNA (Sheet Metal and Air Conditioning Contractors National Association, Inc., 8224 Old Courthouse Road, Vienna, Virginia 22180)

1. *Installation Standards for One and Two Family Dwellings and Multifamily Housing*

Air Diffusers or Supply Registers

Location and selection of the air diffusers and air supply registers can greatly enhance overall system performance.

The importance of using high-quality diffusers and air supply registers rather than one-piece stamped-out registers cannot be overemphasized. At a minimum, both the throw and pressure required should be supplied by the manufacturer.

The selection procedure for supply air diffusers and registers is as follows:

1. The quantity (volume) of air to be handled is determined from the load calculations and equipment selected. That is, the air is divided in direct proportion to each room's load.
2. The diffuser type and location is different for each type of distribution system. Through-the-floor perimeter-type diffusers should be placed near the outer walls. Most small rooms can be adequately served by one diffuser with the exception of corner rooms with windows on each side, Figure 3.17a, and larger rooms like Figures 3.17b and 3.17c. The throw for these diffusers should be about equal to the ceiling height for conventional spreading-vane types.

Convertible diffusers (Table 3.10) are available so that a spreading jet of air can be used in the winter and a straight jet of air can be used in

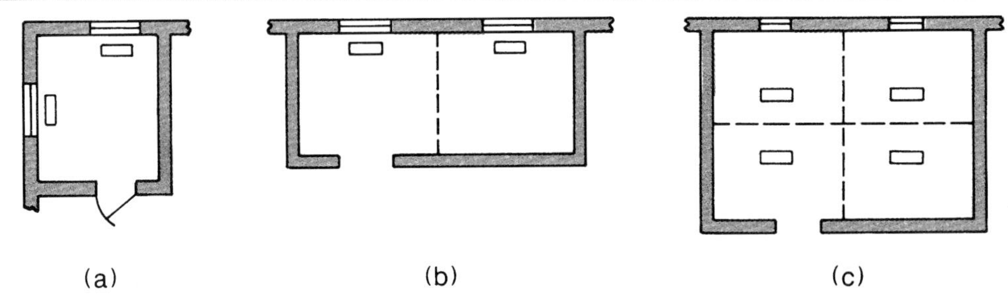

FIGURE 3.17: Diffuser Placement

the summer. These are very effective. For this type of diffuser, the throw should be about 1.2 times the ceiling height for the narrow jet (cooling).

Overhead stepdown diffusers are designed to project air horizontally in all directions as well as in specific directions. Their placement will depend on the type of diffuser selected

TABLE 3.10: Season-Change Floor Diffusers

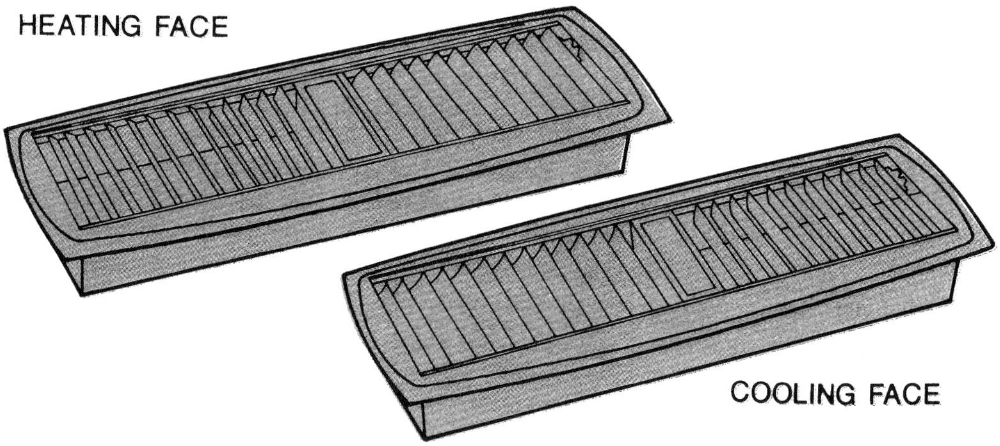

PERFORMANCE DATA — 4" NO. 242 SEASON CHANGER FLOOR DIFFUSERS

SIZE	10" x 4"							12" x 4"							14" x 4"							
Free Area	28.5 sq. in.							35.5 sq. in.							43.0 sq. in.							
	HEATING			COOLING				HEATING			COOLING				HEATING			COOLING				
CFM (Cubic Feet per Minute)	Pressure Loss (in. Water Gage)	Spread (Feet)	Throw (Feet)	Pressure Loss (in. Water Gage)	Spread (Feet)	Throw (Feet)	Avg. Face Velocity (Feet per Minute)	Pressure Loss (in. Water Gage)	Spread (Feet)	Throw (Feet)	Pressure Loss (in. Water Gage)	Spread (Feet)	Throw (Feet)	Avg. Face Velocity (Feet per Minute)	Pressure Loss (in. Water Gage)	Spread (Feet)	Throw (Feet)	Pressure Loss (in. Water Gage)	Spread (Feet)	Throw (Feet)	Avg. Face Velocity (Feet per Minute)
50	.006	10	6.5	.007	5	10	252	.003	9	6	.004	4	9.5	202	.003	8.5	5	.004	3.5	9	167
60	.007	10.5	6.5	.010	5	10	302	.005	9.5	6.5	.006	4	9.5	243	.004	9	5.5	.005	3.5	9	200
70	.012	10.5	7	.013	5	10.5	353	.007	10	7	.008	4.5	10	286	.005	9.5	6	.006	4	9.5	234
80	.013	10.5	7	.015	5.5	10.5	404	.009	10	7	.010	4.5	10	324	.007	9.5	6.5	.008	4	9.5	267
90	.016	11	7.5	.018	5.5	10.5	454	.011	10	7.5	.013	5	10	364	.009	9.5	7	.011	4.5	10	302
100	.020	11	7.5	.022	5.5	11	505	.014	10.5	7.5	.016	5	10.5	405	.012	10	7.5	.014	4.5	10	335
110	.024	11	8	.027	5.5	11	555	.017	10.5	8	.019	5	10.5	445	.014	10	7.5	.016	5	10	368
120	.027	11	8	.031	6	11.5	606	.020	11	8	.023	5.5	11	485	.016	10.5	8	.019	5	10.5	402
130	.032	11.5	8	.036	6	11.5	656	.023	11	8	.026	5.5	11	526	.019	10.5	8	.022	5.5	10.5	435
140	.037	11.5	8.5	.041	6	11.5	707	.026	11	8.5	.030	6	11	567	.021	11	8	.025	5.5	11	468
150	.043	11.5	8.5	.047	6.5	12	757	.031	11.5	8.5	.034	6	11.5	607	.024	11	8.5	.028	5.5	11	502
160	.047	11.5	9	.053	6.5	12	808	.034	11.5	9	.039	6.5	11.5	648	.028	11	8.5	.032	6.0	11	535

and the direction the air is thrown. Perimeter diffusers/registers are designed to throw air in multiple directions. The throw and selection procedures can be obtained from manufacturers' catalogs. Table 3.11 gives performance data for a three-way curved blade diffuser.

3. When selecting diffusers, care should be taken to limit the velocity in the diffuser to prevent unnecessary noise. Grille velocities will depend on the type of air distribution system and its location. Noise will not be a problem if the total velocity through the diffuser is limited to 400 to 600 feet per minute or less. When the diffusers are selected, the total pressure for each one should be recorded. This information is needed to size the duct system.

Duct Layout

When the air diffusers, grilles, and the unit are selected, the duct system can be laid out and sized. Layout of the ducts is an art that is perfected with experience.

Figure 3.18 shows three layouts for the same house, and others can be visualized. The layouts are for a through-the-floor or counterflow perimeter system, but the same general arrangement would apply for an overhead system. Keep in mind that the simple schematic does not show the vertical risers and some elbows, boots, etc.

Figure 3.18a is a preferred situation where both the heat pump and return are centrally located. Central distribution systems such as this are always the easiest to size and generally provide the most even air distribution throughout the home. However, the contractor should still not simply use one duct size for the entire job, but should reduce the trunk size to maintain the proper air velocity.

Duct systems 3.18a and 3.18b are more a function of the home's configuration. Figure 3.18a would be used where the home is long and narrow. If runout lengths do not exceed code requirements, this type of system is suitable. If the width of the house is such that a central supply trunk would not permit reasonable length runouts, then the duct could easily be run near the outer perimeter. Typically, this is done in houses that are close to square to keep the length of the branch ducts as short as possible.

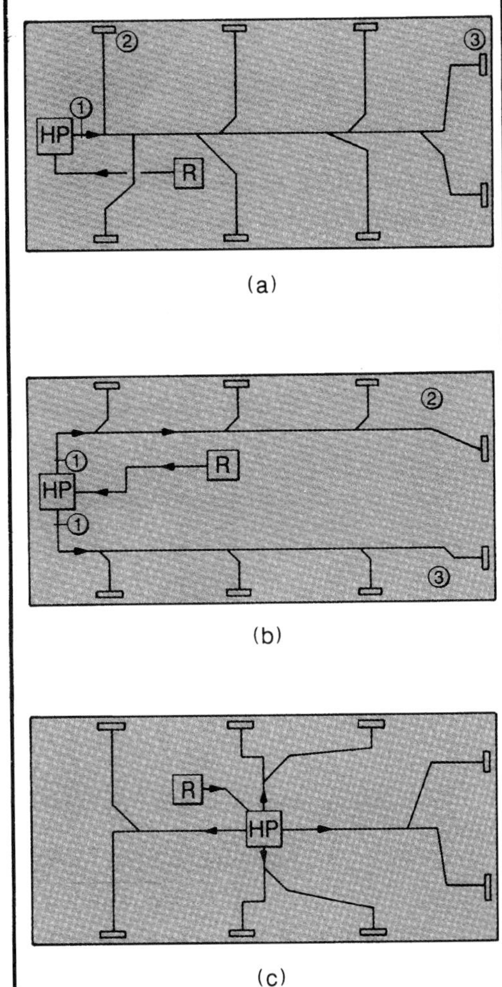

FIGURE 3.18: Alternate Duct Layout

TABLE 3.11: Register Performance

TOTAL CFM = Total Volume All Sides
CFM L/S = Volume Each Side L and S
THROW L/S = Throw in feet. Terminal Velocity 75 FPM
PL = Total Pressure Loss (In. W.G.)
VELOCITY = Jet Velocity Feet per Minute (Average)

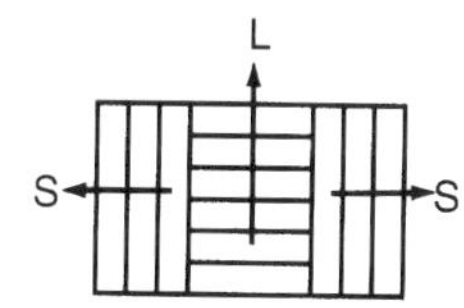

SIZE									
6x6	**TOTAL CFM**	**50**	**75**	**100**	**125**	**150**			
A_k .09	CFM L/S	32/9	47/14	63/18	79/23	95/28			
	THROW L/S	4.5/2.5	6.5/3.5	8.5/4.5	11/5.5	13/7			
	PL	.020	.046	.081	.129	.184			
	VELOCITY	570	860	1140	1440	1720			
8x6	**TOTAL CFM**	**50**	**75**	**100**	**125**	**150**			
A_k .10	CFM L/S	24/13	35/20	46/27	57/34	70/40			
	THROW L/S	3/2.5	4.5/3.5	6/4.5	7.5/6	9/7			
	PL	.015	.034	.061	.096	.140			
	VELOCITY	490	740	990	1240	1500			
10x6	**TOTAL CFM**	**75**	**100**	**125**	**150**	**175**	**200**	**250**	
A_k .15	CFM L/S	27/24	36/32	45/40	54/48	63/56	72/64	90/80	
	THROW L/S	3.5/3.5	5/4.5	6/5.5	7.5/7	8.5/8	10/9	12/11	
	PL	.017	.031	.048	.066	.087	.124	.184	
	VELOCITY	520	700	880	1030	1180	1410	1720	
8x8	**TOTAL CFM**	**75**	**100**	**125**	**150**	**175**	**200**	**250**	
A_k .16	CFM L/S	35/20	46/27	57/34	70/40	80/47	92/54	116/67	
	THROW L/S	4/3	5.5/4	6.5/5	8/6	9/7	11/8	13/10	
	PL	.014	.026	.040	.054	.078	.102	.162	
	VELOCITY	470	640	800	970	1120	1280	1610	
12x6	**TOTAL CFM**	**100**	**125**	**150**	**175**	**200**	**225**	**250**	**300**
A_k .18	CFM L/S	30/35	37/44	44/53	53/61	60/70	67/79	75/88	90/105
	THROW L/S	4/4.5	5/5.5	6/6.5	7/7.5	8/8.5	9/10	10/11	12/13
	PL	.020	.032	.046	.062	.085	.102	.127	.191
	VELOCITY	570	720	860	1000	1170	1280	1430	1750
14x6	**TOTAL CFM**	**100**	**125**	**150**	**175**	**200**	**225**	**250**	**300**
A_k .21	CFM L/S	26/37	32/47	38/56	45/65	52/74	57/84	64/93	76/112
	THROW L/S	3/3.5	3.5/4.5	4.5/5.5	5/6	6/7	6.5/8	7.5/9	9/11
	PL	.014	.023	.032	.044	.057	.073	.090	.129
	VELOCITY	480	600	720	840	960	1080	1200	1440

However, in all cases it is a good idea to locate at least one return near the center of the house.

As the capacity of the heat pump increases, the size of filter return grilles increases. In some cases it is more practical to have two filter grille locations within the home.

While large, complex duct systems require special efforts and detail in sizing, a typical residential system can be sized quite adequately using a few simple modern tools and practical rules of thumb.

One of the handiest tools for the residential duct designer is a device called a "duct-ulator." This is a slide-rule-type device that is available from many manufacturers of HVAC equipment. It allows the user to select duct sizes according to specific airflows and velocities. The designer must make judgments when to downsize a duct to remain within the predetermined velocity and flow rates.

It is always preferred to perform individual room calculations and to determine more precisely the amount of air required in each room. Duct and diffuser sizing can then be done accurately.

Duct Material

In the past, most contractors used metal ducts that were either insulated on the exterior or interior with appropriate weights of duct insulation. Interior duct insulation has the advantage of reducing the noise within a duct system. It is highly recommended that duct liner be used on the first few feet of supply and return systems.

However, interior duct liner often does not have a sufficient thermal resistance to ward off condensation that can be a problem when a duct is located in certain areas (such as crawl spaces of a residential installation). Exterior duct wrap, when properly applied, which includes taping all joints, can provide excellent thermal insulation. The best type of exterior wrap is foil faced. Inexperienced installers tend to stretch plastic-faced wrap tightly over corners and around pipes which greatly reduces insulating value. This is a practice that is important to avoid because stretched two-inch insulation can provide less insulating value than a properly installed half-inch wrap.

Duct board has become a very popular material. It can either be cut in the shop, transported flat and assembled in the field, or simply cut on the job with easy-to-use hand tools. Installers should be cautious of using duct board where it is subject to crushing or abuse. The use of heat-type tapes is highly recommended over the pressure-type material. Another duct board advantage is that it makes for an extremely quiet system. It absorbs any sound being transmitted down a duct that would normally be transmitted down unlined metal duct work.

Designing and Sizing a Duct System

The following is a proven method for designing and sizing a duct system.

It is highly recommended over rule-of-thumb methods.

Some important observations can be made from Figure 3.18.

1. The loss of total pressure from the heat pump discharge through each run of duct into the room must be the same. That is, the change in total pressure (dP_0) from point 1 to point 2 must equal dP_0 from point 1 to point 3 in Figure 3.18a, and the same is true for all the other runs. In Figure 3.18b, dP_0 from 1 to 2 must equal dP_0 from 1 to 3, and in Figure 3.18c, dP_0 for all four branches must be equal. This is true regardless of the duct sizes.
2. The objective is to size each duct so the proper amount of air flows through each diffuser with the total pressures equal as discussed above.
3. The sum of the airflow rates for all diffusers must match the heat pump flow rate, and the change in total static pressure must match the heat pump's total static pressure capability.

There are two basic methods to size the ducts for layouts similar to those shown in Figure 3.18. They are:

1. equal friction method
2. balanced capacity method

These are discussed in detail in McQuiston and Parker, *Heating, Ventilation and Air Conditioning* (Wiley, 1977), and the *ASHRAE Handbook of Fundamentals, 1985*.

The **equal friction method** is the most widely used. It is the method used in ACCA *Manual J*. The commonly used "ductulator" uses this method.

The **balanced capacity method** is accurate because it theoretically produces a balanced system whereas the equal friction method does not. To carry out either procedure for sizing ducts, the following are required:

1. a layout of the duct system showing all lengths, fittings, diffusers, and grilles
2. fan (blower) performance data such as Table 3.12
3. a friction chart such as Figure 3.19 for the type of duct being used
4. equivalent length data for the duct fittings, such as Figure 3.20
5. pressure loss data for the diffusers and grilles, which should be recorded when these devices are selected

The following example briefly describes both methods and compares the results.

TABLE 3.12: Heat Pump Blower Performance Data

Unit Size	Fan Speed	External Static Pressure — Inches of Water									
		0.10	0.15	0.20	0.25	0.30	0.35	0.40	0.50	0.60	0.70
009	Low	320	310	300	290	275					
	High	380	370	360	350	335					
012	Low	350	335	320	305	290					
	High	400	385	365	350	335					
019	Low	560	550	540	530	520					
	High	665	650	535	620	605					
025	Low	845	825	805	780	750					
	High	895	875	850	825	795	765				
033	Low	1,010	980	960	930	890	850				
	High	1,080	1,050	1,010	980	940	900	870			
041	Low	1,215	1,210	1,205	1,200	1,190	1,180	1,170			
	High	1,395	1,385	1,375	1,365	1,350	1,335	1,310	1,260		
049	Low	1,320	1,315	1,310	1,300	1,290	1,275	1,260	1,220		
	High	1,650	1,625	1,600	1,575	1,545	1,510	1,475	1,425	1,340	
059	Low	1,765	1,750	1,735	1,720	1,700	1,680	1,655	1,600	1,550	
	High	1,960	1,930	1,905	1,880	1,855	1,825	1,790	1,730	1,670	1,600

Includes allowances for wet coil and for filter.

Source: WaterFurnace International.[8]

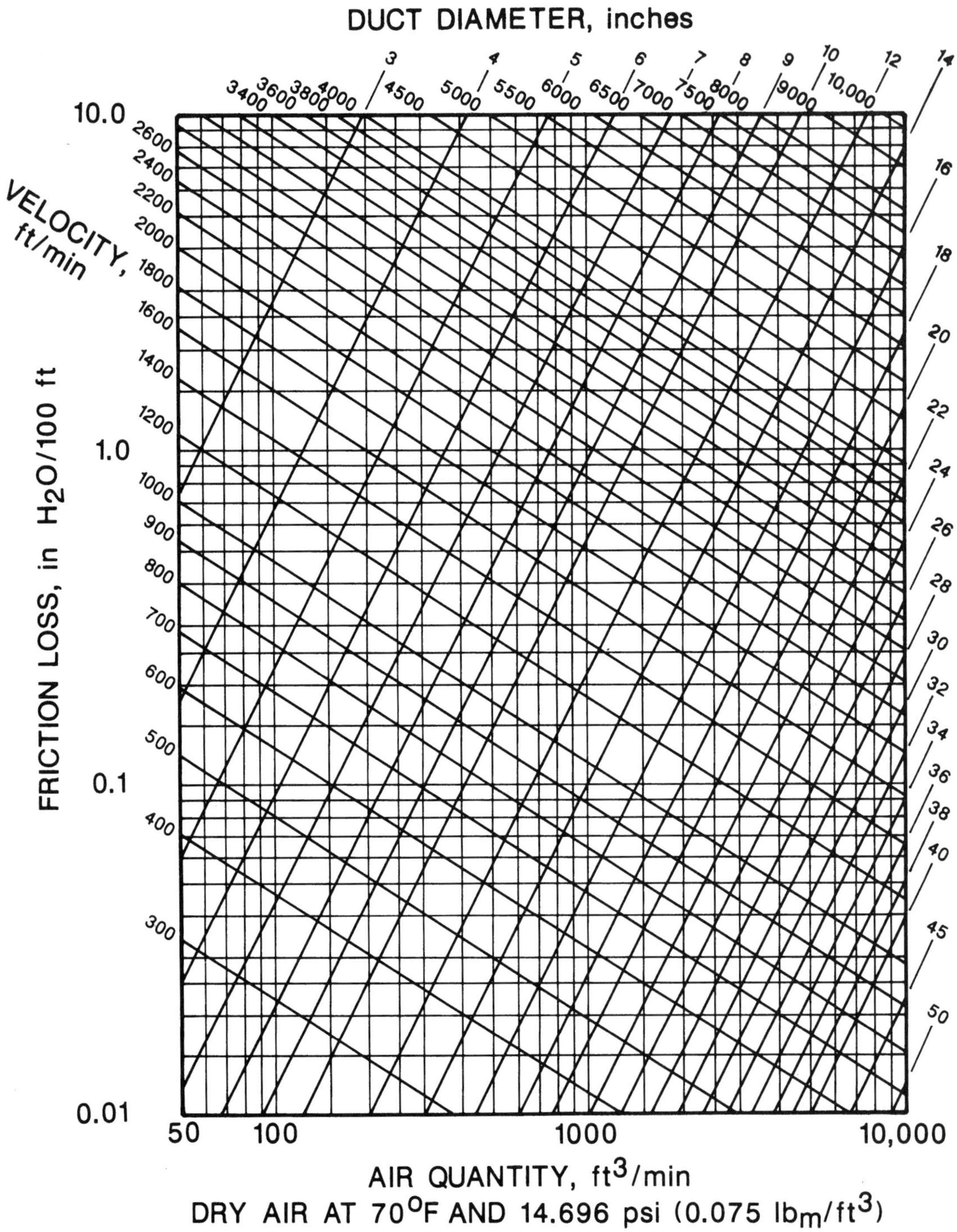

Reprinted by permission from *ASHRAE Handbook of Fundamentals, 1977.*

FIGURE 3.19: Friction Loss in Galvanized Steel Ducts

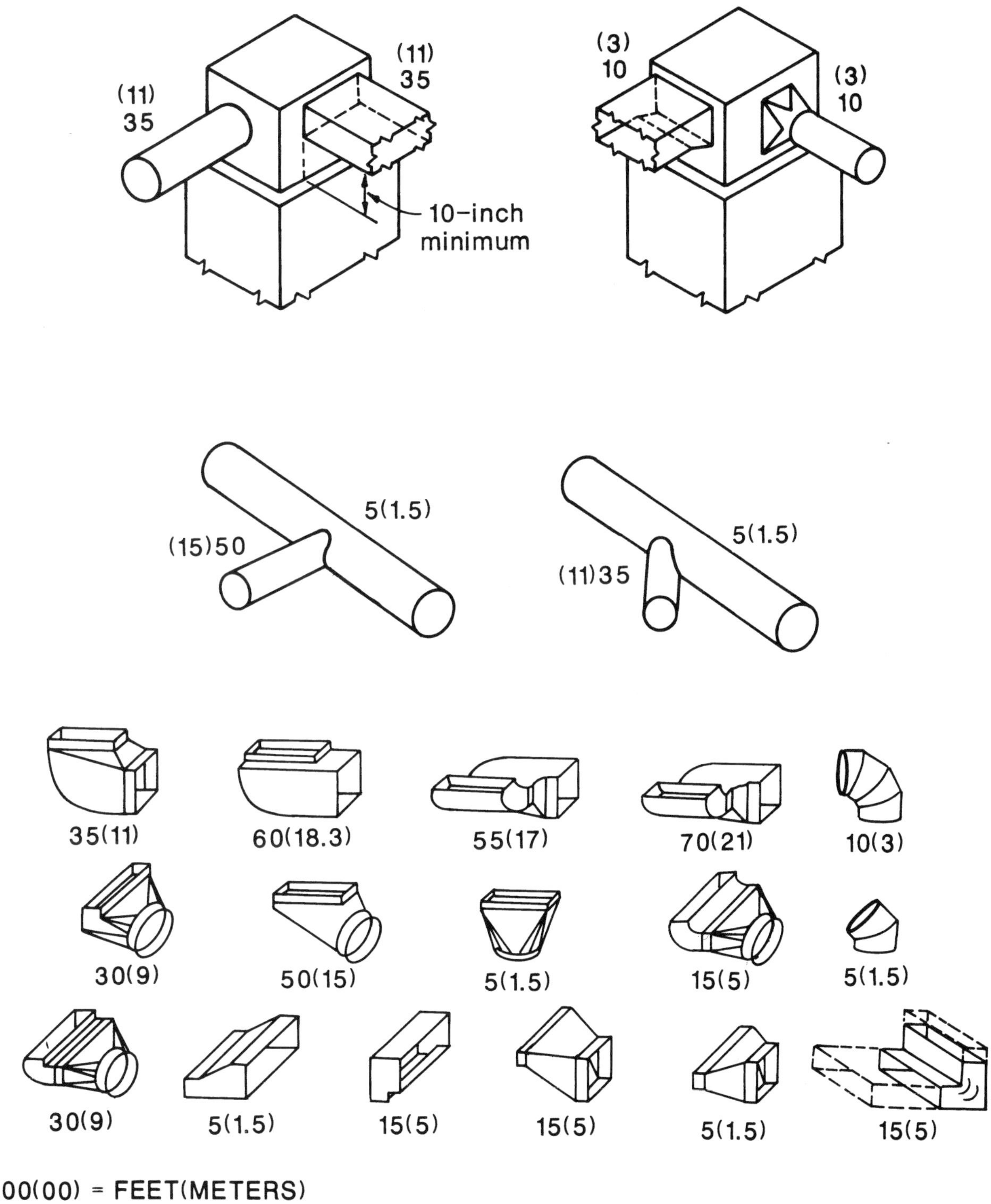

FIGURE 3.20: Equivalent Lengths — Duct Fittings

EXAMPLE: EQUAL FRICTION AND BALANCED CAPACITY METHODS

The following example demonstrates the two methods of duct sizing. Consider the simple layout shown in Figure 3.21 for a heat pump with an overhead distribution system. Note the data for cfm, duct length, and fitting equivalent lengths in Figure 3.21. In an earlier example of selecting a heat pump a nominal airflow rate of 1000 cfm resulted. It is assumed here that this duct system is for that heat pump. Table 3.12 shows the Model 033 unit to produce about 0.20 inches of water at 1010 cfm and 0.25 inches of water at 980 cfm. By interpolation, the pressure will be about 0.215 inches of water at 1000 cfm. The external static pressure is the difference in static pressure between points a and b in Figure 3.21. The air velocities are about equal at a and b; therefore, the external static pressure is equal to the total pressure (dP_0). This total pressure is all that is available for the complete duct system. That is, when the system is running, the combined loss in total pressure for the supply and return ducts must be 0.215 inches of water with a flow rate of 1000 cfm. We see from Figure 3.21 and Table 3.13 that the total equivalent length of the supply and return ducts is 215 feet. This assumes that run 1-2-3-4 is the longest of all the supply duct runs. The return grille on 8 and the diffuser on 4 require total pressure of about 0.015 and 0.03 inches of water, respectively. Therefore, the total pressure remaining for the ducts is:

dP_{0d} = 0.215 − 0.015 − 0.02 = 0.180 inches of water.

Then the lost pressure per unit length of 100 feet is:

dP_0/L = (0.018/215)100 = 0.084 inches of water/100 ft.

Figure 3.19 can then be used to find the duct sizes.

The equal friction and balanced capacity methods are the same to this point for sizing the main run (1-2-3-4) and return (8). After that the methods differ somewhat.

Equal Friction Method

Table 3.14 summarizes the equal friction duct sizing results. The diameter of section no. 1 is determined from Figure 3.19 at dP_0/L = 0.084 and cfm = 1000 as approximately 14.5 inches. If the ducts are to be of a round cross-section, this must be rounded to either 14 or 16 inches. However, a rectangular duct of equivalent capacity and lost pressure may be used.

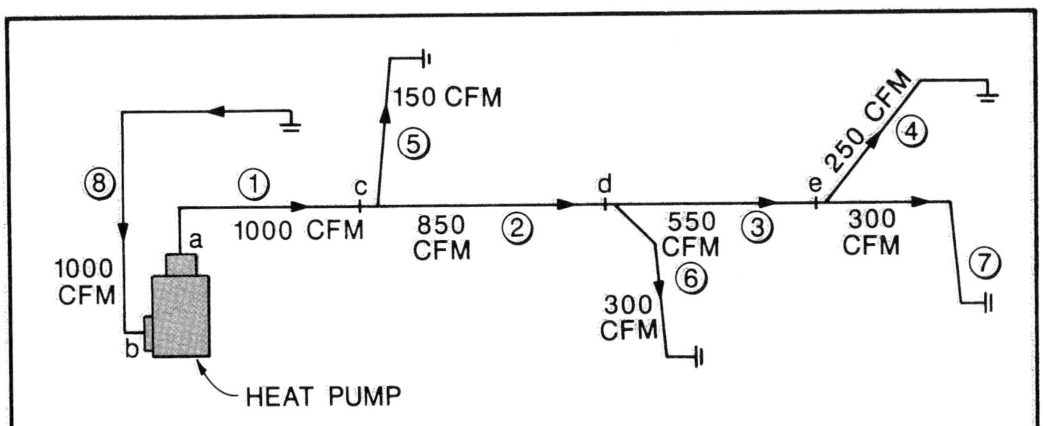

FIGURE 3.21: Duct Layout

TABLE 3.13: Summary Table for Duct Layout

DUCT Sec. No.	CFM	L (ft)	L_E (ft)	L_T (ft)
1	1,000	15	45	60
2	850	10	5	15
3	550	10	5	15
4	250	15	50	65
			TOTAL FOR 1-2-3-4 =	155 (ft)
5	150	17	60	77
6	300	10	50	60
7	300	6	20	26
8	1,000	20	40	60
			TOTAL FOR RUN 8 =	60 (ft)
			GRAND TOTAL =	215 (ft)

TABLE 3.14: Equal Friction Duct Sizing

1	2	3	4	5	6	7
DUCT SEC. NO.	CFM	L_T FEET	dP_0/L (a)	dP_{01} IN WA	DIA INCHES	SIZE WxH IN.
1	1,000	60	0.084	0.050	14.5	15 x 12
2	850	15	0.084	0.013	13.4	15 x 10
3	550	15	0.084	0.013	11.6	12
4	250	65	0.084	0.042	8.7	9
5	150	77	0.084	0.065	7.0	7
6	300	60	0.084	0.050	9.3	9
7	300	26	0.084	0.022	9.3	9
8	1,000	60	0.084	0.050	14.5	15 x 12

Note: (a) Inches of water / 100 feet

Table 3.15 has been developed for this purpose. A 15 x 12-inch duct is one size that would do. The lost pressure for each duct section is shown in column 5.

$$dP_{01} = (d_{P0}/L) \times (L_t/100)$$

then

$$dP_{01} = 0.084 \times (60/100) = 0.050 \text{ inches water}$$

Duct sections 2, 3, 4, and 8 are calculated in a similar way.

The equal friction method simply sizes all the remaining branch ducts using the same dP_0/L used for the main duct runs (1-2-3-4-8). These are shown for sections 5, 6, and 7 in Table 3.14. It is now of interest to compare the lost pressure (dP_0) for each duct run:

Run 1-2-3-4 dP_0 = 0.138 inches of water

Run 1-2-3-7 dP_0 = 0.118 inches of water

Run 1-2-6 dP_0 = 0.133 inches of water

4

Steps in Ground Heat Exchanger Design

The design procedure comprises the following steps:

1. Select a ground heat exchanger configuration:
   a. horizontal or vertical
   b. parallel or series flow arrangement
2. Select plastic pipe, considering:
   a. material (polyethylene and polybutylene)
   b. size (SDR or SCH)
   c. diameter
   d. length
   e. circulating fluid pressure loss
3. Estimate the ground heat exchanger length.
4. Select circulating pump(s).

Ground Heat Exchanger Configuration

Horizontal or Vertical — The choice of whether the system is horizontal or vertical depends on available land, local soil type, and excavation costs. If large land areas without hard rock are available, then the economics of a horizontal system should be studied. Land area requirements have been reduced with the development of multiple-pipe-laying machines.

Vertical systems are used where land area is limited. In many cases, it is the only choice. If hard rock exists, rotary drilling rigs using rock bits have been successful.

The selection of horizontal or vertical can only be determined after a cost study has been made. The true evaluation should be based on a bid by a reliable, experienced contractor.

Parallel or Series Flow — The ground heat exchanger may be either series (Figures 4.2, 4.3, and 4.6) or parallel (Figures 4.4 and 4.5). In the series system there is only one fluid path, whereas in the parallel system the fluid can take two or more paths somewhere in the circuit.

The selection will depend on the installed cost. Series-type systems take larger diameter and more expensive pipe than do parallel systems. For heat pumps with reduced flow requirements (2 GPM/ton and less), the series system should be considered to maintain the fluid velocities in the ground heat exchanger piping necessary to promote good heat transfer.

The principal advantage of the series system is that the fluid path is well defined and trapped air is easily removed by power flushing. There are a number of disadvantages. First, larger diameter pipe is more expensive and requires more antifreeze. Second, the larger pipe is heavier, which increases labor costs. In summary, series systems have the following advantages and disadvantages:

SERIES SYSTEM ADVANTAGES:

1. single flow path and pipe size
2. slightly higher thermal performance per linear foot of pipe since larger diameter pipe is required in the series system

SERIES SYSTEM DISADVANTAGES:

1. larger fluid volume of larger pipe in a series requiring greater antifreeze volumes
2. higher pipe price per unit of performance
3. increased installed labor cost
4. limited capacity (length) due to fluid pressure drop characteristics

Parallel systems are fabricated from smaller diameter pipe which is generally less costly than larger diameter pipe. The disadvantage is that with parallel flow paths special care must be taken in the design and fabrication to ensure all air can be removed with high-velocity water flushing. In addition, each parallel path must be the same length (within 10%) so each loop has the same flow. To ensure that each loop has the same pressure at in diameter. Note that the equal friction method gave a 9-inch duct. Ducts 5 and 6 may be sized the same way:

$dP_{06} = 0.013 + 0.042 = 0.055$ inches of water

$(dP_0/L)_6 = (0.055/60)100 = 0.09$ inches of water/100 ft

$D_6 = 9$ inches

$dP_{05} = 0.013 + 0.013 + 0.042 = 0.068$ inches of water

$(dP_0/L)_5 = (0.068/77)100 = 0.088$ inches of water/100 ft

$D_5 = 7$ inches

If the lost pressure for each run is again checked, the agreement will be much better — perfect in fact:

Run 1-2-3-4 $dP_0 =$ 0.138 inches of water

Run 1-2-3-7 $dP_0 =$ 0.138 inches of water

Run 1-2-6 $dP_0 =$ 0.138 inches of water

Run 1-5 $dP_0 =$ 0.138 inches of water

The system will now balance perfectly, assuming the ability to build it as designed. The system will perform well. The use of a personal computer and/or a worksheet and the balanced capacity method are strongly recommended for most residential and light commercial systems using unit-type equipment.

4 Designing the Ground Heat Exchanger

In This Section: Steps in ground heat exchanger design; ground heat exchanger configuration; plastic pipe selection; ground heat exchanger sizing and design procedure; selecting the circulating pump

The design of ground heat exchangers consists of selecting a ground heat exchanger type and configuration that will result in highest performance for the least cost. Since there is no one type or configuration that is "best," the designer must have some way of choosing between alternatives. Figure 4.1 shows different types of installed vertical and horizontal ground heat exchangers. Figures 4.2, 4.3, 4.4, 4.5, and 4.6 are examples of ground heat exchangers commonly installed.

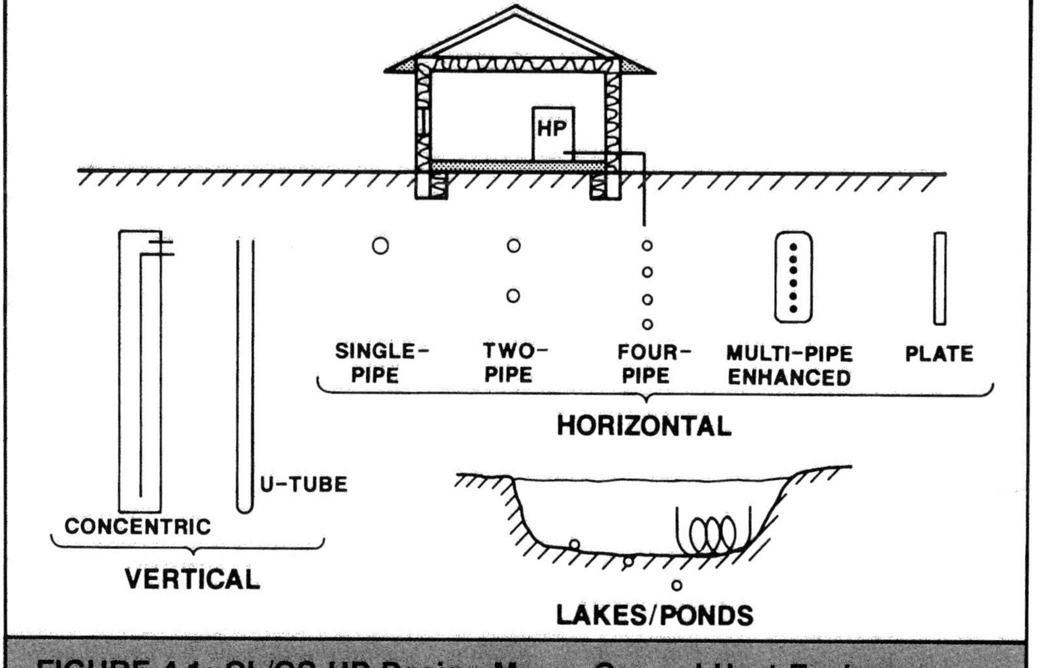

FIGURE 4.1: CL/GS HP Design Menu: Ground Heat Exchanger Options

4

Steps in Ground Heat Exchanger Design

The design procedure comprises the following steps:

1. Select a ground heat exchanger configuration:
   a. horizontal or vertical
   b. parallel or series flow arrangement
2. Select plastic pipe, considering:
   a. material (polyethylene and polybutylene)
   b. size (SDR or SCH)
   c. diameter
   d. length
   e. circulating fluid pressure loss
3. Estimate the ground heat exchanger length.
4. Select circulating pump(s).

Ground Heat Exchanger Configuration

Horizontal or Vertical — The choice of whether the system is horizontal or vertical depends on available land, local soil type, and excavation costs. If large land areas without hard rock are available, then the economics of a horizontal system should be studied. Land area requirements have been reduced with the development of multiple-pipe-laying machines.

Vertical systems are used where land area is limited. In many cases, it is the only choice. If hard rock exists, rotary drilling rigs using rock bits have been successful.

The selection of horizontal or vertical can only be determined after a cost study has been made. The true evaluation should be based on a bid by a reliable, experienced contractor.

Parallel or Series Flow — The ground heat exchanger may be either series (Figures 4.2, 4.3, and 4.6) or parallel (Figures 4.4 and 4.5). In the series system there is only one fluid path, whereas in the parallel system the fluid can take two or more paths somewhere in the circuit.

The selection will depend on the installed cost. Series-type systems take larger diameter and more expensive pipe than do parallel systems. For heat pumps with reduced flow requirements (2 GPM/ton and less), the series system should be considered to maintain the fluid velocities in the ground heat exchanger piping necessary to promote good heat transfer.

The principal advantage of the series system is that the fluid path is well defined and trapped air is easily removed by power flushing. There are a number of disadvantages. First, larger diameter pipe is more expensive and requires more antifreeze. Second, the larger pipe is heavier, which increases labor costs. In summary, series systems have the following advantages and disadvantages:

SERIES SYSTEM ADVANTAGES:

1. single flow path and pipe size
2. slightly higher thermal performance per linear foot of pipe since larger diameter pipe is required in the series system

SERIES SYSTEM DISADVANTAGES:

1. larger fluid volume of larger pipe in a series requiring greater antifreeze volumes
2. higher pipe price per unit of performance
3. increased installed labor cost
4. limited capacity (length) due to fluid pressure drop characteristics

Parallel systems are fabricated from smaller diameter pipe which is generally less costly than larger diameter pipe. The disadvantage is that with parallel flow paths special care must be taken in the design and fabrication to ensure all air can be removed with high-velocity water flushing. In addition, each parallel path must be the same length (within 10%) so each loop has the same flow. To ensure that each loop has the same pressure at

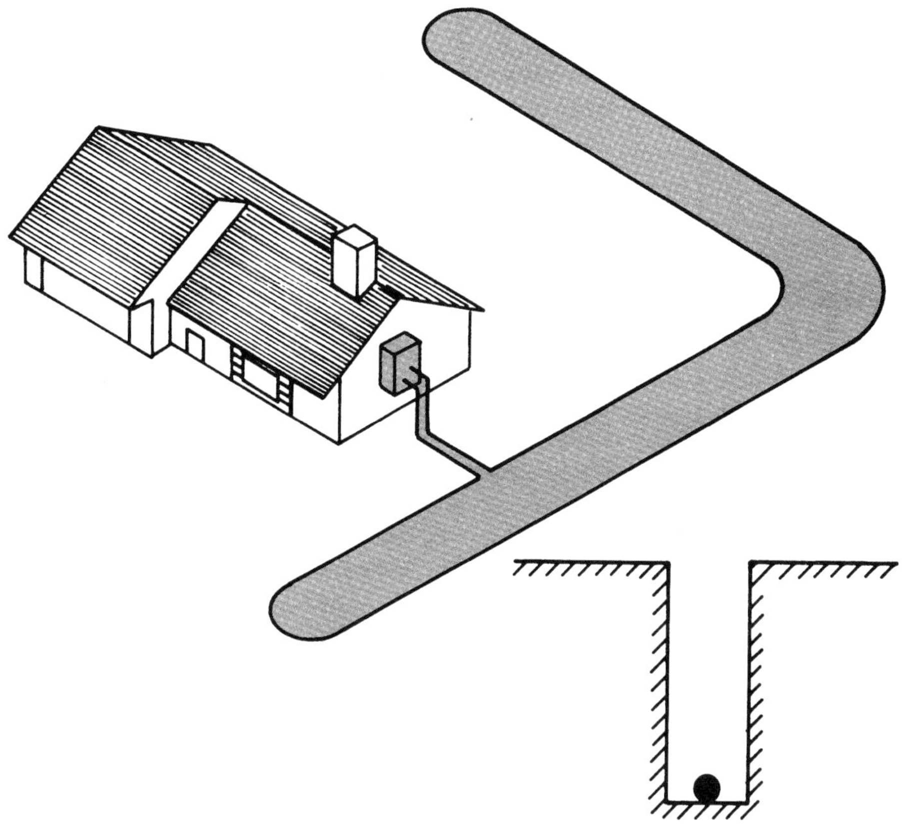

Earth Coil Type:	Horizontal – Single Layer
Water Flow:	Series
Typical Pipe Size:	1 1/2 to 2 inches
Nominal Length:	350 to 500 feet/ton
Burial Depth:	4 to 6 feet

FIGURE 4.2: Single-Pipe Horizontal Ground Heat Exchanger

4

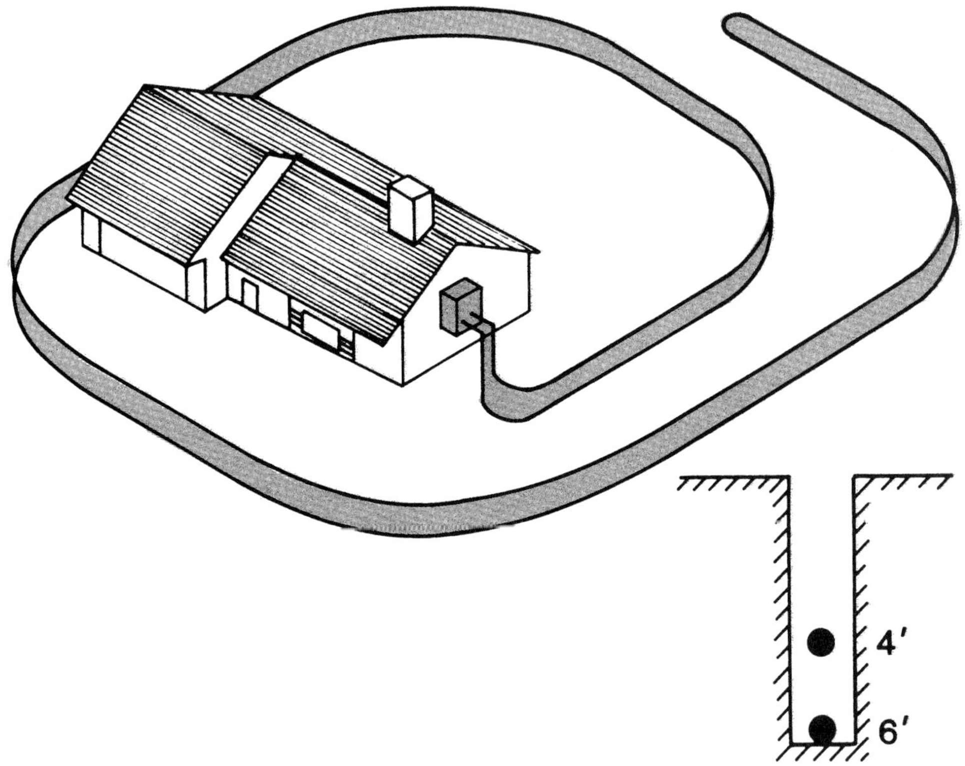

Earth Coil Type:	Horizontal - Two-Layer
Water Flow:	Series
Typical Pipe Size:	1 1/2 to 2 inches
Practical Length:	210 to 300 feet of trench/ton 420 to 600 feet of pipe/ton
Burial Depth:	4 feet and 6 feet

FIGURE 4.3: Two-Pipe Horizontal Ground Heat Exchanger

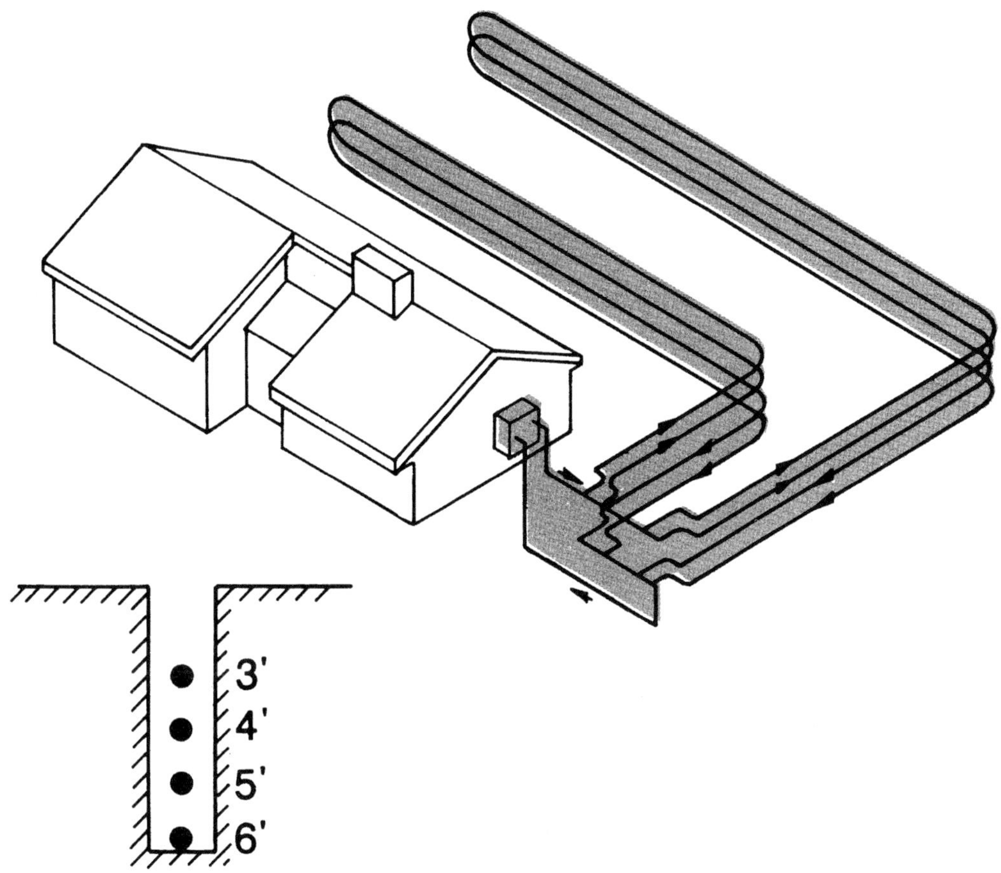

Earth Coil Type:	Horizontal – Four-Layer
Water Flow:	Parallel
Typical Pipe Size:	Parallel loops 3/4 to 1 inch; headers 1 1/2 to 2 inches
Burial Depth:	6 feet, 12-inch spacing

FIGURE 4.4: Four-Pipe Horizontal Ground Heat Exchanger

4

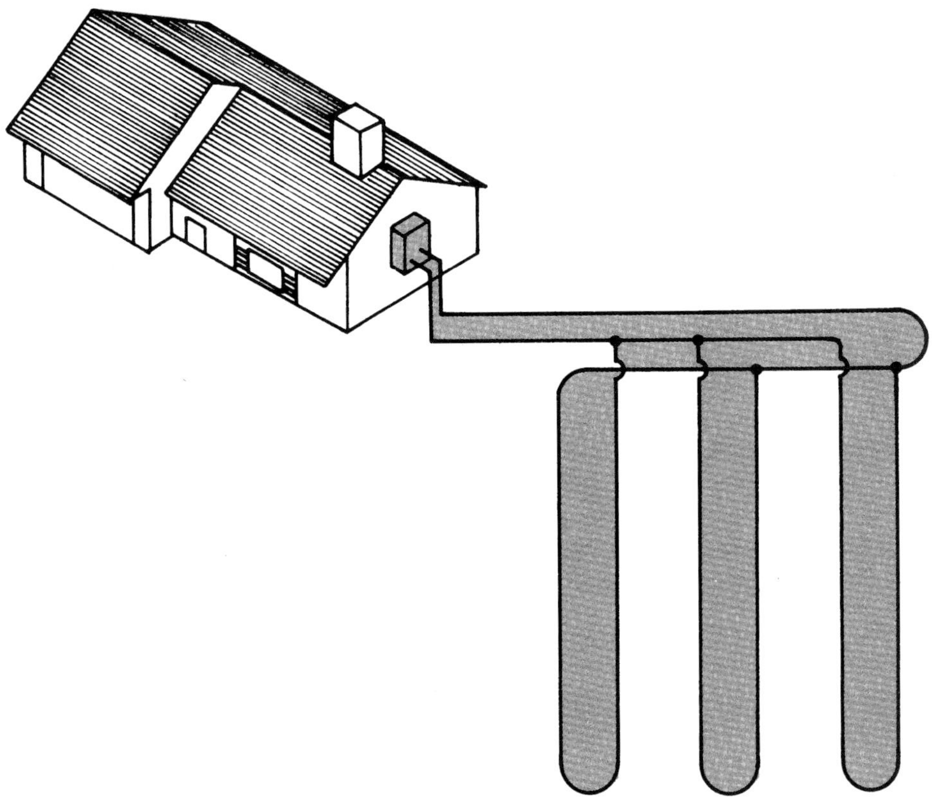

Earth Coil Type:	Vertical – Single U-Bend
Water Flow:	Parallel
Pipe Sizes:	3/4- or 1-inch loops, 1 1/2- or 2-inch headers
Bore Lengths:	175 to 225 feet/ton
Pipe Lengths:	350 to 450 feet/ton

FIGURE 4.5: Parallel Vertical Ground Heat Exchanger

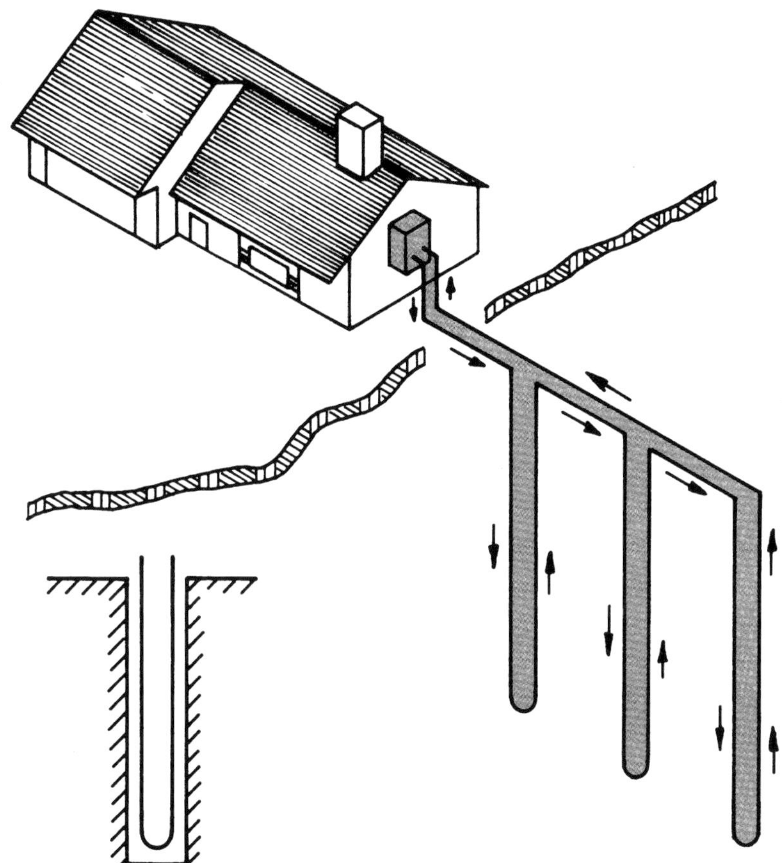

Earth Coil Type:	Vertical – Single U-Bend
Water Flow:	Series
Pipe Sizes:	1, 1 1/2, & 2 inches
Bore Length:	165 to 200 feet/ton
Pipe Length:	230 to 400 feet/ton

FIGURE 4.6: Series Vertical Ground Heat Exchanger

the inlet and outlet, large-diameter headers are used.

In summary, parallel systems have the following advantages and disadvantages:

PARALLEL SYSTEM ADVANTAGES:

1. lower cost pipe because smaller diameter
2. less antifreeze required
3. lower installation labor cost

PARALLEL SYSTEM DISADVANTAGES:

1. special attention to assure air removal
2. flow balancing between each parallel path to result in equal length loops

Figures 4.7 and 4.8 describe some basic elements of the horizontal and vertical ground heat exchangers in series and parallel flow arrangements.

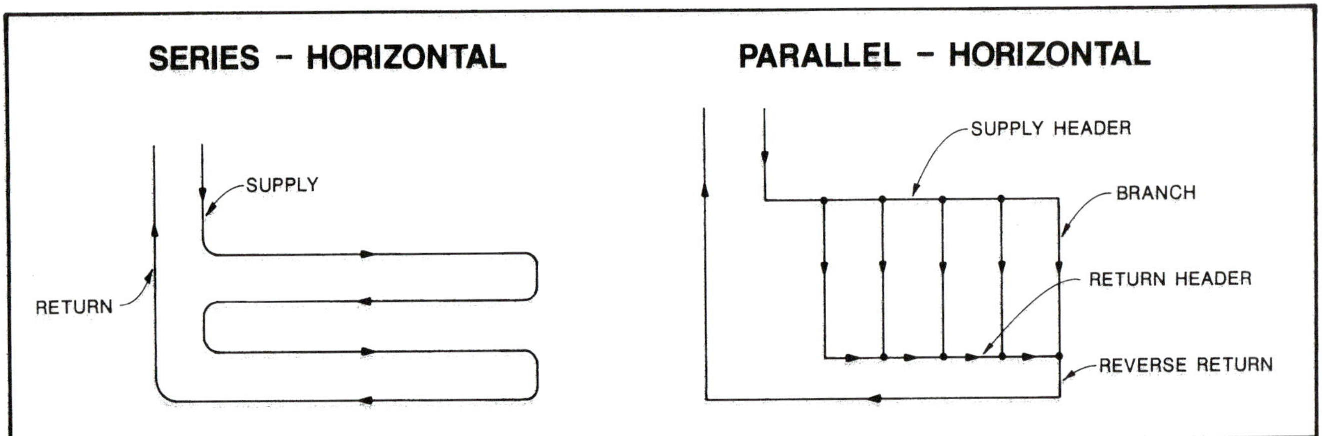

FIGURE 4.7: Series and Parallel Horizontal Systems

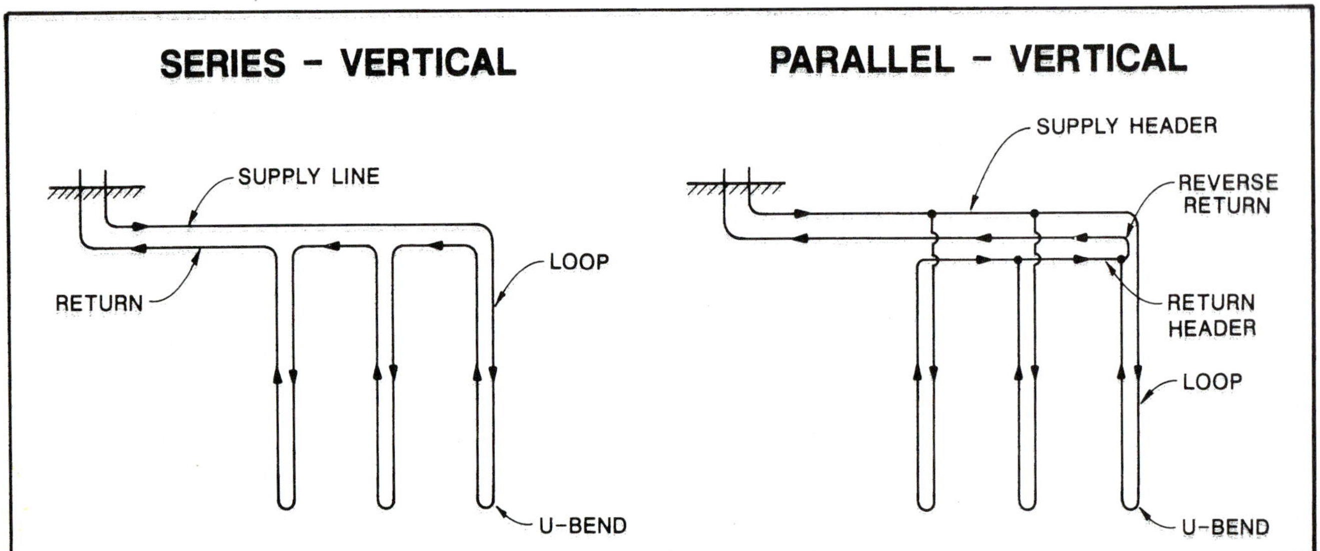

FIGURE 4.8: Series and Parallel Vertical Systems

4

The major components in these heat exchangers are:

1. *SUPPLY AND RETURN HEADERS* — Supply and return headers are the fluid supply and return lines from the heat pump to the parallel loops of a ground heat exchanger system. They carry the total system flow of the heat pump system. Headers are constructed from large-diameter pipe to minimize fluid flow pressure drop down the length of the pipe run.
2. *LOOPS* — Pipes extending from the supply header down a bore hole or trench, returned in the same trench or bore hole, and joined to the return header.
3. *REVERSE RETURN* — A piping arrangement that allows each loop of a parallel system to have the same inlet and outlet pressure. It is used to cancel the effects of pressure loss along the header lines.
4. *U-BEND* — A 180° fitting in a loop-type heat exchanger used at the bottom of a bore hole or at the end of a trench to return the fluid.

Plastic Pipe Selection

Plastic Pipe Material — Polyethylene and polybutylene are the most common pipe materials found in ground heat exchangers. These pipe materials are flexible and can be heat fused to form joints stronger than the pipe itself.

PVC is not recommended for the buried portion of the ground heat exchanger.

Plastic Pipe Size — Table 4.1 gives the most common pipe sizes used in ground heat exchanger designs. Wall thickness and consequently pipe strength is defined in terms of Schedule Rating (SCH) or a Size Dimension Ratio (SDR). For ground heat exchangers the following two polybutylene pipes are recommended:

1. polybutylene PB 2110, SDR 13.5
2. polybutylene PB 2110, SDR 17

The recommended polyethylene pipes are:

1. polyethylene PE 3408, SDR 11
2. polyethylene PE 3408, SCH 40

For the polyethylene pipes, the designer should select the pipe with the thinnest wall for loops. Headers should be constructed with the higher wall thickness pipe in Table 4.1 to provide structural strength.

Based on local material and labor costs, the increased thermal performance and increased cost of larger diameter pipe is seldom justified over less expensive, smaller diameter pipe. Consequently, many systems are fabricated in a parallel flow configuration which allows the use of small-diameter pipe (¾- to 1-inch) loops connected to a common header.

Plastic Pipe Diameter — There are two engineering guidelines that must be followed before a particular pipe diameter can be selected. The pipe must:

1. be large enough to keep the pumping power small
2. be small enough to cause turbulence inside the pipe to help support good heat transfer between the circulating fluid and the inside pipe wall

As a result of these two requirements, pipe diameter will be selected based on a compromise between fluid pressure drop and thermal performance. The designer then selects among the available pipe sizes, and selects a pipe material and pipe size that results in the lowest installation cost.

Table 4.2 gives the minimum fluid flow in ground heat exchanger pipes to ensure turbulence (Reynolds number greater than 2500). Data is presented for a number of common pipe sizes and circulating fluids. The designer must check to make sure that the above conditions are met in

4

TABLE 4.1: Plastic Pipe Sizes and Pipe Thermal Resistance Values

No.	Description	Nominal Size	Outside Diameter (OD)	Inside Diameter (ID)	Pipe Resistance R_S	Single Loop N = 2 R_{PE}	Single Loop N = 2 D_{OE}	Double Loop N = 4 R_{PE}	Double Loop N = 4 D_{OE}
Polyethylene									
1	SDR-11	¾	1.050	0.860	0.141	0.096	1.48	0.067	2.10
2		1	1.315	1.077	0.141	0.096	1.86	0.067	2.63
3		1¼	1.660	1.358	0.141	0.096	2.35	0.067	3.32
4		1½	1.900	1.554	0.141	0.097	2.69	0.067	3.80
5		2	2.375	1.943	0.141	0.097	3.36	0.067	4.75
6	SCH 40	¾	1.050	0.824	0.170	0.116	1.48	0.080	2.10
7		1	1.315	1.049	0.159	0.109	1.86	0.075	2.63
8		1¼	1.660	1.380	0.130	0.089	2.35	0.062	3.32
9		1½	1.900	1.610	0.117	0.080	2.69	0.056	3.80
10		2	2.375	2.067	0.098	0.068	3.36	0.047	4.75
Polybutylene									
11	SDR-17	1½	1.900	1.676	0.16	0.110	2.69	0.077	3.80
12	IPS	2	2.375	2.095	0.16	0.110	3.36	0.077	4.75
13	SDR-13.5	1	1.125	0.957	0.20	0.140	1.59	0.098	2.25
14	CTS	1¼	1.375	1.171	0.20	0.141	1.94	0.098	2.75
15		1½	1.625	1.385	0.20	0.140	2.30	0.098	3.25
16		2	2.125	1.811	0.20	0.140	3.00	0.098	4.25

All dimensions are in inches
Pipe resistance, R_{PE}: hr-ft-°F/Btu
Number of pipes, N

all sections of the ground heat exchanger. The Reynolds number is defined in Table 4.3.

Table 4.4 gives the fluid properties used to calculate the above minimum flow requirements for turbulence. Appendix A gives complete fluid properties for these and other antifreeze solutions. Table 4.3 gives the equations used to calculate the values given in Table 4.2.

REMEMBER: Large-diameter pipe is more expensive, takes more antifreeze (if required), and is more difficult to handle and install than smaller-diameter pipe.

A majority of ground heat exchangers are constructed from ¾-, 1-, 1¼-, 1½-, and 2-inch pipe. Pipe loops used in parallel configurations will be constructed from the smaller pipe sizes and headers carrying the fluid to and from the building will be constructed from the larger pipe sizes. A typical design for a 3-ton ground heat exchanger may use 200-foot headers fabricated from 1½-inch pipe with 400 feet of ¾- or 1-inch pipe in each parallel loop connected to the supply and return headers.

Pipe Length — The length of any pipe run will depend on the flow rate and the corresponding allowable pressure drop. If the pressure drop in the ground heat exchanger becomes too large for efficient

4

TABLE 4.2: Minimum Flow in Pipe for Turbulence (GPM)

Nominal Pipe Size	(Pipe ID)	Water at 40°F	20% Calcium Chloride at 25°F	Propylene Glycol-20% 25°F	Methanol 20% 25°F
PE (SDR-11)					
¾"	(0.86)	1.1	2.3	3.4	2.4
1"	(1.077)	1.3	2.9	4.4	3.1
1¼"	(1.385)	1.7	3.6	5.5	3.9
1½"	(1.554)	1.9	4.1	6.3	4.4
2"	(1.943)	2.4	5.2	7.9	5.5
PE (SCH 40)					
¾"	(0.824)	1.0	2.2	3.3	2.3
1"	(1.049)	1.3	2.8	4.2	3.0
1¼"	(1.380)	1.7	3.7	5.6	3.9
1½"	(1.610)	2.0	4.3	6.5	4.6
2"	(2.067)	2.5	5.5	8.4	5.9
PB (SDR-17, IPS)					
1½"	(1.676)	2.1	4.5	6.8	4.8
2"	(2.095)	2.6	5.6	8.5	5.9
PB (SDR-13.5, CTS)					
1"	(0.957)	1.2	2.5	3.9	2.7
1¼"	(1.171)	1.4	3.1	4.7	4.9
1½"	(1.385)	1.7	3.7	5.6	3.9
2"	(1.811)	2.2	4.8	7.3	5.1

TABLE 4.3: Pressure Drop Design Equations

PRESSURE DROP AND FLOW (SMOOTH PIPE)

- DARCY'S $P_f = \left\{0.000175 + \frac{0.0155}{Re^{0.32}}\right\} \frac{Q^2 L}{D^5}$

- HAZEN–WILLIAMS $P_f = \frac{0.0004\ Q^{1.85}\ L}{D_i^{\ 4.86}}$

REYNOLDS NUMBER

- $Re = 122.6 \frac{Q\rho}{D_i \mu}$

- $Re = 50.7 \frac{Q\rho}{D_i \nu}$

FLUID VELOCITY

- $V = \frac{0.4085\ Q}{D_i^{\ 2}}$

Q = gal/min	μ (lb/ft•hr)
ρ = lb/ft³	ν (centipoise)
P_f = Head, ft	L = feet
D_i = inches	V = ft/sec

4

TABLE 4.4: Assumed Fluid Properties for Table 4.2

Solution	Temperature (°F)	Viscosity (centipoise)	Density (lb/ft³)
Water	40	1.55	62.4
Calcium Chloride	25	4.0	74.3
Propylene Glycol	25	5.2	63.6
Methanol	25	3.5	60.8

pumping, then one of the following should be used:

1. a shorter pipe run
2. a larger diameter pipe
3. a parallel fluid path

As a general rule, the pressure drop through the heat pump water heat exchanger will be approximately equal to the pressure drop through the ground heat exchanger and the associated piping. While this is not a requirement, a number of designs fall within this range. Pipe diameters and length then can be chosen to fit within this broad guideline.

The equations given in Table 4.3 can be used to calculate a given fluid's pressure drop, flow rate, and pipe diameter. Table 4.5 gives the piping feet of head loss at different flow rates per 100 feet for a number of common pipe sizes and materials commonly found in cl/gs heat pump systems. These values are for water at 40°F. For fluids other than water, use the equations given in Table 4.3 and the fluid properties given in Appendix A. The following rule of thumb may be used for estimating piping head loss for water and antifreeze solutions:

RULE OF THUMB #1

To adjust the total earth loop piping head loss for antifreeze and water solutions at 25°F, multiply the pressure loss values in Table 4.5 by:

FLUID	MULTIPLIER
20% Propylene Glycol	1.36
20% Calcium Chloride	1.23
20% Methanol Alcohol	1.25

Pipe Cost — Generally stated, smaller diameter pipe is more economical and can be installed at a lower initial cost.

Antifreeze Cost — Smaller diameter piping systems will require less antifreeze.

Circulating Fluid Pressure Drop — The pipe's inside diameter and its maximum length will depend on fluid system pressure drop.

TABLE 4.5: Piping Feet of Head Loss at Different Flow Rates per 100 Feet

PIPE SIZE AND MATERIAL		ID	G.P.M. FLOW RATE									
			1	2	3	4	5	6	8	10	12	14
CONNECTION HOSE 1"		1.050	*	*	*	1.33	1.95	2.68	4.43	6.53	8.99	11.77
PVC 3/4" - 200 PSI			*	*	*	3.7	5.7	*	*	*	*	*
PVC 1" - 200 PSI			*	*	*	1.0	1.9	2.7	4.2	6.3	8.9	11.8
COPPER 3/4"			*	*	*	4.3	6.3	*	*	*	*	*
COPPER 1"			*	*	*	1.5	1.9	2.7	4.5	6.9	9.6	12.8
PE3408 (Polyethylene)		ID										
1. SDR-11	3/4	0.860	0.20	1.03	2.07	3.41	5.03	*	*	*	*	*
2. SDR-11	1	1.077	0.07	0.36	0.71	1.18	1.73	2.38	3.92	*	*	*
3. SDR-11	1-1/4	1.358	*	0.12	0.24	0.39	0.58	0.79	1.31	1.93	2.65	3.47
4. SDR-11	1-1/2	1.554	*	*	0.13	0.21	0.31	0.42	0.69	1.02	1.40	1.83
5. SDR-11	2	1.943	*	*	*	0.07	0.11	0.15	0.24	0.35	0.48	0.63
6. SCH 40	3/4	0.824	0.25	1.26	2.54	4.18	6.16	8.46	*	*	*	*
7. SCH 40	1	1.049	0.08	0.40	0.81	1.33	1.96	2.69	4.45	*	*	*
8. SCH 40	1-1/4	1.380	*	0.11	0.22	0.36	0.54	0.74	1.21	1.79	2.46	3.21
9. SCH 40	1-1/2	1.610	*	*	0.11	0.18	0.26	0.35	0.58	0.86	1.18	1.55
10. SCH 40	2	2.067	*	*	*	*	0.08	0.11	0.18	0.26	0.36	0.47
PB2110 (Polybutylene)		ID										
11. SDR-17, IPS	1-1/2	1.676	*	*	0.09	0.15	0.21	0.29	0.48	0.71	0.98	1.28
12. SDR-17, IPS	2	2.095	*	*	*	0.05	0.07	0.10	0.17	0.25	0.34	0.44
13. SDR-13.5, Cts	1	0.957	0.11	0.62	1.25	2.06	3.03	4.16	*	*	*	*
14. SDR-13.5, Cts	1-1/4	1.171	*	0.24	0.48	0.79	1.17	1.60	2.64	*	*	*
15. SDR-13.5, Cts	1-1/2	1.385	*	0.11	0.22	0.36	0.53	0.72	1.19	1.76	2.41	3.20
16. SDR-13.5, Cts	2	1.811	*	*	0.06	0.10	0.15	0.20	0.33	0.49	0.68	0.88

Notes:
1. These head losses are for water at 40°F temperature.
2. Count each elbow, tee, reducer, air scoop, flow meter, etc., as 3 feet of equivalent pipe length and add to actual measured pipe length for total length.
3. To adjust the piping head loss for fluids other than water, multiply the total head loss for water by the following:
   a) 20% Propylene Glycol 1.36 b) 20% Calcium Chloride 1.23 c) 20% Methanol Alcohol 1.25

4

Ground Heat Exchanger Sizing and Design Procedure

The following inputs should be considered in sizing and designing the ground heat exchanger:

1. heat pump water flow rate
2. circulating fluid type
3. minimum entering fluid temperature expected

The ground heat exchanger design process which follows is recommended:

STEP 1: Select Ground Heat Exchanger Layout

The designer selects a ground heat exchanger configuration and layout that falls within the property boundaries of the building site. The designer has a choice between various types of ground heat exchanger layouts but the most common are:

HORIZONTAL

1. single pipe (Figure 4.2)
2. two-pipe (Figure 4.3)
3. four-pipe (Figure 4.4)

VERTICAL

1. vertical single U-bend loop (Figures 4.5 and 4.6)
2. vertical double U-bend loop

There are certain general design guides that can help in the layout of the ground heat exchanger. These design guides and the configurations contained in them have evolved from field experience and represent an industry design standard.

DESIGN GUIDE 1: SINGLE VERSUS MULTIPLE PIPE — HORIZONTAL SYSTEMS

Experience has shown that trenching costs are higher than pipe costs per linear foot. The result has been many horizontal systems using multiple pipes in the same trench. New multi-pipe-laying machines which place and backfill pipes automatically have brought increased interest in the four-pipe system.

Since pipes buried close to each other have lower thermal performance than pipes separated four to six feet, the following rule of thumb is helpful in initially laying out the ground heat exchanger:

RULE OF THUMB #2

Trench length is reduced in the following proportion when multiple pipes are installed in a single trench:

NUMBER OF PIPES	TRENCH/PIPE (FEET)	DEPTHS (FT)
1 (Figure 4.2)	500/500	5
2 (Figure 4.3)	275/550	4, 6
3 (Figure 4.4)	225/900	3, 4, 5, 6

Trench spacing must be increased as the number of pipes per trench increases. Trench spacing for a single pipe in a trench is 4 feet; for two pipes, 6 feet; and for four pipes, 12 feet. This is based on a 5% decrease in performance using a single pipe in a trench as the standard.

DESIGN GUIDE 2:

In determining the maximum length of any pipe run for series and parallel systems while maintaining reasonable pumping power, Rule of Thumb #3 is recommended.

DESIGN GUIDE 3:

The guideline for the layout of the number of loops in a system for residential-size ground heat exchangers is given in Rule of Thumb #4.

The one loop per ton allows a standardized header design to be developed which

RULE OF THUMB #3

Maximum recommended pipe length in series:
(Pressure loss, water @ 40°F)

NOMINAL SIZE	DESCRIPTION FLOW, LENGTH	HEAD (FEET)
¾"	SDR-11 (0.86 inch ID)	
	3 GPM, 500 ft.	10.3
1"	SDR-11 (1.077 inch ID)	
	3 GPM, 1,500 ft.	10.4
	4.5 GPM, 800 ft.	11.4
1¼"	SCH 40 (1.380 inch ID)	
	7.5 GPM, 1,200 ft.	12.9
1½"	SCH 40 (1.61 inch ID)	
	9 GPM, 1,500 ft.	10.7
2"	SCH 40 (2.067 inch ID)	
	15 GPM, 2,500 ft.	10.7

NOTE: Fluid pressure drop dictates the length of pipe run in any ground heat exchanger design.

RULE OF THUMB #4

For parallel systems, use one loop for each one ton (12,000 Btu/hr) of heat pump capacity. For example, a three-ton heat pump would have three parallel loops. Headers are constructed from 1½- and 2-inch pipe, and loops are constructed from ¾- and 1-inch pipe.

details pipe sizing for air removal and pressure loss.

STEP 2:
Check Minimum Fluid Velocity

Check Table 4.2 to determine if the minimum flow rate exists in all fluid paths for the specific fluid chosen. If not, select a new piping arrangement that gives higher flow in all paths.

The minimum fluid velocity is different for each type of fluid. For fluid properties different from those given in Table 4.3, see Appendix A.

STEP 3:
Design for Air Removal

Be sure that all potential air traps are designed out of the system. This includes items such as "pigtails" on the ends of headers which should be replaced by reducing elbows. In all cases, the design should be specific on how the system is to be fabricated. Refer to Section 7, Figure 7.2, for header designs that have been created for easy air purging and debris removal.

STEP 4:
Size the Ground Heat Exchanger Length

The design will incorporate commercially available pipe widely used in this industry. The process requires the ground heat exchanger configuration be defined. It also requires a knowledge of earth tempera-

4

tures and thermal conductivities at specified depths. The substeps are as follows:

1. Define buried pipe placement.
2. Determine underground soil temperatures.
3. Determine maximum and minimum earth temperatures.
4. Determine heat pump maximum and minimum entering fluid temperatures.
5. Calculate the earth-to-loop temperature difference.
6. Determine the pipe resistance to heat flow.
7. Determine the soil/field resistance to heat flow.
8. Calculate the heating and cooling run fractions.
9. Calculate the ground heat exchanger length:
   a. the length of pipe for a horizontal ground heat exchanger, or
   b. the feet of bore hole for a vertical system.

Many of the above values have been calculated and are provided herein in tabular form. In cases where special or unique designs are selected, it may be necessary to use the references at the end of this volume.

STEP 4.1: Define Buried Pipe Placement

Pipe depth and spacing in each trench and trench spacing must be defined for horizontal ground heat exchangers. If one of the ground heat exchangers defined in Figures 4.9 through 4.14 or Table 4.6 is selected, the values of soil/field resistance needed in Step 4.7 can be found in those illustrations. If these are insufficient, the designer is referred to the *ASHRAE Design/Data Manual for Closed-Loop Ground-Coupled Heat Pump Systems* or the *Oklahoma State University Earth Coil Design Manual* (International Ground Source Heat Pump Association — IGSHPA) for a calculation procedure. Software programs specifically developed

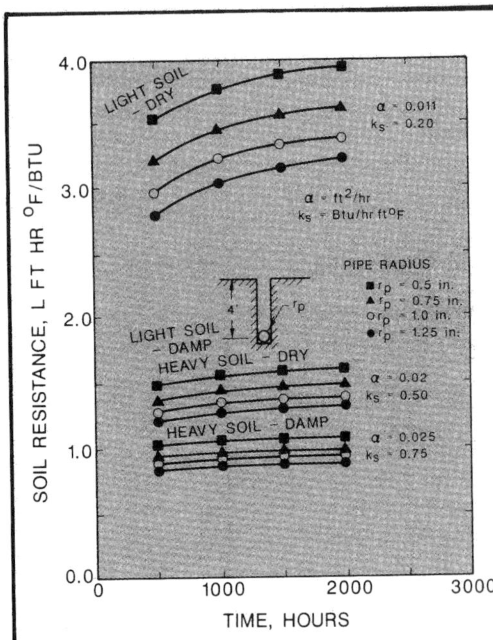

FIGURE 4.9: Soil/Field Resistance — Single-Pipe Horizontal Ground Heat Exchanger

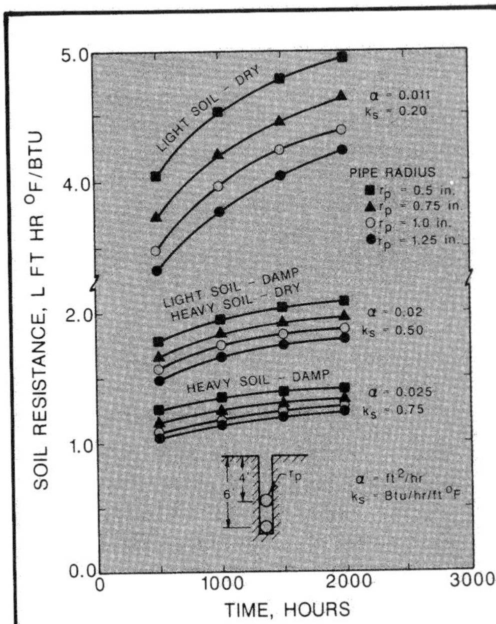

FIGURE 4.10: Soil/Field Resistance — Two-Pipe (2 x 1) Horizontal Ground Heat Exchanger

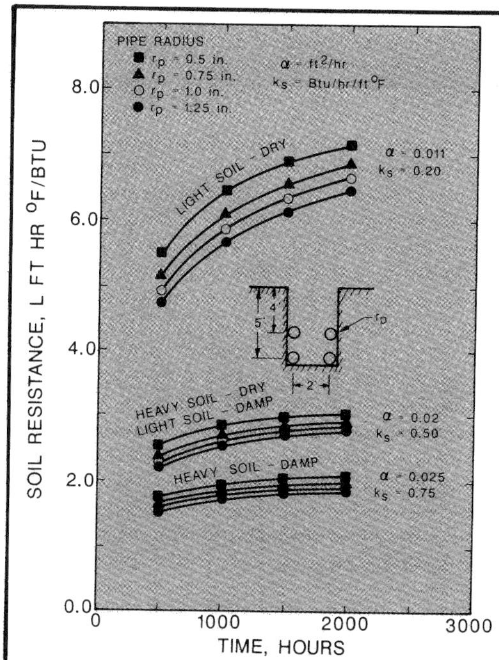

FIGURE 4.11: Soil/Field Resistance — Four-Pipe (2 x 2) Horizontal Ground Heat Exchanger

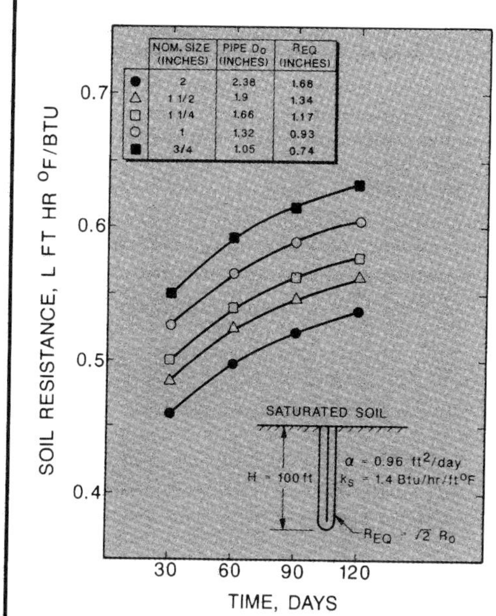

FIGURE 4.13: Soil/Field Resistance — Single-Loop Vertical Ground Heat Exchanger

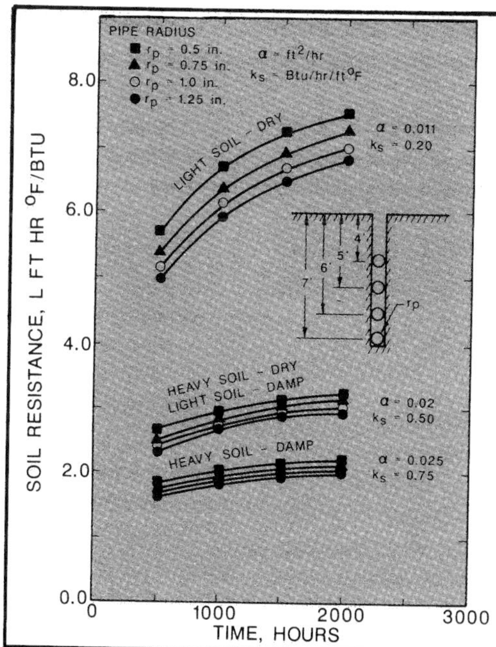

FIGURE 4.12: Soil/Field Resistance — Four-Pipe (4 x 1) Horizontal Ground Heat Exchanger

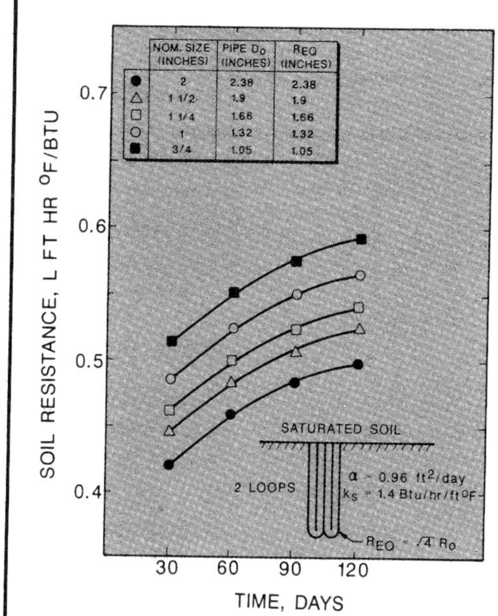

FIGURE 4.14: Soil/Field Resistance — Double-Loop (Single-Bore) Vertical Ground Heat Exchanger

4

TABLE 4.6: Pipe and Soil/Field Resistance Values

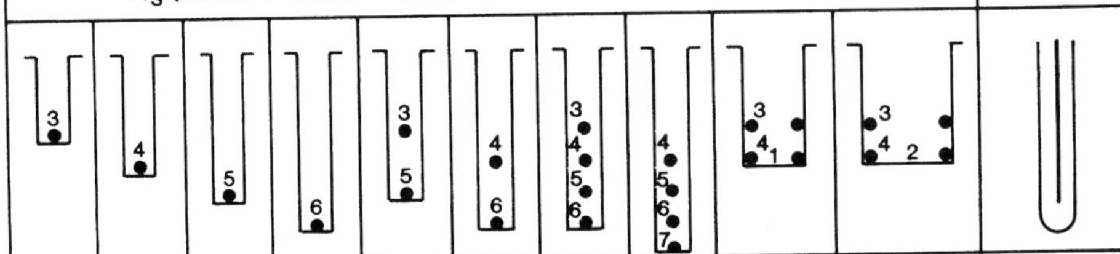

Pipe Size	PE SCH 40 (R_p/R_{pe})	PE SDR-11 (R_p/R_{pe})	PB SDR-17 (R_p/R_{pe})	PB SDR-13.5 (R_p/R_{pe})	R_s(HEAVY SOIL - DAMP) / R_s(HEAVY SOIL - DRY OR LIGHT SOIL - DAMP)										R_s(ROCK) / R_s(HS-DAMP)
3/4	.17/.116	↑	↑	↑	1.02/1.38	1.06/1.44	1.09/1.47	1.11/1.49	1.31/1.77	1.37/1.84	2.05/2.75	2.15/2.86	2.11/2.85	1.88/2.53	0.60/1.06
1	.159/.109				0.97/1.32	1.02/1.37	1.04/1.40	1.06/1.42	1.26/1.70	1.32/1.77	2.00/2.68	2.10/2.79	2.07/2.78	1.84/2.47	0.57/1.01
1 1/4	.130/.089	.141/.096	.16/.11	.20/.14	0.92/1.25	0.97/1.31	0.99/1.34	1.01/1.36	1.22/1.63	1.27/1.70	1.96/2.61	2.05/2.72	2.02/2.71	1.79/2.40	0.54/0.96
1 1/2	.117/.080				0.89/1.21	0.94/1.27	0.97/1.30	0.98/1.32	1.19/1.59	1.25/1.66	1.92/2.57	2.02/2.68	1.99/2.67	1.76/2.36	0.53/0.94
2	.098/.068	↓	↓	↓	0.85/1.15	0.89/1.20	0.92/1.24	0.94/1.26	1.14/1.53	1.20/1.60	1.88/2.51	1.98/2.62	1.94/2.61	1.71/2.29	0.50/0.89

T = 1500 hours

for this calculation process are available and are recommended since the process is long and tedious.

For vertical systems, the designer must define the depth of each bore hole, the number of U-bend loops, and the spacing of the bore holes. Figure 4.15 defines the equivalent diameter for both single- and double-loop pipes. Table 4.6 gives calculated values.

STEP 4.2: Determine Underground Soil Temperatures

Annual air temperatures, moisture content, and soil type, and vegetative cover, all have an effect on underground soil temperatures.[9, 10, 11, 12] Several sources can be used to determine underground soil temperatures:

1. a knowledge of local soil temperature conditions based on experience or measured data
2. design charts (funnel curve) as shown in Figure 4.16 for determining soil maximum and minimum temperatures as a function of soil type and depth
3. design tables as given in Table 4.7 to determine maximum and minimum soil temperatures for different values of annual soil swing (A_S)
4. analytical equations as given in Table 4.8 to calculate the soil temperature at any depth and time of year

For multiple-pipe systems, the soil temperature can be determined by taking an average of the temperatures at each depth.

The values of the following can be obtained from tabular data as given in Table 4.9 and in maps as presented in Figures 4.17, 4.18, and 4.19, respectively:

T_M — mean earth temperature (°F)

A_S — annual soil swing temperature (°F)

T_O — phase shift at the surface (days)

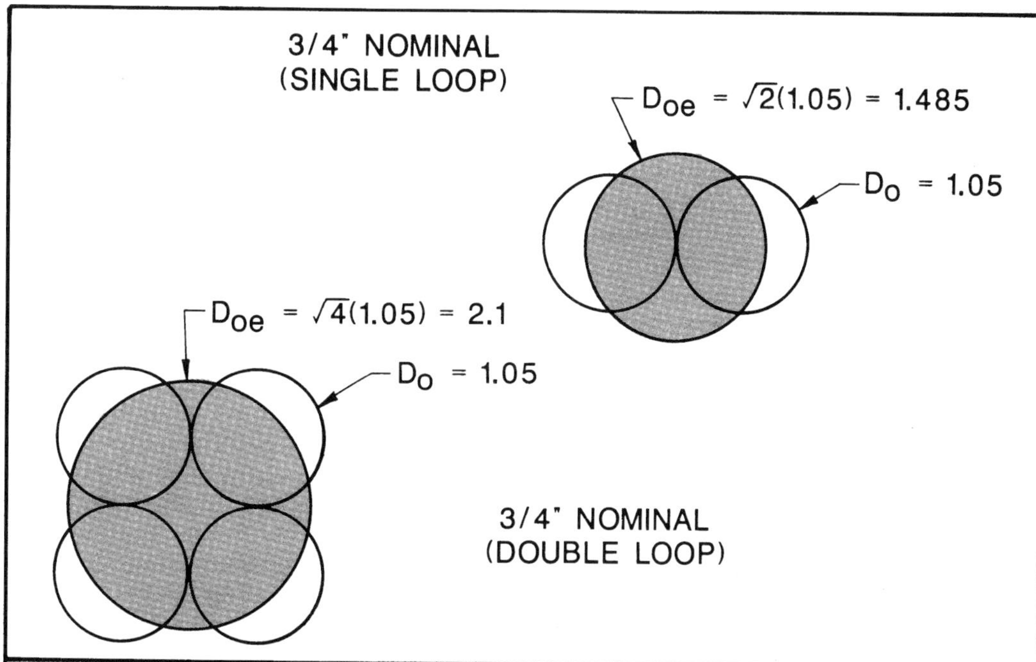

FIGURE 4.15: Equivalent Pipe Diameters for Loop Heat Exchangers

α — soil thermal diffusivity (ft^2/day)

For vertical ground heat exchangers a constant ground temperature value can be assumed to be equal to T_M.

Plug-in Values for T_M, A_S, and T_O

The mean earth temperature (T_M) can be assumed to be constant for all depths up to 200 feet. Below that depth the increase per 100 feet of depth is small. T_M can be assumed equal to the groundwater temperature or the annual air temperature plus 2°F. In areas with geothermal effects, the local groundwater temperature would be a better estimate.

The annual swing (A_S), yearly temperature amplitude of daily means, ranges from 19° to 26°F for most sodded surfaces. Its value depends on location, soil type, and moisture content. Values of A_S can be found in the *ASHRAE Handbook of Fundamentals, 1985* in Chapter 25, page 25.6, Figure 4.

The phase constant (T_O) references the annual surface temperature swing to the calendar year. Kusuda's[10] analysis gives this a typical value of 35 ± 10 days or the minimum soil temperature at the surface occurring within a week and a half of February 4. Consequently, the maximum will occur within about a week and a half of August 4.

Values of soil thermal diffusivity (α) depend on soil type and moisture content. Representative values for typical soils are given in Table 4.10 with field identification descriptions given in Table 4.11. A typical value is 0.6 ft^2/day, but extreme caution should be taken in using representative values. See the work by Salomone.[13]

STEP 4.3: Determine Maximum and Minimum Earth Temperatures

From the earth cycle calculated in Step 4.2, obtain the maximum (T_H) and minimum (T_L) earth temperatures that occur during the yearly cycle or the temperature on any day of the year for any depth. Analytical equations for T_H and T_L

4

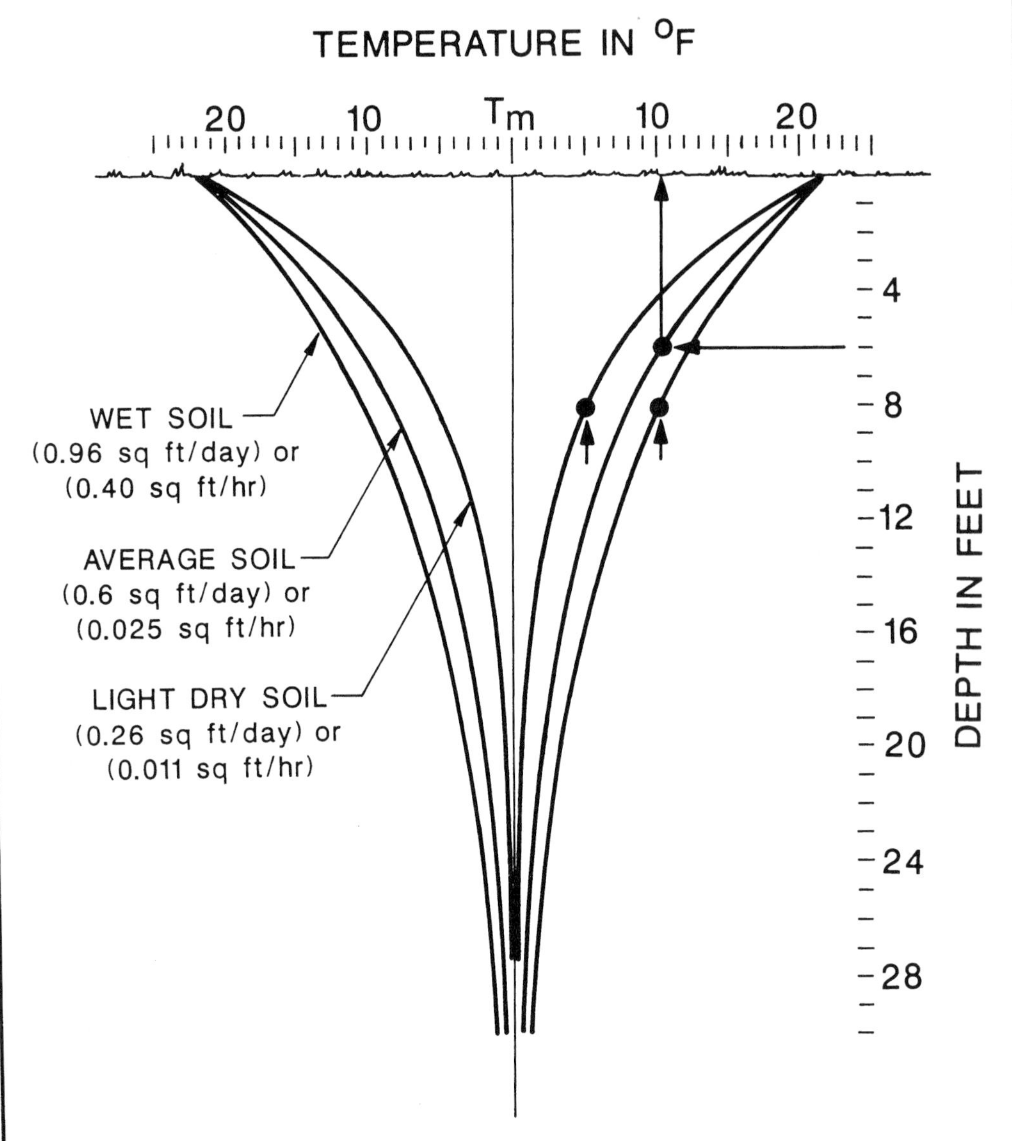

FIGURE 4.16: Soil Temperature Variations

are given in Table 4.8 and tabular values are given in Table 4.7. In all cases, the mean earth temperature (T_M), the annual soil temperature swing at the surface (A_S), and the soil phase shift (T_O) at the surface can be selected from tabular data for specific geographic locations.

STEP 4.4: Determine Heat Pump Maximum and Minimum Entering Water Temperatures

Performance data for water-source heat pumps is available from the manufac-

TABLE 4.7: Soil Temperature Variations for Different Values of Annual Surface Swing

Depth (feet)	Annual Swing = 10°F Wet	Average	Dry	Annual Swing = 15°F Wet	Average	Dry
1	9.0	8.8	8.3	13.6	13.3	12.5
2	8.2	7.8	6.9	12.4	11.8	10.4
3	7.5	6.9	5.7	11.2	10.4	8.6
4	6.8	6.1	4.8	10.2	9.2	7.2
5	6.2	5.4	4.0	9.3	8.2	6.0
6	5.6	4.8	3.3	8.4	7.3	5.0
7	5.1	4.3	2.7	7.7	6.4	4.1
8	4.6	3.8	2.3	7.0	5.7	3.4
	Annual Swing = 20°F			Annual Swing = 25°F		
1	18.1	17.7	16.6	22.7	22.1	20.8
2	16.5	15.7	13.8	20.6	19.6	17.3
3	15.0	13.9	11.5	18.8	17.4	14.4
4	13.6	12.3	9.6	17.1	15.4	12.0
5	12.4	10.9	8.0	15.5	13.7	10.0
6	11.3	9.7	6.7	14.1	12.1	8.3
7	10.3	8.6	5.5	12.8	10.8	6.9
8	9.3	7.6	4.6	11.7	9.5	5.8
	Annual Swing = 30°F			Annual Swing = 35°F		
1	27.2	26.6	25.0	31.8	31.0	29.1
2	24.8	23.6	20.8	28.9	27.5	24.3
3	22.5	20.9	17.3	26.3	24.4	20.2
4	20.5	18.5	14.4	23.9	21.6	16.9
5	18.6	16.4	12.0	21.8	19.2	14.0
6	16.9	14.6	10.0	19.8	17.0	11.7
7	15.4	12.9	8.3	18.0	15.1	9.7
8	14.0	11.5	6.9	16.4	13.4	8.1

Soil Type	Thermal Diffusivity (ft^2/day)
Wet	0.96
Average	0.60
Dry Soil	0.26

turer. The range of temperatures of available heat pumps designed for cl/gs heat pumps is from 25° to 110°F. Not all heat pumps have this broad a temperature operating range.

A design minimum EWT (T_{MIN}) value of 30° to 40°F above the coldest outdoor temperature at a given location is a good first estimate when selecting heat pump capacity. For cooling loads, assume 100°F as the first approximation for the maximum EWT (T_{MAX}) in all locations except the deep South (T_M greater than 65°F) where 105°F should be used.

4

TABLE 4.8: Analytical Equations for Estimating Earth Temperatures

$$T(X_S, t) = T_M - A_S \times EXP\left\{- X_S \left(\frac{\pi}{365\,\alpha}\right)^{1/2}\right\} \times$$

$$\cos\left\{\frac{2\pi}{365}\left[t - t_o - \frac{X_S}{2}\left(\frac{365}{\pi\,\alpha}\right)^{1/2}\right]\right\}$$

$T(X_S, t)$ — Earth temperature at soil depth X_S and time of year, t
T_M — Mean earth temperature, °F
A_S — Annual surface soil temperature swing, °F
α — Soil thermal diffusivity, $k_S/(\rho\, c)$, ft^2/day
k_s — Soil thermal conductivity, Btu/hr-ft-°F
ρ — Soil density, lb/ft^3
c — Soil specific heat, Btu/lb-°F
t_o — Phase constant, day of minimum soil surface temperature, days
X_s — Soil depth, feet
t — Time, days

and:

$$T_L = T_M - A_S \times EXP\left\{- X_s \left(\frac{\pi}{365\,\alpha}\right)^{1/2}\right\}$$

$$T_H = T_M + A_S \times EXP\left\{- X_s \left(\frac{\pi}{365\,\alpha}\right)^{1/2}\right\}$$

where:

T_L — Minimum annual soil temperature @ depth X_S
T_H — Maximum annual soil temperature @ depth X_S

STEP 4.5: Calculate the Earth-to-Loop Temperature Difference

Calculate the temperature difference between the earth temperature minimum (T_L) and heat pump minimum entering water (or fluid) temperature (T_{MIN}) for winter heating. For summer cooling cycles, calculate the difference in the heat pump maximum (T_{MAX}) entering water temperature and the earth maximum temperature (T_H). Figure 4.20 describes graphically the temperature difference defined as:

$T_{HD} = T_L - T_{MIN}$ = Temperature difference for heating °F

$T_{CD} = T_{MAX} - T_H$ = Temperature difference for cooling °F

For example, assume the maximum and minimum earth temperatures at a 4-foot depth were 78° and 41°F, respectively, and that the heat pump maximum water temperature is 100°F for cooling and the

TABLE 4.9: Earth Temperature Data for Selected U.S. Cities

State	City	T_M(°F)	A_S(°F)	T_o (Days)
AL	Birmingham	65	19	31
	Montgomery	67	18	31
AZ	Phoenix	73	23	33
	Tucson	70	18	34
AR	Little Rock	64	21	32
CA	Los Angeles	64	7	54
	Merced	68	25	33
	San Diego	64	7	54
CO	Colo.Springs	51	21	36
	Denver	52	22	37
	Grand Junc.	55	25	32
DC	Washington	57	22	36
FL	Appalachico.	70	15	36
	Jacksonville	71	14	32
GA	Atlanta	62	19	32
	Augusta	65	18	30
	Macon	67	17	33
ID	Boise	53	21	34
	Idaho Falls	46	23	34
IL	Chicago	51	25	37
	E. St. Louis	57	24	34
	Urbana	53	26	35
IN	Fort Wayne	53	24	35
	Indianapolis	55	24	34
	South Bend	52	25	37
IA	Des Moines	52	28	35
	Sioux City	51	29	34
KS	Dodge City	57	25	35
	Topeka	56	26	35
KY	Louisville	60	22	33
LA	Lake Charles	70	16	32
	New Orleans	70	15	32
	Shreveport	66	19	32
MA	Portland	48	22	39
MS	Plymouth	51	21	43
MI	Battle Creek	50	24	35
	Detroit	50	25	39
	Sau St. Marie	42	26	40
MN	Duluth	41	28	37
	Int. Falls	39	31	34
	Minneapolis	47	29	35
MS	Biloxi	70	17	32
	Columbus	65	19	32
	Jackson	67	18	31
MO	Columbia	57	24	35
	Kansas City	58	26	35
	Springfield	58	23	34
MT	Billings	49	23	37
	Great Falls	48	23	36
	Missoula	46	21	32
NB	Grand Island	52	27	35
	Lincoln	53	28	34
	N. Platte	51	26	35
NV	Ely	47	22	35
	Las Vegas	69	23	32
	Winnemucca	52	22	33
NJ	Trenton	55	22	38
NM	Albuquerque	59	22	31
	Roswell	63	22	30
NY	Albany	50	25	38
	Binghamton	48	24	38
	Niagara Falls	50	24	24
	Syracuse	50	24	38
NC	New Bern	65	17	35
	Greensboro	60	20	31
ND	Bismark	44	31	33
	Grand Forks	42	33	35
	Williston	45	29	34
OH	Akron	52	23	37
	Columbus	55	22	34
	Dayton	56	24	35
	Toledo	51	25	36
OK	Altus	65	24	33
	Ok. City	62	23	34
	Tulsa	62	23	34
OR	Astoria	53	9	45
	Medford	55	17	34
	Portland	54	13	37
PA	Middletown	55	23	35
	Philadelphia	55	22	34
	Pittsburg	52	23	36
	Wilkes-Barre	52	23	36
SC	Charleston	66	16	32
	Greenville	62	19	33
	Sumpter	65	18	32
SD	Huron	47	30	35
	Rapid City	50	25	38
TN	Bristol	59	20	32
	Knoxville	61	21	31
	Memphis	63	21	32
	Nashville	60	21	32
TX	El Paso	66	20	28
	Ft. Worth	68	21	34
	Houston	71	16	33
	San Antonio	72	16	32
UT	Salt Lake	53	24	35
VT	Burlington	46	26	37
VA	Norfolk	61	20	37
	Richmond	60	19	33
	Roanoke	59	20	33
WA	Moses Lake	53	23	29
	Seattle	53	12	36
	Spokane	49	21	32
WV	Charleston	58	20	33
	Elkins	52	20	32
WI	Green Bay	46	26	37
	Madison	49	27	36
WY	Casper	49	24	37
	Cheyenne	48	21	39
	Lander	46	24	38
	Sheridan	48	24	35

minimum is 30°F for heating. The temperature difference for heating would be:

$$T_{HD} = 41 - 30 = 11°F$$

and the temperature difference for cooling would be:

$$T_{CD} = 100 - 78 = 22°F$$

4

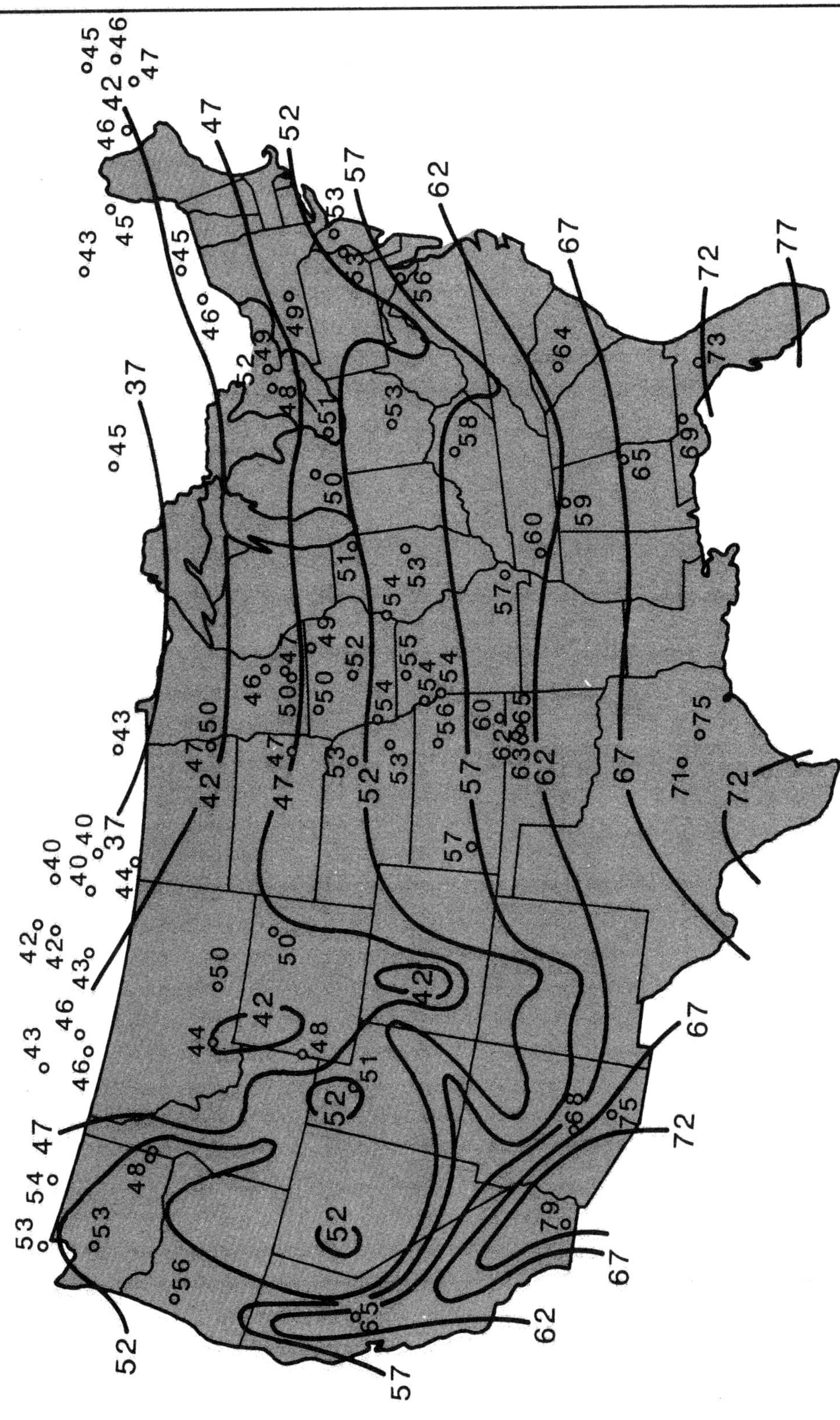

VALUES OF MEAN ANNUAL EARTH TEMPERATURE FROM OBSERVATIONS AT INDIVIDUAL STATIONS ARE SUPERIMPOSED UPON COLLINS' ESTIMATED (1925) WELL-WATER ISOTHERMS.

FIGURE 4.17: Mean Earth Temperature, T_M (°F)

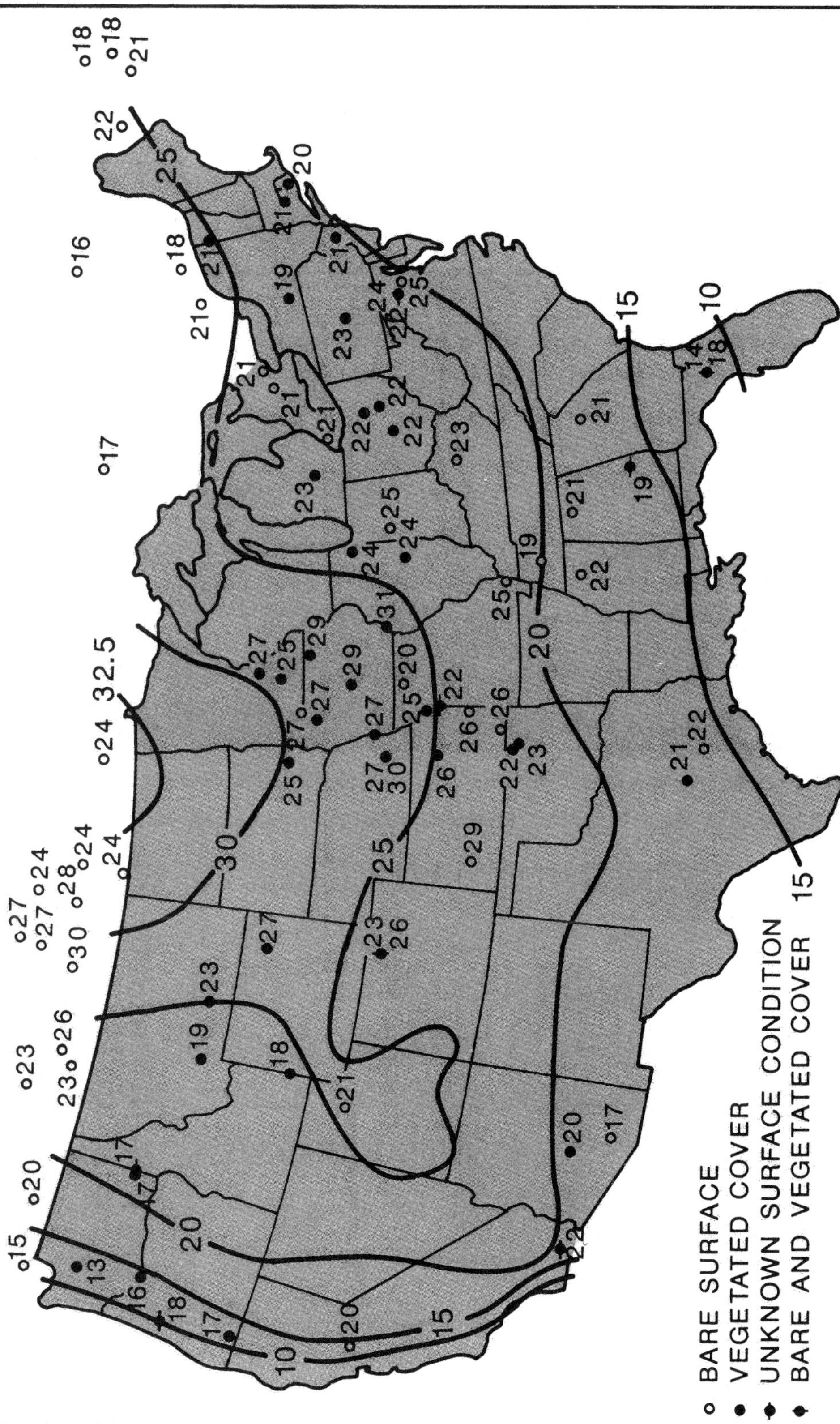

EARTH SURFACE TEMPERATURE AMPLITUDE DETERMINED THROUGH LEAST SQUARES ANALYSIS BY KUSUDA AND ACHENBACH OF OBSERVED DATA IS SUPERIMPOSED ON MAP OF ANNUAL AIR TEMPERATURE AMPLITUDE. CANADIAN DATA OBTAINED BY TAKING ONE-HALF THE RANGE OF ANNUAL TEMPERATURE AT THE 10mm DEPTH.

FIGURE 4.18: Earth Surface Temperature Amplitude, A_s (°F)

4

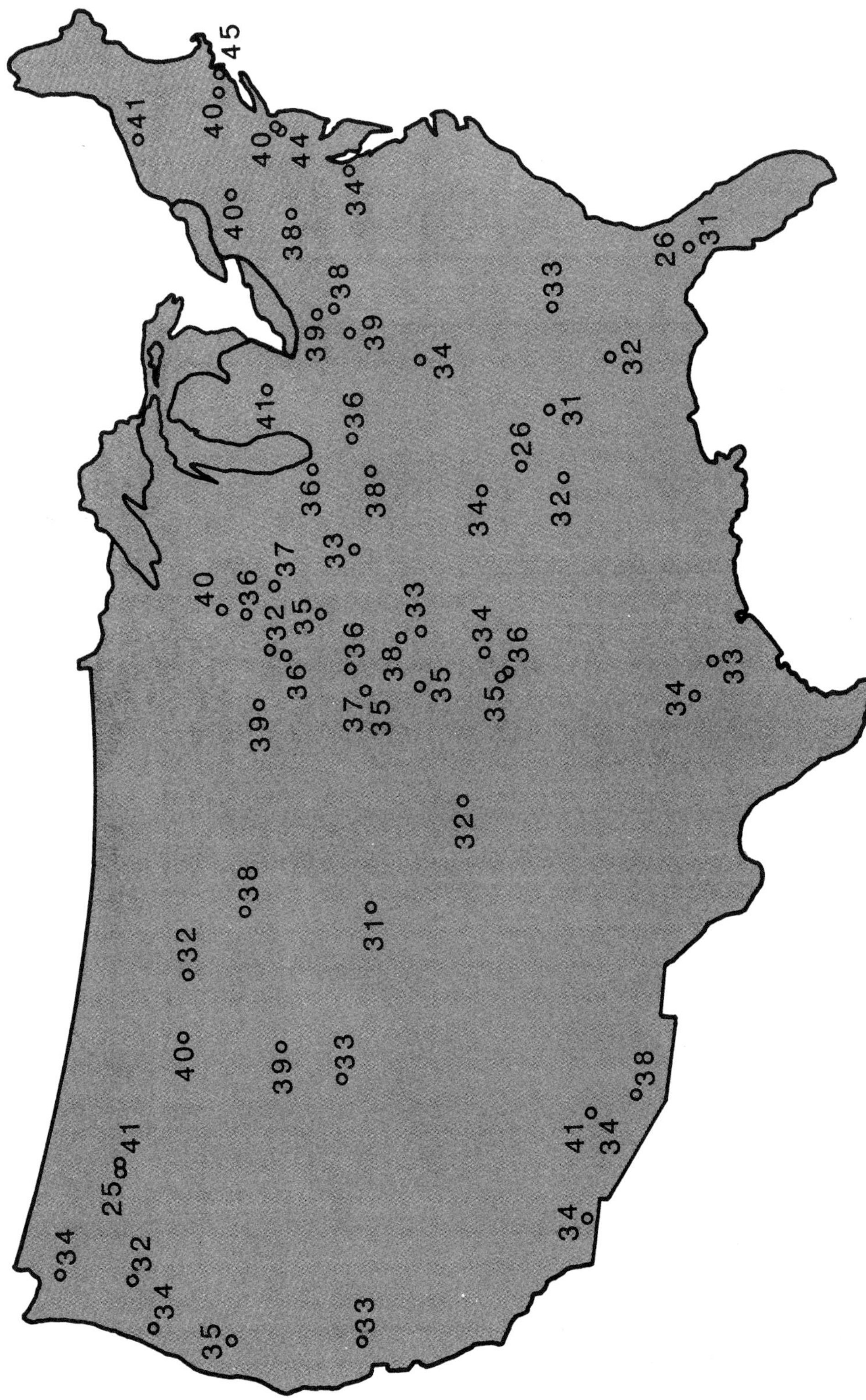

PHASE CONSTANTS OF INDIVIDUAL EARTH TEMPERATURE STATIONS, DETERMINED FROM LEAST SQUARES ANALYSIS PERFORMED BY KUSUDA AND ACHENBACH.

FIGURE 4.19: Earth Surface Phase Constant, T_o (Days)

4

TABLE 4.10: Thermal Properties of Soils, Rocks, and Concrete

Material Descriptor	Thermal Conductivity (Btu/hr-ft-°F)	Thermal Diffusivity (ft^2/day)	Density (lb/ft^3)	Specific Heat (Btu/lb-°F)
Dense rock	2.00	1.20	200	0.20
Average rock	1.40	0.96	175	0.20
Dense concrete	1.00	0.79	150	0.20
Heavy soil, saturated	1.40	0.84	200	0.20
Solid masonry			143	0.21
Heavy soil, damp	0.75	0.60	131	0.23
Heavy soil, dry			125	0.20
Light soil, damp	0.50	0.48	100	0.25
Light soil, dry	0.20	0.26	90	0.20

Source: Modified from *ASHRAE Handbook of Fundamentals, 1985.*

TABLE 4.11: Field Identification of Inorganic Soil Textural Types

	General Characteristics				
Textural Type	Main Fractions*	Appearance, Dry Conditions	Rubbed Between Fingers, Dry Conditions	Squeezed in Palm of Hand, Moist Conditions at Field Capacity	Rolled into Ribbon, Moist Conditions
Sandy soil	At least 85% sand particles	Crumbly with no clods or lumps; individual soil grains visible to naked eye	Gritty; soil grains readily felt	Can form cast; crumbles with least amount of handling	Cannot form ribbon
Sandy loam soil	At least 50% sand particles; not more than 20% clay	Mainly crumbly and loose; grains readily seen and felt	Gritty: soil grains readily felt	Can form cast that will bear careful handling	Cannot form ribbon
Loam soil	At least 80% sand silt in about equal proportion; not more than 20% clay	Mainly crumbly; some clods or lumps	Fairly smooth but some gritty feeling; lumps easily broken	Can form cast that can be handled freely	Cannot form ribbon
Silt loam soil	At leat 50% silt; not more than 20% clay	Quite cloddy but some crumbly materials	Lumps easily broken and easily pulverized; there-after floury texture and soft feel	Can form cast that can be handled freely; wet soil runs together and puddles	Cannot form perfect ribbon; has broken surface, cracks appear
Clay loam soil	20 to 30% clay	Fine-textured soil; quite cloddy but some crumbly material	Lumps hard; not easily broken	Can form cast that can be handled freely; soil plastic	Can form perfect ribbon but breaks easily
Clay soil	30 to 100% clay	Fine-textured soil; breaks into very hard clods	Lumps very hard; difficult, if not impossible, to break	Can form cast that can be handled freely; soil plastic	Can form ribbon that will support its own weight
Silt soil	At least 80% silt				

*Based on the textural classification of the U.S. Bureau of Public Roads.
Refer to text for: shaking (dilatancy) test, shine test, dry-strength test.

Source: *Soil Thermal Characteristics in Relation to Underground Power Cables.*[14]

4

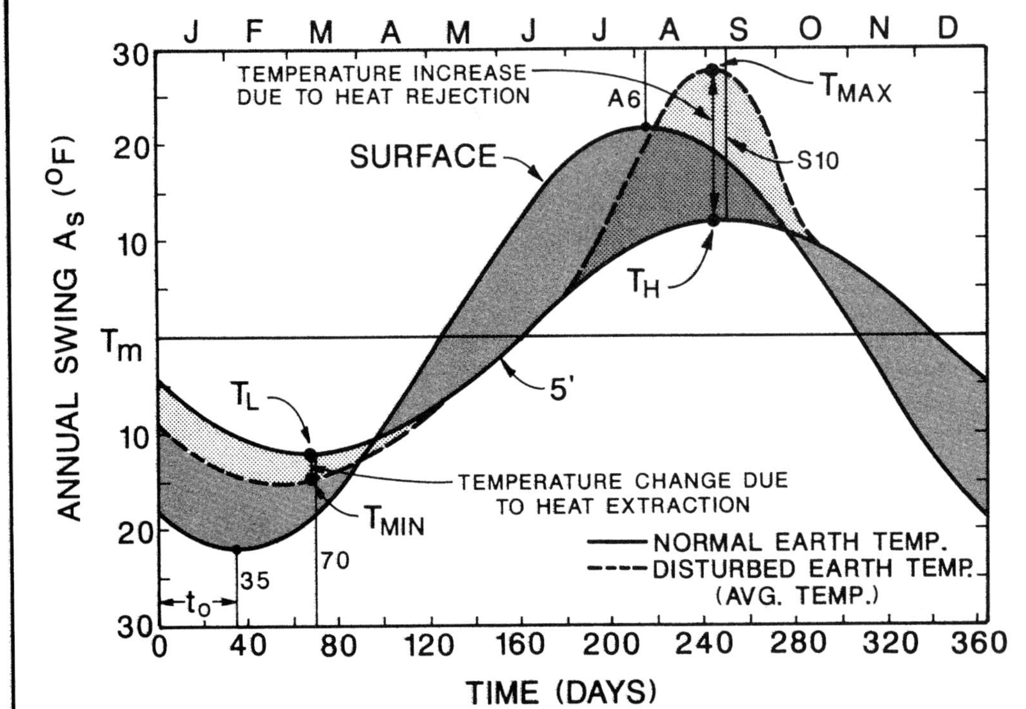

FIGURE 4.20: Horizontal Earth Coil Temperature Variations

STEP 4.6: Calculate the Pipe Resistance to Heat Flow

Select from Table 4.1 or calculate the pipe resistance for either single-pipe or loop-type heat exchangers. For single pipes:

$$R_{PE} = \frac{1}{2\pi k_p} \; L_N \left(\frac{D_o}{D_I}\right)$$

For two pipes in contact (vertical heat exchangers, Figure 4.15), the equivalent pipe resistance is given by:

$$R_{PE} = \frac{1}{2\pi k_p} \; L_N \left(\frac{D_{OE}}{D_{OE} - (D_o - D_I)}\right)$$

D_O = Pipe outside diameter, inches
D_I = Pipe inside diameter, inches
k_P = Pipe material thermal conductivity, Btu/hr-ft-°F
L_N = Natural logarithm, base e
D_{OE} = $(N)^{1/2}$ x D_O
N = Number of pipes in loop
π = 3.14159

STEP 4.7: Calculate the Soil/Field Resistance

Calculation of the soil/field resistance is a time-consuming process. This manual provides the data in table form for different types of soils and heat exchanger configurations. Figures 4.9 through 4.14 and Table 4.6 give data for a number of different types of soils and a number of horizontal and vertical ground heat exchangers. Tables 4.10 and 4.11 give soil identification and property values.

The resistance values given in Table 4.6 and Figures 4.9 through 4.14 are based on the line source theory which assumes that heat pump operation is continuous. The values given in Table 4.6 are for continuous operation for 1,500 hours (approximately two months), which would result in higher soil resistance values than would be found in actual operation. A run fraction as defined in Step 4.8 gives an average value for a design month.

STEP 4.8: Calculate the Heating and Cooling Run Fractions

HEATING RUN FRACTION (F_H) CALCULATION PROCEDURE

The procedure is as follows (see Figure 4.21 for nomenclature and definitions):

1. Estimate the building's heat load (Btu/hr) as given in Section 3.
2. Calculate the heat loss per degree difference.
3. Assume a balance temperature. Recommended values are:
   Average Construction------------ 60°F
   Energy-Efficient Construction - 55°F
   Super-Insulated Construction - 50°F
4. Calculate the building's load for each bin temperature for the heating design month of January.
5. Select heat pump capacity at the design entering water temperature for the design month of January.
6. Check building demand with heat pump capacity. Calculate the heat pump run hours for each temperature bin (run hours = bin hours x heat pump run time for the bin temperature range). When building demand exceeds heat pump capacity, run hours equal bin hours.
7. Sum run hours for all temperature bins for the design month.
8. Calculate run fraction from:
   F_H = (run hours) / (31 x 24)
   where: 31 days in January
   24 hours per day

EXAMPLE – Heating Run Fraction

1. Building load using ASHRAE *GRP-158:*
   Location: Oklahoma City, Oklahoma
   Design Heat Loss: 55,000 Btu/hr
   Outdoor Design Temperature: 13°F
   Indoor Design Temperature: 72°F
2. Heat loss per degree Fahrenheit (°F):
   55,000 / (72 – 13) = 932.2 Btu/hr/°F

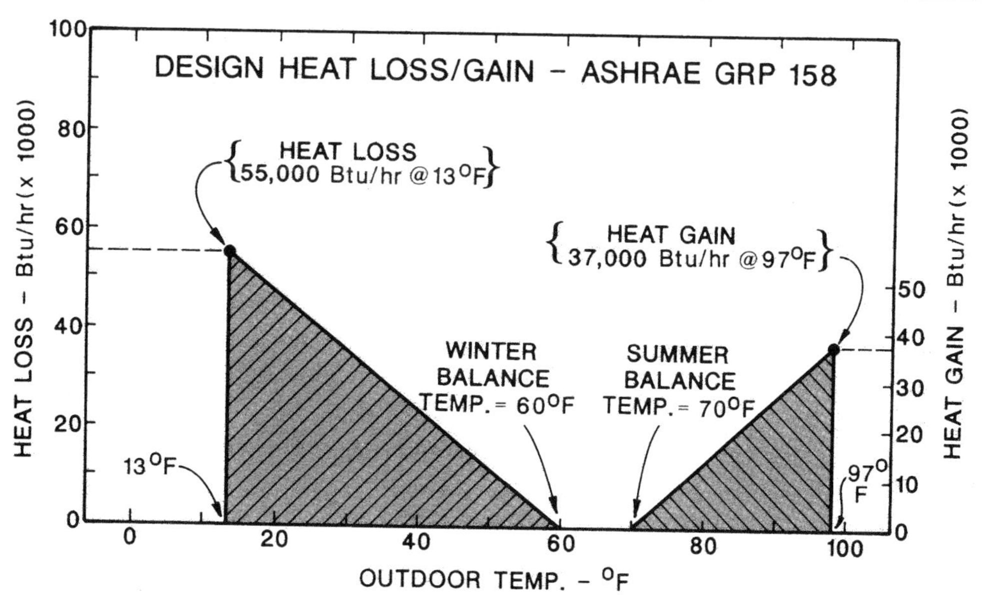

FIGURE 4.21: Design Heat Loss/Gain

3. Assume a balance temperature of 60°F.
4. For bin temperature range 45°/49°: 932.2 x (60 – 0.5(45+49)) = 12,118 Btu/hr
5. Heat pump capacity @ 35°F = 30,000 Btu/hr (from manufacturer's data)
6. Run hours = 69 x (12,118 / 30,000) = 27.9 hrs.
7. Repeat for all bin temperature ranges.
8. Sum all run hours and calculate the run fraction for January.

Figure 4.22 gives the heating run fraction (F_H) as a function of heat pump sizing, as a percent of the total load, and as a function of the local degree days.

COOLING RUN FRACTION (F_C) CALCULATION PROCEDURE

The calculation procedure for estimating the cooling run fraction is:

1. Estimate the heat gain (ASHRAE *GRP-158*).
2. Calculate the heat gain per degree F.
3. Assume a balance temperature (70°F).
4. Calculate the building's load for each bin temperature.
5. Check the building's demand with heat pump capacity. Calculate heat pump run hours for each bin temperature:
6. Sum run hours for each bin temperature and calculate run fraction:

$$F_H = \frac{\text{Monthly Run Hours}}{31 \times 24}$$

where: 31 days in July
24 hours per day

TABLE 4.12: Summary Table for Calculating Heating Run Fraction

Design Month - - - - - - January
Location - - - - - - Oklahoma City, Oklahoma
Building Heat Loss - - - - - - 55,000 Btu/hr @ 13°F
Heat Pump Capacity - - - - - - 30,000 Btu/hr @ 35°F

Temp Range °F	Temp °F	Bin Hours	Build Load Btu/hr	Heat Pump Hours
55/59	57	32	2,796	2
50/54	52	55	7,457	13
45/49	47	69	12,118	27
44/44	42	94	16,779	52
35/39	37	103	21,440	73
30/34	32	119	26,101	103
25/29	27	91	30,762	91
20/24	22	61	35,423	61
15/19	17	42	40,084	42
10/14	12	30	44,745	30
5/9	7	12	49,406	12
0/4	2	8	54,067	8
-5/-1	-3	1	58,728	1
				515

Heating run fraction = 515/24 x 31 = 0.69
Resistant heat required = (58,728 – 30,000)/3412 = 8.5 kW

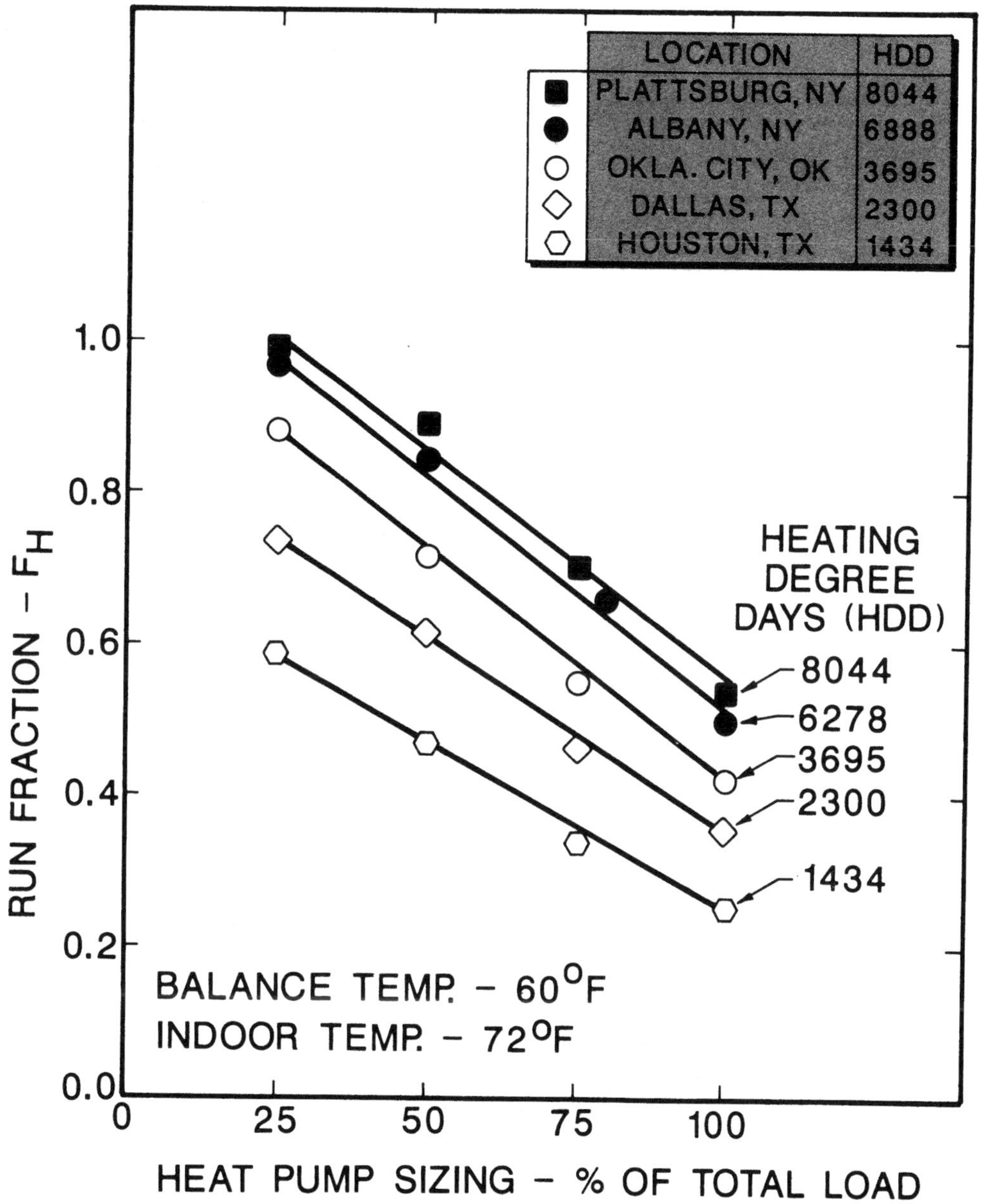

FIGURE 4.22: Heating Run Fraction — F_H

EXAMPLE –
Cooling Run Fraction

1. Building heat gain (ASHRAE *GRP-158*):
   Heat Gain: 37,000 Btu/hr
   Outdoor Temperature: 97°F
   (Oklahoma City, Oklahoma)
   Indoor Temperature: 78°F
2. Heat gain per degree F:
   37,000 / (97° – 78°) = 1947 Btu/hr/°F
3. Balance temperature = 70°F
4. For bin temperature range 100°/104°

4

TABLE 4.13: Summary Table for Calculating Cooling Run Fraction

Month - July
Location - Oklahoma City, Oklahoma
Building Heat Gain - - - - - - - - - - - 37,000 Btu/hr @ 97°F ODT, 78° IDT
Heat Pump Capacity - - - - - - - - - - - - - 36,000 Btu/hr @ 100°F EWT

Temp Range °F	Temp °F	Bin Hours	Building Load Btu/hr	Heat Pump Hours
100/104	102	9	62,315	15
95/99	97	51	52,578	74
90/94	92	91	42,842	108
85/89	87	103	33,105	94
80/84	82	142	23,368	92
75/79	77	171	13,631	64
70/74	72	123	3,894	13
				460

Run fraction = 460/(24 x 31) = 0.62

Avg temp = 0.5(100° + 104°) = 102°F
Building load = 1947.4 x (102°–70°) = 62,315 Btu/hr

5. Heat pump capacity = 36,000 Btu/hr @ 100°F EWT
62,315/36,000 x 9 hrs. = 15 hours @ 100°/104°F bin range
6. Repeat for all temperature ranges.

Figure 4.23 gives the cooling run fraction (F_C) versus cooling degree days for a number of locations.

STEP 4.9: Calculate the Ground Heat Exchanger Length

There are many design approaches that can be selected to yield similar results. There is one straightforward method that has found general acceptance:

1. Use the soil resistance values as given in Table 4.6. Adjust these values by the run time (run fraction for heating (F_H) and for cooling (F_H)) in both heating and cooling during the design months of January and July. Procedures for calculating the heating and cooling run fractions are given in Step 4.8. Figures 4.22 and 4.23 give typical values of run fractions based on heating and cooling degree days.
2. Assume the following design equations:

HEATING:

$$L_H \text{ (ft/ton)} = \frac{12\,000\,((COP_H - 1)/COP_H) \times (R_P + R_S \times F_H)}{(T_L - T_{MIN})}$$

COOLING:

$$L_C \text{ (ft/ton)} = \frac{12{,}000\,((COP_C + 1)/COP_C) \times (R_P + R_S \times F_C)}{(T_{MAX} - T_H)}$$

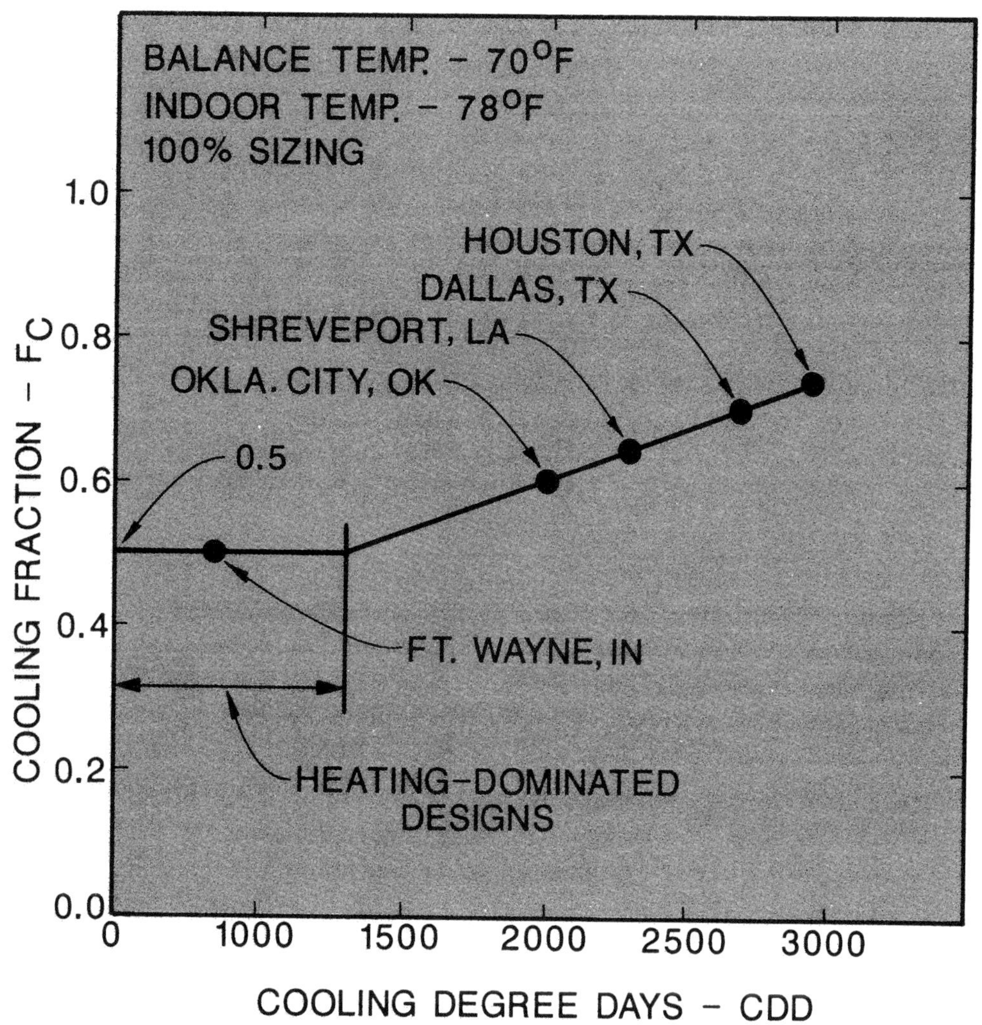

FIGURE 4.23: Cooling Run Fraction — F_C

Where:

L_H (ft/ton) = The length of heat exchanger required for each ton (12,000 Btu/hr) of heat pump heating capacity @ T_{MIN}

COP_H = Heat pump heating coefficient of performance @ T_{MIN}

L_C (ft/ton) = The length of heat exchanger required for each ton (12,000 Btu/hr) of heat pump cooling capacity @ T_{MAX}

COP_C = Heat pump cooling coefficient of performance @ T_{MAX}
= EER/3.412

R_P = Pipe resistance from Table 4.1 or 4.6

R_S = Soil/field resistance from Figures 4.9 through 4.14 or Table 4.6.

T_{MAX} = Defined in STEP 4.3 (designer's choice)

4

T_{MIN} = Defined in STEP 4.3 (designer's choice)

T_L = Defined in STEP 4.2 (set by location and heat exchanger depth)

T_H = Defined in STEP 4.2 (set by location and heat exchanger depth)

F_H = Heating run fraction, the heat pump run time during the design month of January expressed as a number less than 1 (Figure 4.22)

F_C = Cooling run fraction, the heat pump run time during the design month of July expressed as a number less than 1 (Figure 4.23)

T_{MAX} is the average maximum for the design month of July. It is not the highest water temperature that can occur during the cooling season.

T_{MIN} is the average minimum for the design month of January. It is not the lowest water temperature that can occur during the heating season.

An estimate of the highest and lowest temperatures that could be expected is given by:

$$T_{HIGH} = T_H + \frac{12{,}000\ \{[COP_C + 1]/COP_C\}\ \{R_P + R_S\}}{L_H\ (ft/ton)}$$

$$T_{LOW} = T_L - \frac{12{,}000\ \{[COP_H - 1]/COP_H\}\ \{R_P + R_S\}}{L_C\ (ft/ton)}$$

Recall that both the capacity and efficiency (COP_H and COP_C (EER/3.412)) will change at the new lower and higher temperatures.

Example — Ground Heat Exchanger Design Length

The following procedure is recommended for designing a ground heat exchanger:

1. Determine the local maximum and minimum air temperatures.
2. Determine the building heat loss and heat gain.
3. Select a heat pump based on the cooling load.

   NOTES:

   a. Oversizing of the heat pump by 25% of the cooling load is acceptable. See Section 3 for specific details.

   b. Initial estimates of the maximum heat pump entering water temperature (EWT) for sizing purposes for cooling are 10°F below the warmest outdoor air temperature. This value is used as the first estimate of T_{MAX}.

   T_{MAX} cannot be greater than the highest EWT given by the heat pump manufacturer in his data.

   At the heat pump EWT of T_{MAX} record:

   1. entering air temperatures (°F) ______ DB, ______ WB
   2. flow ______ GPM
   3. heat pump cooling capacity (CAP_C) ______ Btu/hr
   4. EER ______

   c. Initial estimates of the minimum heat pump entering water temperature (EWT) for sizing purposes for heating is 40°F warmer than the coldest air temperature during heating. This value is used as the first estimate of T_{MIN}.

   T_{MIN} cannot be lower than the minimum EWT specified by the manufacturer.

   At the heat pump EWT of T_{MIN} record:

   1. heat pump entering air temp (EA) ______ °F
   2. heating capacity (CAP_H) ______ Btu/hr
   3. COP_H ______
   4. water pressure drop ______ feet

4. Select the ground heat exchanger type: vertical or horizontal.
5. Estimate the ground temperatures T_L and T_H at the ground heat exchanger burial depth. Use Table 4.8 or Figure 4.16.
   a. For multiple horizontal buried pipe systems, use the average of the ground temperatures at each depth.
   b. For vertical systems, use $T_L = T_H = T_M$.
6. Select the cooling run fraction (F_C) (based on 100% sizing of the cooling load) from Figure 4.23.
7. Calculate the heating sizing factor:

$$SF(\%) = \frac{\text{Building heat loss (Btu/hr)}}{\text{Heat pump heating capacity (CAP}_H\text{, Btu/hr)}}$$

8. Select the heating run fraction (F_H) from Figure 4.22.
9. Select the pipe resistance from Table 4.6:
   a. Use R_P for single pipes.
   b. Use R_{PE} for loops (two pipes).
10. Select the soil/field resistance from Table 4.6.
11. Calculate the ground heat exchanger length using the following two equations given in Step 4.9. The equations have been modified from a length of feet/ton to a total length in feet for the specified heat pump at a heating capacity of CAP_H and a cooling capacity of CAP_C:

HEATING LENGTH (FEET):

$$L_H \text{ (feet)} = \frac{CAP_H\,((COP_H - 1)/COP_H) \times (R_P + R_S \times F_H)}{(T_L - T_{MIN})}$$

COOLING LENGTH (FEET):

$$L_C \text{ (feet)} = \frac{CAP_C\,((EER + 3.412)/EER) \times (R_P + R_S \times F_C)}{(T_{MAX} - T_H)}$$

12. If the ground heat exchanger length is unacceptable for any reason, select a new set of parameters and repeat the process.

EXAMPLE: Single-Pipe Horizontal Ground Heat Exchanger

1. Location: Stillwater, Oklahoma
   minimum outdoor air temperature $= -2°F$
   maximum outdoor air temperature $= 112°F$
2. Building heat loss: 55,000 Btu/hr
   Building heat gain: 40,000 Btu/hr
3. Heat pump selected: Model 049
   Trial value: $T_{MIN} = 40 + -2 = 38°F$; use 40°F
   Trial value: $T_{MAX} = 112 - 10 = 102°F$; use 100°F
   a. Size based on cooling load:
      Cooling: $EWT = T_{MAX} = 100°F$
         Entering air temp. = 75°F DB and 63°F WB
         Flow = 9 GPM
         Cooling capacity (CAP_C) = 40,100 Btu/hr
         EER = 7.8
      Heating: $EWT = T_{MIN} = 40°F$
         Entering air temp. = 70°F
         Heating capacity (CAP_H) = 45,800 Btu/hr
         $COP_H = 3.24$
         Water pressure drop = 13.1 feet of head

4

COOLING PERFORMANCE DATA							
EWT	GPM	EA	TC	SC	KW	HR	EER
90	9	75/63	42.6	31.9	4.84	59.1	8.8
100	9	75/63	40.1	30.8	5.17	57.7	7.8
110	9	75/63	37.6	29.7	5.50	56.3	6.8

EWT = Entering water temperature, °F
GPM = Gallons per minute
EA = Entering air temperature, DB/WB, °F
TC = Total capacity, Btu/hr
SC = Sensible capacity, Btu/hr
KW = Kilowatts
HR = Heat rejected, Btu/hr
EER = Energy Efficiency Ratio, Btu/Watt

HEATING PERFORMANCE DATA							
EWT	GPM	WPD	EA	HC	KW	HE	COP
30	9	14.2	70	38.9	3.74	26.1	3.04
40	9	13.1	70	45.8	4.13	31.7	3.24
50	9	12.0	70	52.7	4.51	37.3	3.43

WPD = Water pressure drop, feet of head
EA = Entering air temperature, DB, °F
HC = Heating capacity, Btu/hr
HE = Heat extracted, Btu/hr
COP = Coefficient of Performance

4. Horizontal single-pipe system, 1½-inch pipe
5. Estimate T_H and T_L:
   Assumed soil properties for Stillwater, Oklahoma:
   a. mean earth temperature: T_M = 62°F
   b. soil swing: A_S = 23°F
   c. phase shift: T_O = 35 days (not needed for maximum and minimum soil temperatures)
   d. thermal diffusivity: = 0.60 ft²/day

   The soil temperatures for Stillwater, Oklahoma, can be obtained in either of two principal ways:

   a. from the following two equations of Table 4.8:

   $T_L = T_M - A_S \times EXP(-X\ (\pi/(365 \times \alpha))^{1/2})$

   $62-23 \times EXP(-4\ (3.14/(365 \times 0.6))^{1/2}) = 47.8°F$

   $T_H = T_M + A_S \times EXP(-X\ (\pi/(365\ \alpha))^{1/2})$

   $62 + 23 \times EXP(-4\ (3.14/(365 \times 0.6))^{1/2}) = 76.2°F$

   b. from the soil temperature variation curve of Figure 4.16 which gives the following temperatures at depths of:
   X = 2, 3...9 feet:

T_L/T_H			
DEPTH, X (FEET)	SATURATED (α = 0.96) °F	MOIST (α = 0.60) °F	DRY (α = 0.26) °F
2	42/81	43/80	46/77
3	44/79	45/78	48/75
4	46/77	47.8/76.2	51/72
5	47/76	49/74	52/72
6	48/75	50/73	54/69
7	50/73	52/71	55/68
8	51/72	53/70	56/67
9	52/71	54/69	57/66

T_L and T_H are estimated from the above table to be:

T_L = 47.8°F; use 48°F

T_H = 76.2°F; use 76°F

6. The 100% sizing to the cooling load gives a cooling run fraction of (from Figure 4.23):

   $F_C = 0.6$

7. From Figure 4.22 the heating run fraction for Stillwater, Oklahoma, with 3600 DD and a sizing of 45,800/55,000 = 0.83 or 83%:

   $F_H = 0.5$

9. The pipe resistance (Table 4.6) for a single 1½-inch (SCH 40) pipe is:

   R_P = 0.117 hr-ft-°F/Btu

10. The soil/field resistance value for a horizontal ground heat exchanger with a single 1½-inch (SCH 40) pipe in a heavy soil (damp) at 4 feet is (from Table 4.6):

    R_S = 0.94 hr-ft-°F/Btu

11. Ground heat exchanger length (feet):

$$L_H = \frac{45{,}800\ \{(3.24-1)/3.24\}\times[0.117+0.5\times0.94]}{(48-40)} = 2{,}323 \text{ feet}$$

$$L_C = \frac{40{,}100\ [(7.8+3.412)/7.8]\times[0.117+0.6\times0.94]}{(100-76)} = 1{,}636 \text{ feet}$$

The following table gives ground heat exchanger lengths for other entering water temperatures (EWT):

HORIZONTAL SINGLE-PIPE GROUND HEAT EXCHANGER

COOLING

EWT (T_{MAX})	CAP_C	F_C	EER	PIPE (Ft/Ton)	L_C (Ft)
90	42,600	0.57	8.8	776	2,756
100	40,100	0.60	7.8	489	1,636
110	37,600	0.64	6.8	380	1,193

HEATING

EWT (T_{MIN})	CAP_H	F_H	COP_H	SF%	PIPE (Ft/Ton)	L_H (Ft)
30	38,900	0.58	3.04	71	296	960
40	45,800	0.50	3.24	83	608	2,323
50	52,700	0.44	3.43	96	—	—

4

The designer has a wide range of choices in the length of the ground heat exchanger. Using the shorter length of about 1,200 feet will result in a temperature swing from 30° to 110°F from winter to summer. This may be the most economic choice from the owner's standpoint.

The longer length of 2,500 feet would result in a swing in loop temperature from 40° to 90°F and an increase in heat pump capacity and efficiency. There is a slight demand reduction using the longer length, but, perhaps more importantly, there is a margin of safety built in for possible soil drying in the summer cooling periods.

During heating, the heat pump cannot run at an EWT of 50°F since the ground temperature minimum is 48°F. A deeper depth is not feasible.

EXAMPLE: Horizontal Two-Pipe Ground Heat Exchanger

The ground heat exchanger length for two 1-inch pipes buried in the same trench at depths of 3 and 5 feet would be the same as above except that:

1. the soil/field resistance value would be changed from 0.94 to 1.26 and
2. the pipe resistance would change from .117 to .141 for PE SDR-11 pipe.

Also, since the average depth is (3 + 5)/2 = 4 feet, the soil temperatures remain the same as in the previous example.

TWO-PIPE HORIZONTAL GROUND HEAT EXCHANGER

EWT (T_{MAX}/T_{MIN})	PIPE (Ft/Ton)	TRENCH (Ft/Ton)	PIPE (Feet)	TRENCH (Feet)
		COOLING		
90	1,022	511	3,628	1,814
100	644	322	2,152	1,076
110	502	226	1,513	786
		HEATING		
30	390	185	1,264	632
40	800	400	1,526	3,052
50	—	—	—	—

A trench length of 1,500 feet looks to be a good compromise between the heating and cooling requirements.

EXAMPLE: Vertical Ground Heat Exchanger

The following inputs will change for a vertical ground heat exchanger:

1. $T_L = T_H = T_M = 62°F$
2. Pipe resistance R_{PE} for a ¾-inch PE SCH-40 pipe in a vertical loop is (from Table 4.6):
   R_{PE} = 0.116 ft–°F–hr/Btu
3. The soil/field resistance (from Table 4.6) will be assumed to be 0.65. This assumes some rock layers and saturated soil in the bore hole.
   R_S = 0.65 ft–°F–hr/Btu

A system bore hole length of 800 feet would result in an EWT of 40°F+ in the winter at design and 100°F– in the summer. Additional length for demand reduction is not warranted because of the additional cost. For example, to go from an EWT of 40° to 50°F would require a doubling of the bore hole length.

VERTICAL GROUND HEAT EXCHANGER			
EWT (T_{MAX}/T_{MIN})	BORE LENGTH (Feet/Ton)	BORE HOLE (Feet)	PIPE (Feet)
COOLING			
90	289	1,025	2,050
100	230	768	1,537
110	200	626	1,253
HEATING			
30	124	401	802
40	166	633	1,266
50	284	1,247	2,494

Selecting the Circulating Pump

STEP 1: Check Fluid System Pressure Drop and Select Circulating Pump

The circulating pump must have sufficient power to overcome the total back pressure from flow through the:

1. ground heat exchanger
2. heat pump water heat exchanger
3. equipment room piping including all fittings, hoses, etc.

The following example details a typical system design and layout which can be used as a basis for determining system pressure drop and circulating pump selection.

EXAMPLE – System Design and Layout

Figure 4.24 gives an overall layout and details of a typical heat pump system. The system can be described as follows:

GROUND HEAT EXCHANGER

1. The system uses a parallel ground heat exchanger with 1½-inch headers and ¾-inch PE or 1-inch PB loops.
2. Supply and return headers are manifolded together with a reverse return system.
3. Bell reducers are used to eliminate all air traps that could exist in the buried ground heat exchanger.
4. All underground piping is heat fused.

CIRCULATING SYSTEM

1. A circulating manifold is detailed in Figure 4.25, and Figure 4.26 shows the service valves connected at the point where the ground heat exchanger penetrates the building floor. Figure 4.27 details the use of 4-inch PVC pipe for stubbing out the supply and return headers. This allows the supply and return headers to be installed after the slab/floor has been poured.

   Figure 4.28 shows the equipment room hookup and the necessary equipment to service the connected system. Two service valves are included to fill, flush, and pressurize the system.

   Several manufacturers have combined the function of the circulating manifold and service valves into a single package called a "pump kit" or a "pump module" as shown in Figure 4.29. These prefabricated units save time and eliminate a number of potential mistakes such as improper pipe joining, incorrect and incompatible materials, and improper circulating pump and pipe sizing.

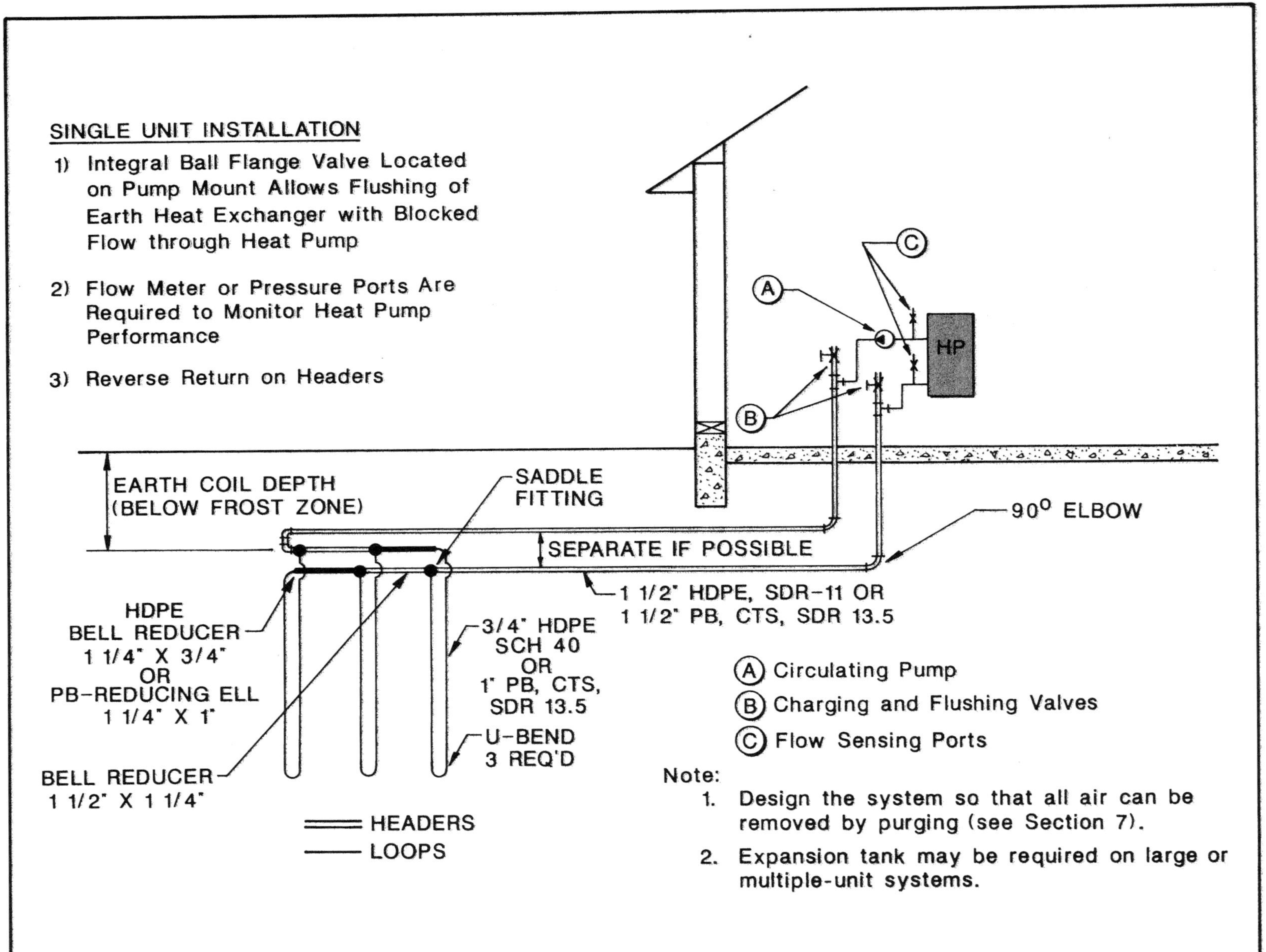

FIGURE 4.24: Layout and Details of a Typical CL/GS Heat Pump System

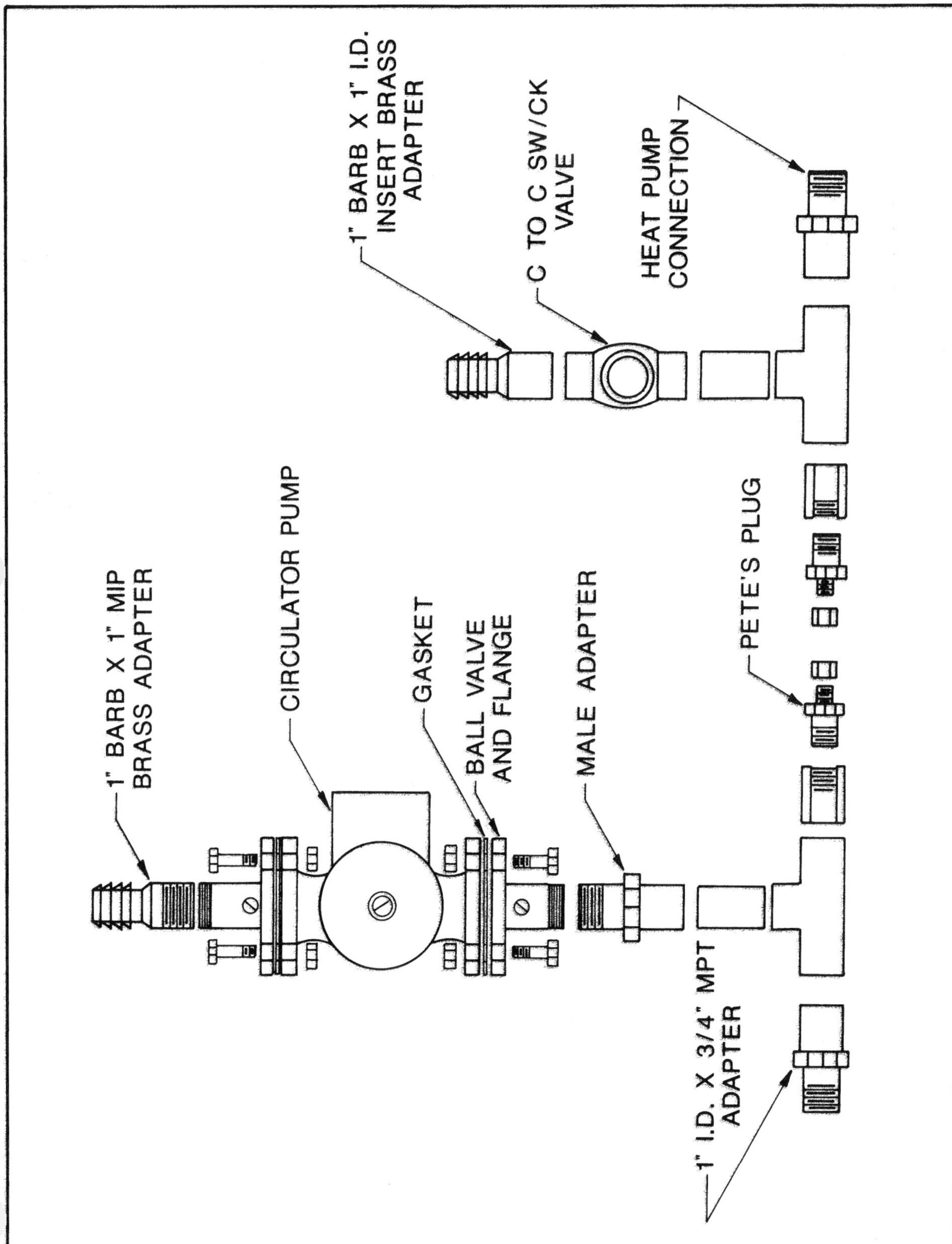

FIGURE 4.25: Circulating Manifold Detail

4

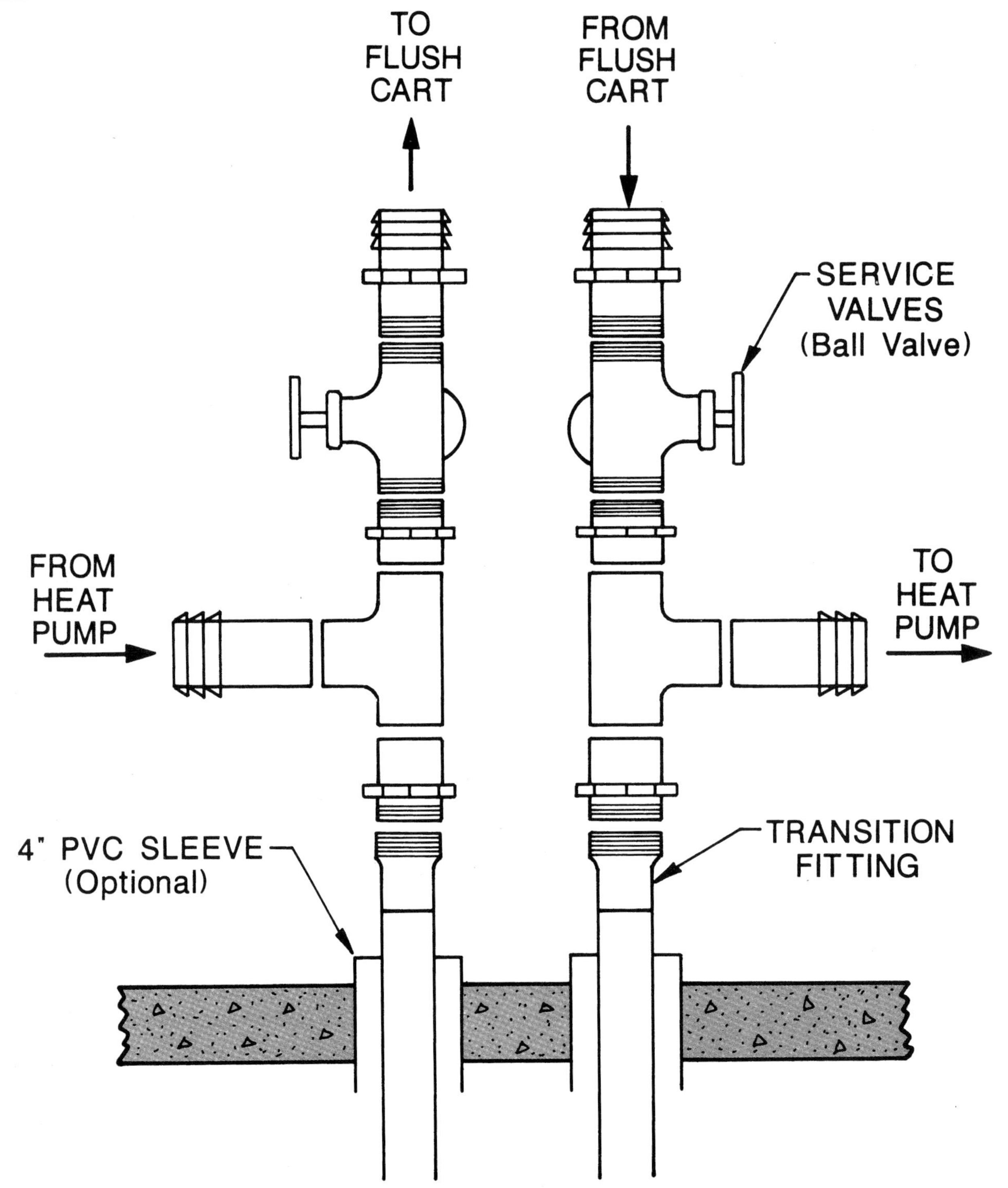

FIGURE 4.26: Loop Servicing System

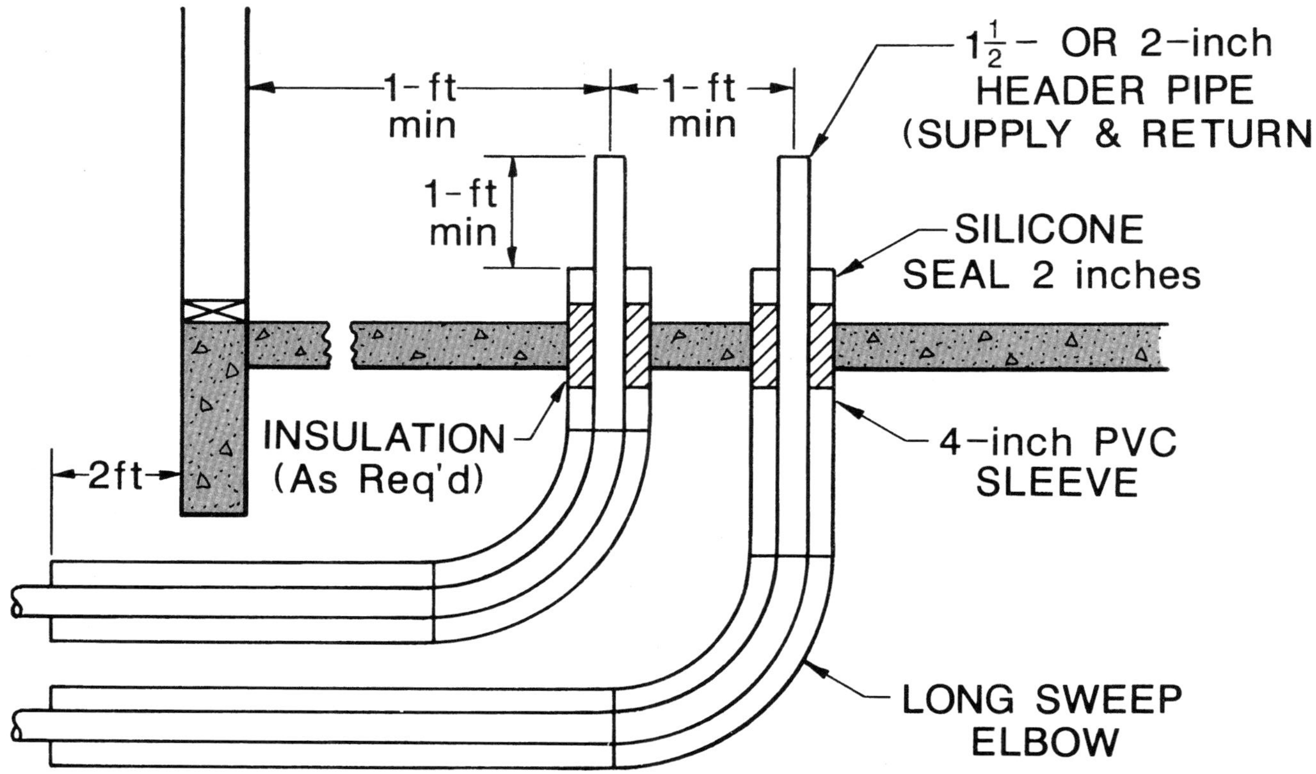

Note:

(1) Ground heat exchanger stub outs should be spaced a minimum of 1 foot from any wall or obstruction.

(2) Long sweep elbows are required for $1\frac{1}{2}$- or 2-inch header pipes.

FIGURE 4.27: Ground Heat Exchanger Building Stub Out

4

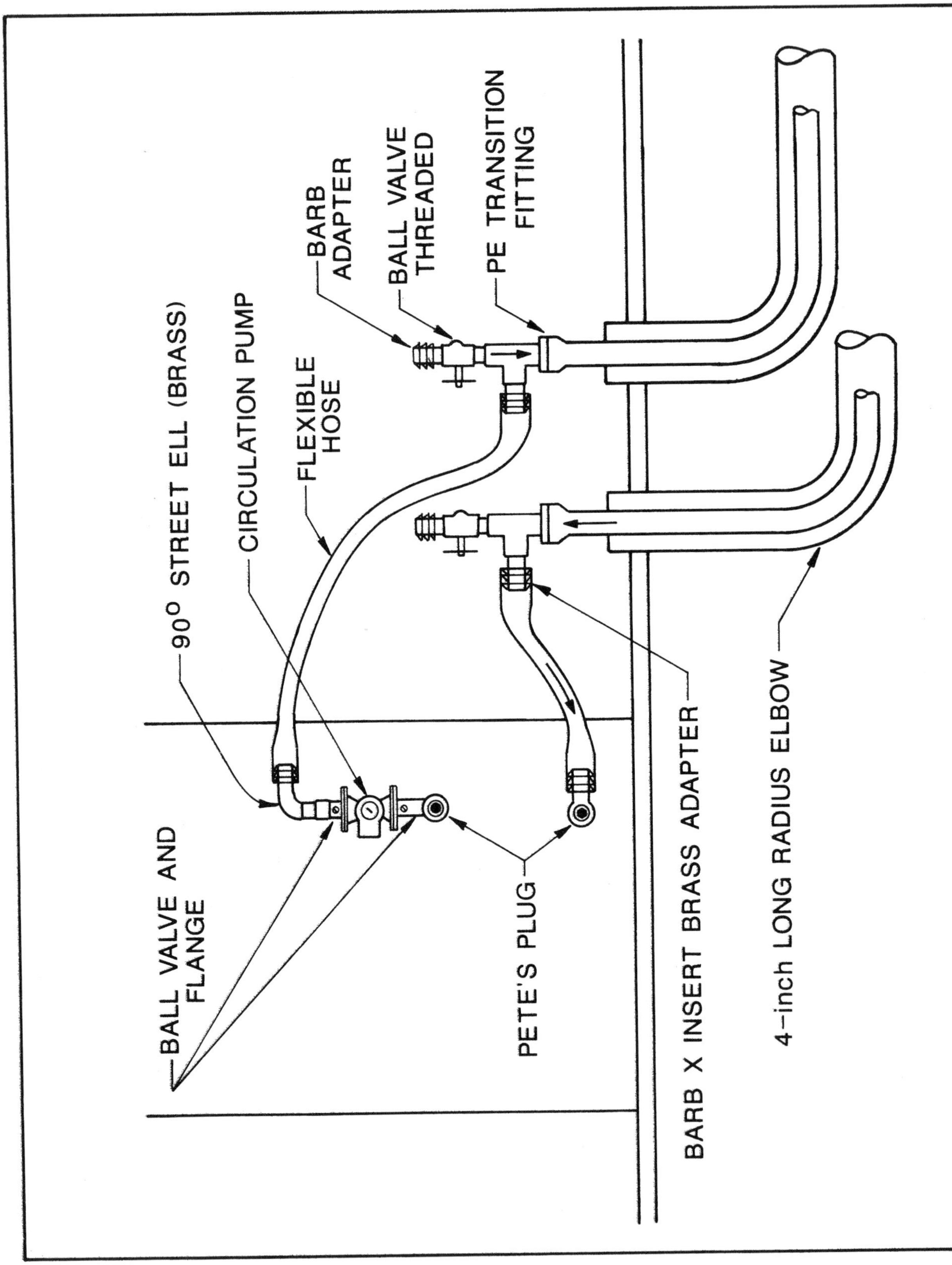

FIGURE 4.28: Heat Pump and Ground Heat Exchanger Hookup

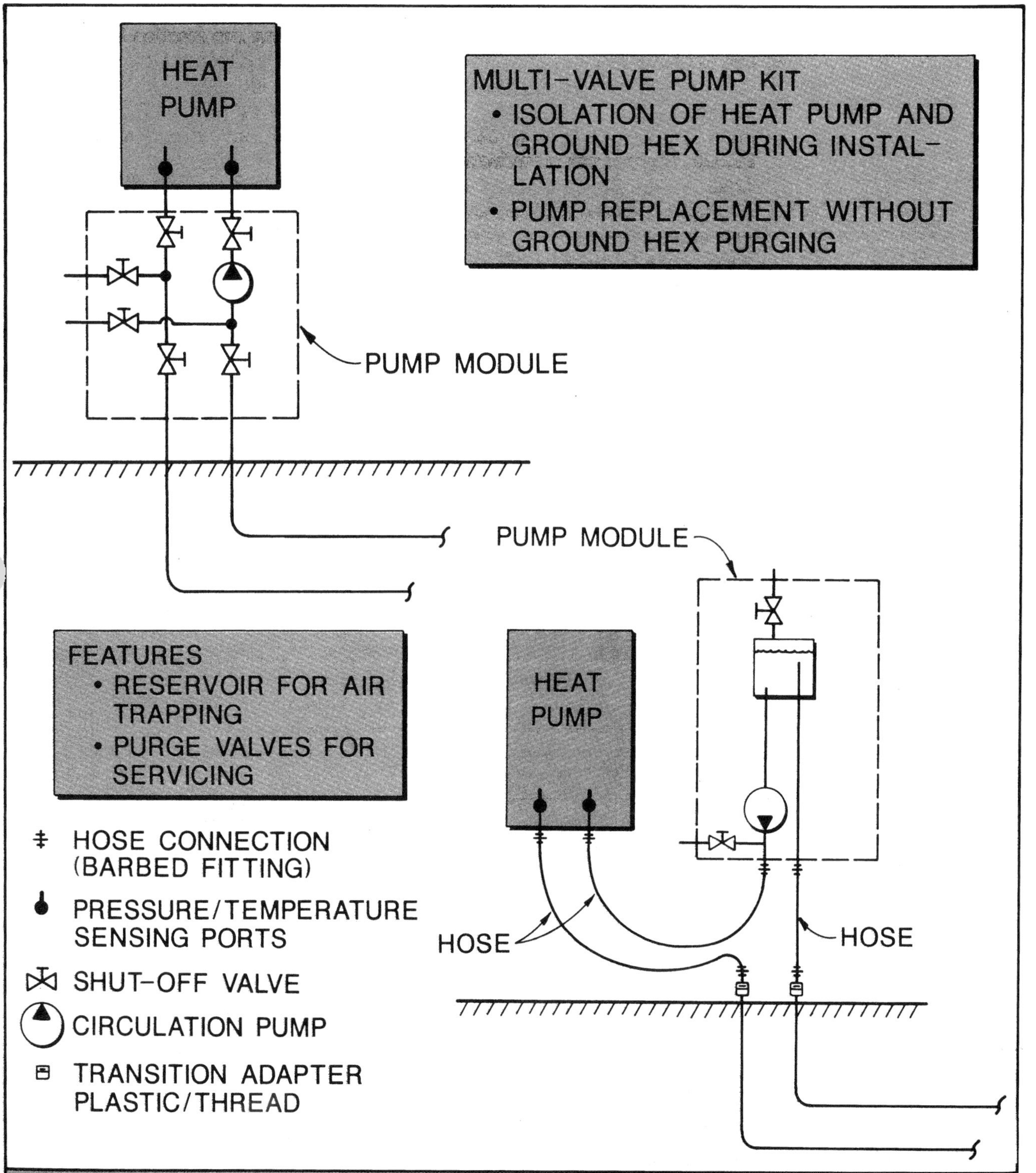

FIGURE 4.29: Pump Module and Hookup

4

2. Pressure and temperature ports are included on the inlet and outlet (Figure 4.28) of the water-source heat pump. Figure 4.30 shows some of the equipment that can be used to check the heat pump's performance. Also shown (Figure 4.30) is a special adapter that reduces the number of fittings necessary to install the service ports. These ports are located as close to the heat pump unit as practical. A pressure gauge and temperature probe adapter can be used to measure the fluid conditions at the inlet and outlet of the water heat exchanger on the heat pump.
3. Ball valves are integral to the pump flange valves and are used during air purging or pump replacement. These valves will be fully open during heat pump operation and have no useful purpose beyond system startup and maintenance (if required).
4. A check valve (Figure 4.25) is required where multiple units are connected to a single ground heat exchanger (Figure 4.31). Without the check valve, it is possible for one heat pump to circulate through a second heat pump that is not running, thus bypassing the ground heat exchanger.

Caution should be used when more than three heat pumps are connected as shown in Figure 4.31. A flow balancing problem can exist when attempts are made to connect heat pumps of different sizes and flow requirements.

5. Also shown is a method of floor penetration (Figure 4.27). This allows the ground heat exchanger system to be installed after the concrete slab has been poured. Service valves are included to allow the flushing of de-

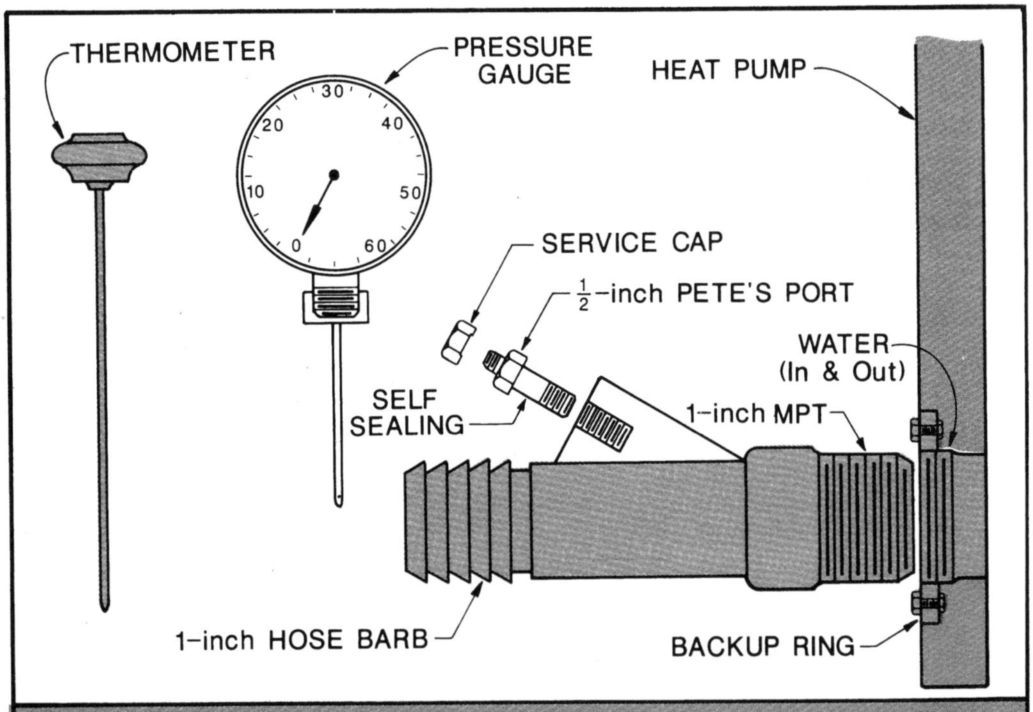

FIGURE 4.30: Pressure and Temperature Sensing Adapter and Components

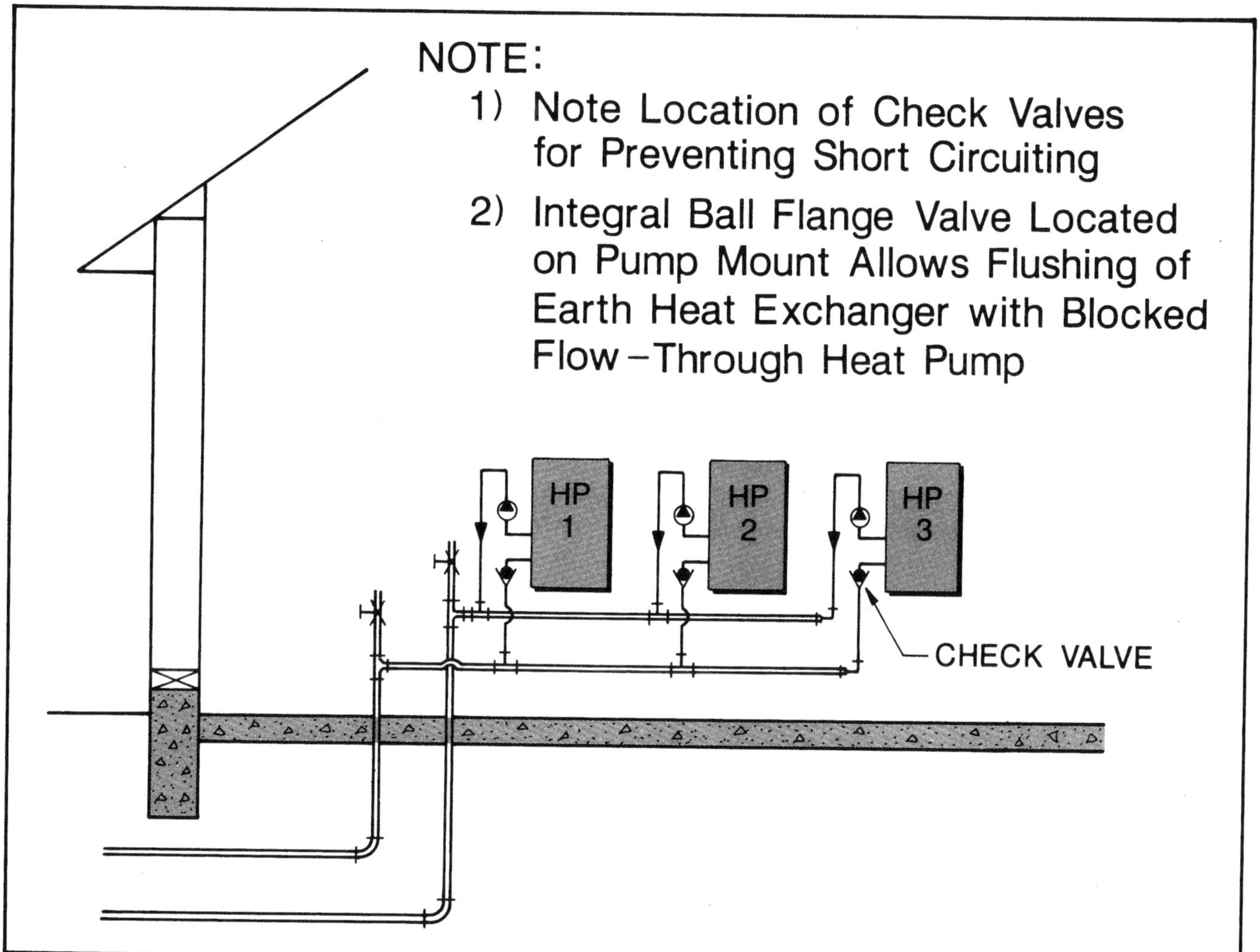

FIGURE 4.31: Multiple Unit Installation

bris, air purging, and the addition of antifreeze if needed.

6. Barbed hose fittings are used to connect the ground heat exchanger, the circulating manifold, and the floor penetration valve assembly.

Figure 4.32 gives pressure/flow performance curves for the ground heat exchanger (loop), system (loop plus heat pump), and the circulating pump. These curves help in understanding the design steps in matching the heat pump, circulating pump, and ground heat exchanger.

1. The loop pressure (ΔP) gives the pressure in feet of head as a function of flow rate Q through the ground heat exchanger.
2. The system ΔP (loop plus heat pump) gives the pressure in feet of head as a function of flow rate Q (GPM) through the ground heat exchanger, circulating system, elbows, fittings, hose, etc.
3. The two dashed curves give these same pressure drops but with 20% calcium chloride at a design

4

minimum entering liquid temperature of 30°F.

4. The manufacturer's data for recommended flow is 5 to 11 GPM. This limits or sets the maximum flow that can be utilized in the system.
5. Two circulating pumps' (UP 26-96 F or CIR 26-96 and UP 26-64 or CIR 26-64, Figures 4.32 and 4.33) performance curves are plotted on the system (loop plus heat pump) curve. The designer should choose the particular unit based on circulating power input and the corresponding heat pump output for the given flow.
6. The intersection of the pump curve and the system ΔP curve gives the system flow as 9 GPM if CIR 26-96 is used.
7. In this case, the larger pump (CIR 26-96) is recommended since the capacity of the heat pump increases from approximately 32,500 to 33,500 Btu/hr (Table 4.14) when the flow is increased from 7 to 9 GPM.
8. The minimum flow rate for this system is 6.9 GPM (3 loops x 2.3 GPM/loop from Table 4.2) since a lower flow rate would result in laminar flow in the ¾-inch pipe loops of the ground heat exchanger.
9. A larger circulating pump could be chosen to increase the flow up to the manufacturer's maximum flow rate of 9 GPM. Before this is done, the extra cost of the circulating pump and the increase in heat pump performance should be ascertained.

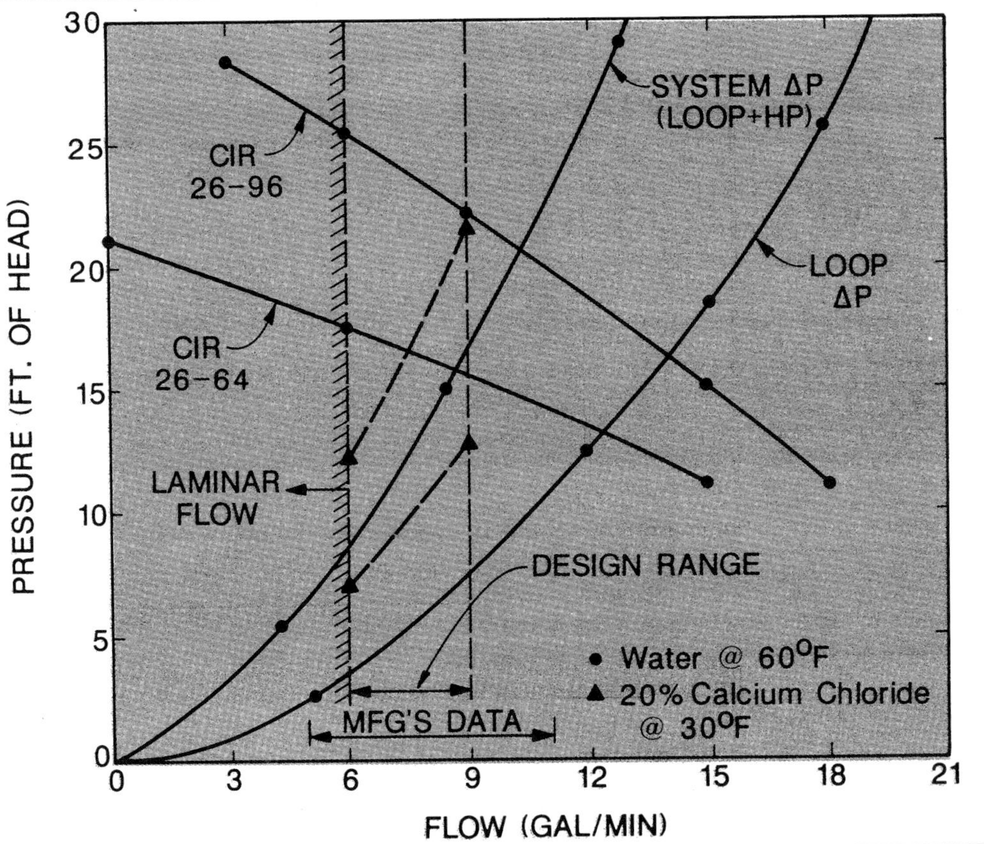

Figure 4.32: System Pressure Drop Curves

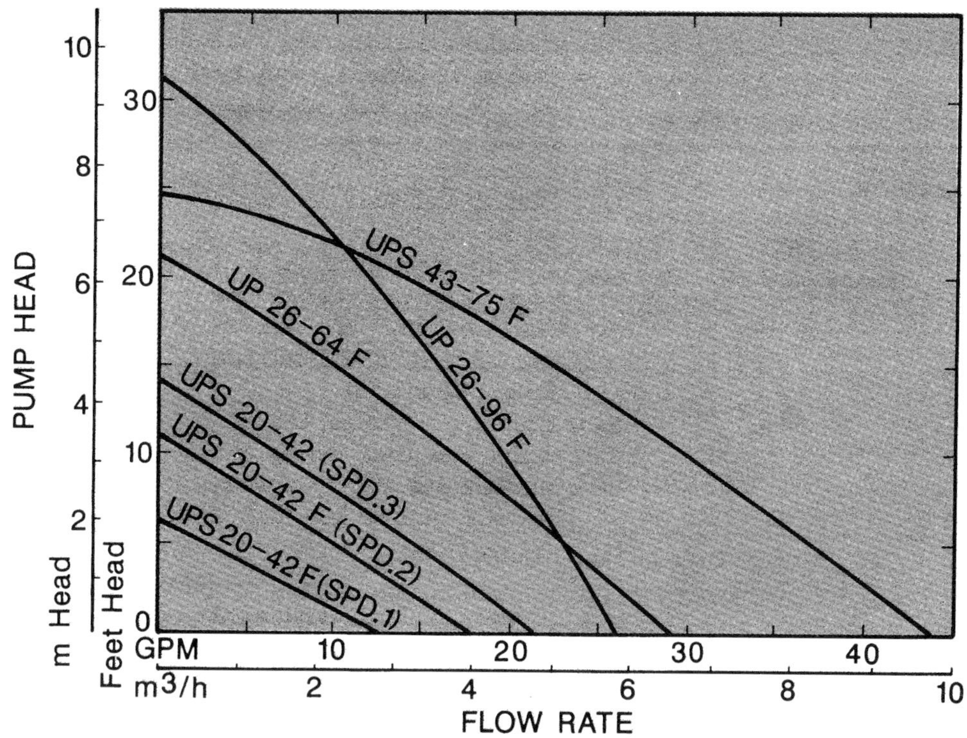

FIGURE 4.33: Typical Circulating Pump Performance Curves

TABLE 4.14: Heat Pump Performance Data (EWT = 30°F, EA = 70°F)

FLOW RATE (GPM)	WPD (Feet)	CAPACITY (Btu/hr)	COP
4.5	3.5	23,200	2.94
7.0	8.4	24.500	3.00
9.0	13.5	25,400	3.02

Table 4.15 is a worksheet which organizes the steps in sizing and selecting a circulating pump. Table 4.16 or Figure 4.33 can be used to pick a particular circulating pump to meet the requirements of the system detailed in Figure 4.32. The 6 GPM at 13 feet of head can be satisfied by pump model 26-64. At 9 GPM and 23 feet of head, pump model 26-96 can be used.

EXAMPLE – Circulating Pump Selection

The sample worksheet (Table 4.17) details the selection of the circulating pump for the system described in Figure 4.24. The system consists of:

1. supply and return headers of 150 feet each or 300 feet total including 8 elbows, tees, etc.

TABLE 4.15: Circulating Pump Worksheet

1. HEAT PUMP MODEL NUMBER ________, Flow ______ GPM, HEAD ______ (FEET) @ ______ °F
2. Count each elbow, tee, reducer, air scoop, flowmeter, etc., as 3 FEET OF PIPE EQUIVALENT. Add the equivalent feet of pipe to the actual feet of pipe used. The TOTAL LENGTH is used to determine the PIPING HEAD LOSS below.

NOTE: For parallel loops, divide the heat pump GPM by the number of loops to determine the flow through each loop.

	Pipe type and size	No. elbows, tees, etc.		Equiv. feet of pipe		Actual pipe used		Total pipe length
a.	______	______	x3	______	+	______	=	______
b.	______	______	x3	______	+	______	=	______
c.	______	______	x3	______	+	______	=	______
d.	______	______	x3	______	+	______	=	______

3. Piping head loss for different types of pipe at GPM flow rate of water-source heat pump.
NOTE: For parallel earth-loop system, figure for only one loop.

	Pipe type and size	Flow GPM	Total pipe length		Piping head loss/100 ft.		
a.	______	______	(/100)	x	( )	=	______ ft. head
b.	______	______	/100)	x	( )	=	______ ft. head
c.	______	______	(/100)	x	( )	=	______ ft. head
d.	______	______	(/100)	x	( )	=	______ ft. head
e.	Heat Pump	______					______ ft. head
					Subtotal		______ ft. head
					Total		______ ft. head

For antifreeze solutions, multiply the subtotal by the following constants:

( ) a. 1.36 — Propylene glycol 20% @ 25°F
( ) b. 1.25 — Calcium Chloride 20% @ 25°F
( ) c. 1.25 — Methanol Alcohol 20% @ 25°F

4. Pump Selection: Use Table 4.16 and select pump output which is larger or equal to the total pressure drop. (______ Cir. Pump Model) — (______ No. of Pumps)

Source: Adapted from Bard Manufacturing Co.'s *Earth Coupled Loop System Design.*[16]

2. vertical loops of 200-foot bore holes or 400 feet of pipe in each loop including 6 elbows, tees, etc.
3. heat pump connection with 10 feet of rubber hose
4. heat pump liquid flow of 9 GPM
5. methanol alcohol (20%) as an antifreeze solution

The total feet of head required to overcome the friction in the system is 30.6 feet of head. From Table 4.16, two pumps will be required to meet the pressure loss.

TABLE 4.16: Selection of Circulating Pump or Pumps

		Water Flow Rate Required in GPM						
Pump Models (Grundfos)	**No. of Pumps**	**4**	**5**	**6**	**8**	**10**	**12**	**14**
20–42	1	11.5	11.0	—	—	—	—	—
26–64	1	18.8	18.0	17.5	16.0	14.5	13.5	12.0
26–64	2	37.6	36.0	35.0	32.0	29.0	27.0	24.0
40–75	1	23.8	23.5	23.0	22.5	21.5	21.0	19.5
40–75	2	47.6	47.0	46.0	45.0	43.0	42.0	39.0
26–96	1	27.5	27.0	26.0	23.5	21.5	19.0	16.0
26–96	2	55.0	54.0	52.0	47.0	43.0	38.0	32.0

Model numbers 26-64, 40-75, or 26-96 could be used. Price and/or availability of each particular model could be the determining factor.

The information given in Figure 4.34 is helpful in determining when two pumps are required. In most cases, this system will require two pumps when an antifreeze solution is required.

Table 4.2 gives the minimum flow rate for turbulence in ¾-inch (0.86-inch) pipe. The system in Figure 4.24 with three loops will require a system (or header) flow rate of three times the individual loop minimum values to ensure adequate heat transfer in the ground loop.

A single pump can be used if the fluid used in the system is water. The total system flow required to ensure turbulence in each loop is only 3.3 GPM when water is used. The system flow with a single pump and water (from Figure 4.34) is approximately 8.5 GPM. The use of a second pump would result in a system flow of 12 GPM and an increase of only 1200 Btu/hr in heat pump capacity.

However, water-only systems can only be used in southern locations.

The success in using methanol and calcium chloride and a single pump is marginal. In this case, two pumps should be used of the type detailed in Table 4.16.

Propylene glycol requires the use of two pumps. It is important that the level of antifreeze protection is not exceeded since viscosity increases very rapidly with propylene glycol.

FLUID	SINGLE LOOP (GPM)	SYSTEM FLOW (THREE LOOPS) (GPM)
Water	1.1	3.3
Calcium Chloride	2.3	6.9
Methanol	2.4	7.2
Propylene Glycol	3.4	10.2

4

TABLE 4.17: Circulating Pump Worksheet — Sample

1. HEAT PUMP MODEL NUMBER 049, Flow 9 GPM, HEAD 13.1 (FEET) @ 40 °F
2. Count each elbow, tee, reducer, air scoop, flowmeter, etc., as 3 FEET OF PIPE EQUIVALENT. Add the equivalent feet of pipe to the actual feet of pipe used. The TOTAL LENGTH is used to determine the PIPING HEAD LOSS below.

NOTE: For parallel loops, divide the heat pump GPM by the number of loops to determine the flow through each loop.

	Pipe type and size	No. elbows, tees, etc.		Equiv. feet of pipe		Actual pipe used		Total pipe length
a.	1½-inch SCH 40	8	x 3	24	+	300	=	324
b.	¾-inch SDR-11	6	x 3	18	+	400	=	418
c.	1-inch hose	—	x 3	—	+	10	=	10
d.			x 3		+		=	

3. Piping head loss for different types of pipe at GPM flow rate of water-source heat pump.
NOTE: For parallel earth-loop system, figure for only one loop.

	Pipe type and size	Flow GPM	Total pipe length		Piping head loss/100 ft.		
a.	1½-inch SCH 40	9	(324/100)	x	(0.7)	=	2.26 ft. head
b.	¾-inch SDR-11	3	418/100)	x	(2.07)	=	8.65 ft. head
c.	1-inch hose	9	(10/100)	x	(5.0)	=	0.5 ft. head
d.			(/100)	x	( )	=	ft. head
e.	Heat Pump	9					13.1 ft. head
					Subtotal		24.51 ft. head
					Total		30.6 ft. head

For antifreeze solutions, multiply the subtotal by the following constants:

( ) a. 1.36 — Propylene glycol 20% @ 25°F
( ) b. 1.25 — Calcium Chloride 20% @ 25°F
(x) c. 1.25 — Methanol Alcohol 20% @ 25°F

4. Pump Selection: Use Table 4.16 and select pump output which is larger or equal to the total pressure drop. (26–96) – (2)
Cir. Pump Model — No. of Pumps

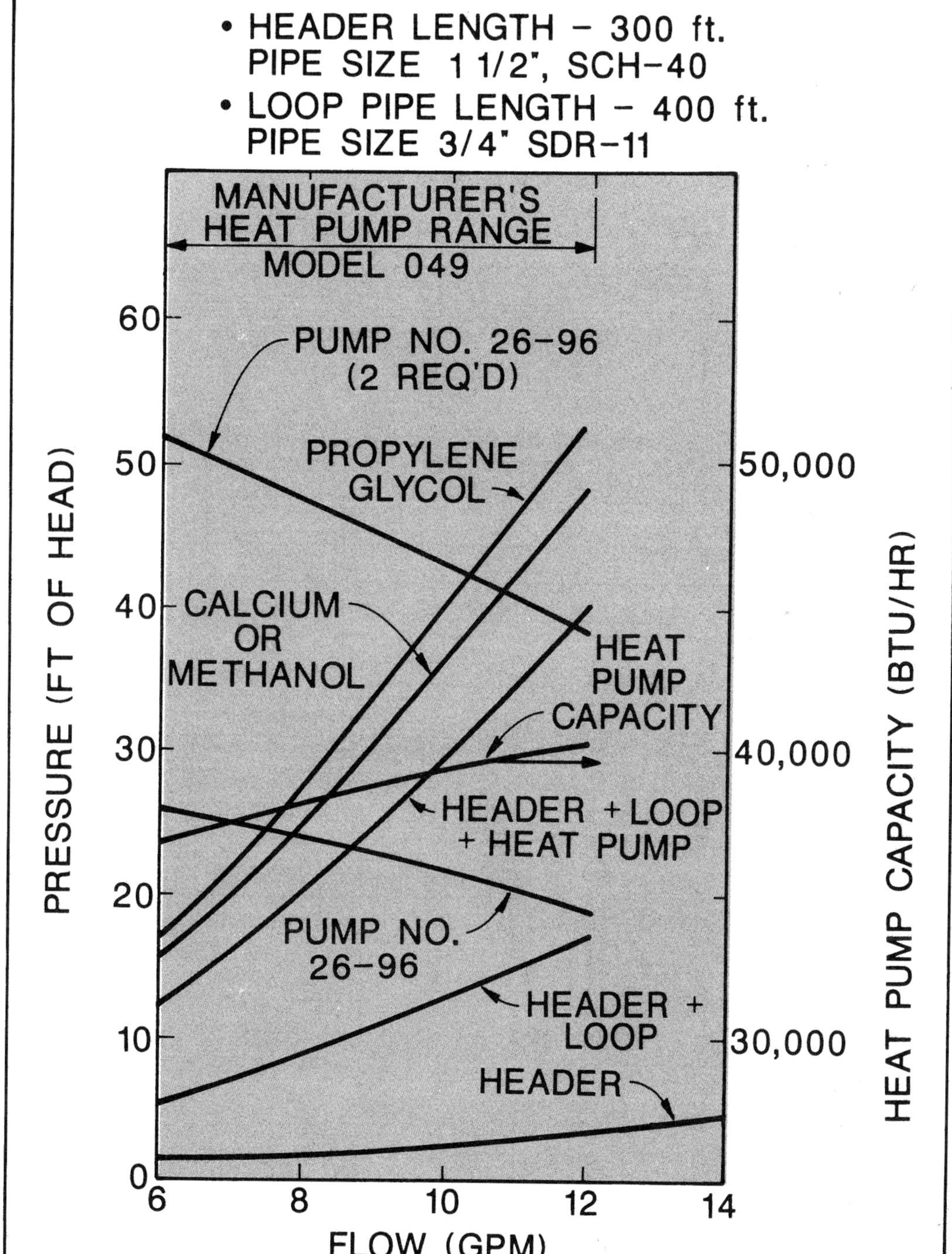

FIGURE 4.34: System Pressure Drop Curves for Sample System

5 Pipe Joining Methods

In This Section: Butt fusion procedure; sidewall fusion joining; socket fusion

Heat Fusion: Heat fusion is the process in which plastic pipe materials are aligned, cleaned or trimmed, heated to their melting point, brought together, and allowed to cool to form a homogeneous material. The appropriate American Society for Testing and Materials (ASTM) standards are:

1. D 2610 "Specifications for Butt Fusion Polyethylene (PE) Plastic Pipe Fittings, Schedule 40"
2. D 2683 "Specification for Socket-Type Polyethylene Fittings for Outside Diameter Controlled Polyethylene Pipe"
3. D 2657 "Practice for Heat Joining of Polyolefin Thermoplastic Pipe and Fittings"

For reliability, all underground piping joints must be thermally fused rather than mechanically coupled. The arguments for this are:

1. Heat fusion joining results in a joint which is stronger than the pipe itself. The pipe wall at this point has a larger cross section which gives the joint the added strength.
2. The connection or joint is all plastic, eliminating corrosion problems.
3. Industry standards (ASTM, PPI) in their respective written guidelines recommend thermal fusion for proper joining.

Training programs in the correct methods are available from the manufacturers of pipe and fusion machines.

Heat Fusion Methods: The industry has accepted two heat fusion techniques, butt and socket. The fusion method chosen will depend on the pipe manufacturer's requirements and recomendations and the ease with which field personnel can be trained to reliably fabricate ground heat exchangers.

Socket Fusion Joining: In the socket fusion method, the two pipe ends are joined by fusing each pipe end to a socket fitting (Figure 5.1). This requires two heat fusion procedures for each joint.

Butt Fusion: The second type of heat fusion joining is the butt fusion procedure where the two pipe ends are simultaneously heated to a plastic state by a heater plate and brought together to form the heat-fused joint (Figure 5.2). A single heat fusion process is required to form the joint between the two plastic pipe ends. The butt fusion process is performed by using specially designed machines (available from several manufacturers) which provide for securely holding the two pieces to be fused, aligning them, trimming and squaring their ends, heating the surfaces

5

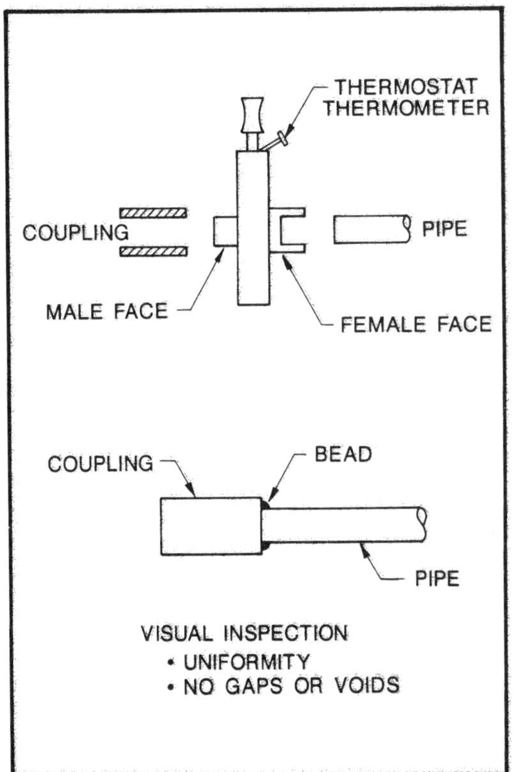

FIGURE 5.1: Socket Fusion

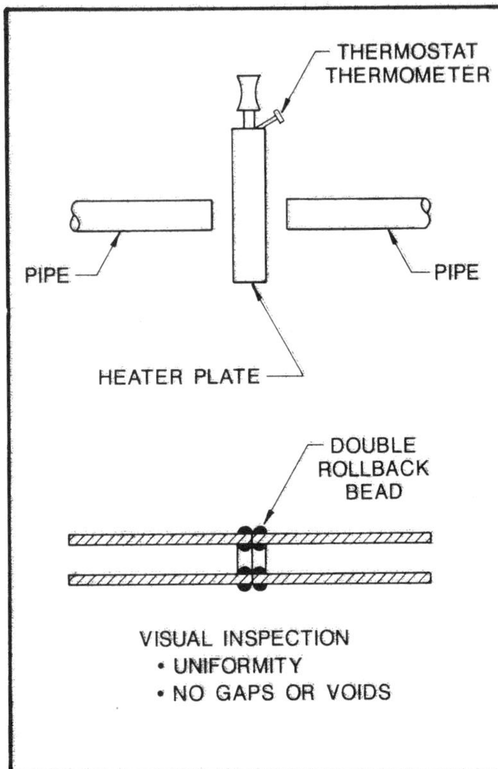

FIGURE 5.2: Butt Fusion

to be joined with a heater plate, and butting them together while they remain in a plastic state which produces a double rollback bead.

Sidewall Butt Fusion Joining: In addition to joining the flat ends of pipe or fittings, the butt fusion process can also be used when joining the concave surface of a service saddle to the convex surface on the sidewall of a pipe (Figure 5.3). This is done by heating these surfaces with appropriate convex- and concave-shaped heaters and then butting these surfaces together. This is known as sidewall fusion joining. Figure 5.4 shows a cross section of a completed sidewall fusion joint with the triple bead that is formed during the fusion process.

Selection of Heat Fusion Method: Polyethylene is heat fused and joined using both butt and socket procedures. Material grade, density, etc., will determine if the particular grade can be fused with either method. Some high-density polyethylene materials cannot be socket fused.

Polybutylene in the sizes used in ground heat exchangers is normally socket fused.

Both fusion procedures when properly done yield highly reliable joints that are stronger than the pipe itself. Both procedures are described below.

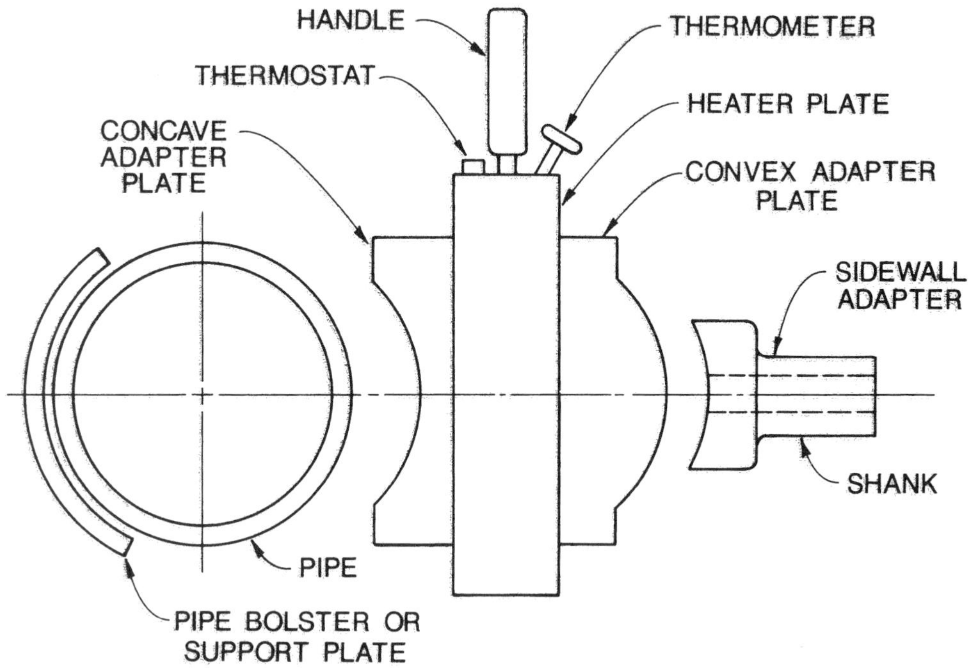

FIGURE 5.3: Sidewall Fusion

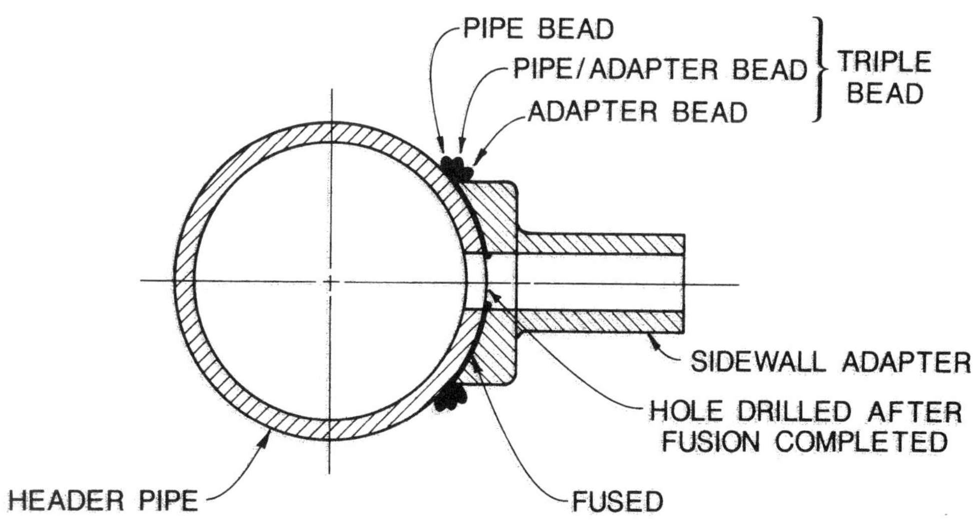

FIGURE 5.4: Sidewall Fusion Beads

5

Butt Fusion Procedure*

1. ASSEMBLE THE REQUIRED EQUIPMENT AND TOOLS:
   a. appropriate power supply
   b. heater plate with appropriate adapters
   c. tubing cutters
   d. clean, dry cotton rags
   e. pipe facer (hand or electrical)

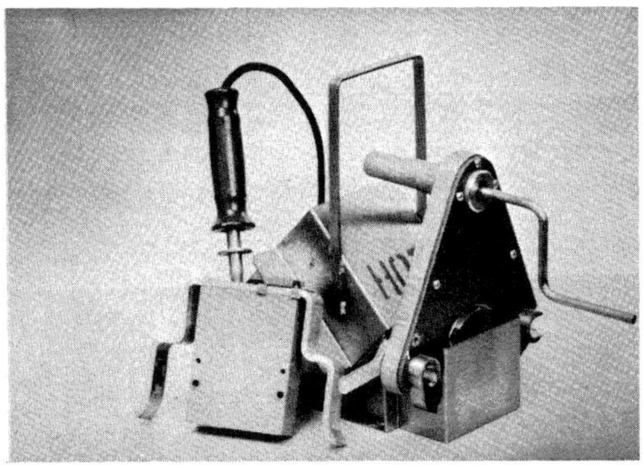

Equipment and Tools

2. PREPARE THE FUSION MACHINE CORRECTLY.
   Do's
   a. Install appropriate pipe inserts into the upper and lower fixed and movable jaws.
   b. Plug in the heater and place it in a storage box or storage container to allow it to come to temperature.
   c. Use a power supply not in use with other tools.

   Dont's
   a. Do not use an inadequate power supply or inadequate power cord which could cause loss of amperage and result in improper heating.

3. CLEAN AND HEAT THE HEATER TO THE PROPER TEMPERATURE.
   Do's
   a. Clean the heater plate with a clean dry, non-synthetic rag. Heaters are coated with an anti-stick coating to facilitate cleaning.
   b. Set heater temperature using the adjustable thermostatic control.
   c. Periodically check surface temperature with a pyrometer or Tempil Stik™ and adjust accordingly.
   d. Under cold or windy conditions, increase temperature to ensure adequate surface temperature if necessary.

   Dont's
   a. *CAUTION:* The heater plate is not explosion proof. If fusion is to take place under hazardous conditions, bring the heater plate up to temperature in a safe environment and unplug it before entering the ditch or explosive atmosphere.

---

*As adapted from *McElroy No. 2 CU Fusion Manual.*[17]

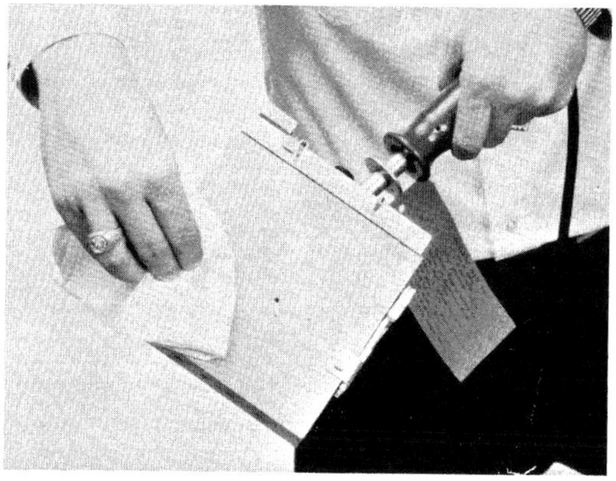

Clean Heater Plate

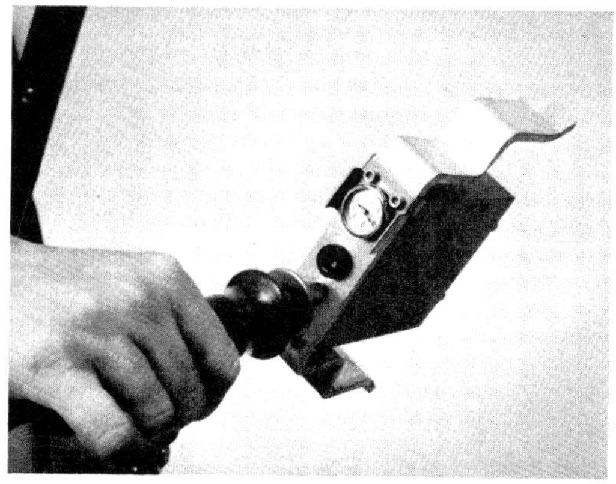

Check Heater Temperature

4. PREPARE THE PIPE ENDS PROPERLY FOR FUSION.
   a. Inspect pipe ends for out-of-roundness or damage due to shipping. Remove the out-of-round or damaged portion of pipe by using tubing cutters or shears.
   b. Clean and dry the pipe ends before installing in the machine. Dirty pipe can cause joint contamination, damage to the facer blades, and pipe slippage during fusion.
   c. Install the pipe in the fixed and movable jaws.
   d. Position the top jaws over the pipe and clamp loosely so that the pipe will slip through the jaws.
   e. Place the facer into position.
   f. Lock the facer on the guide rod.
   g. To adjust for face-off, use the hand lever to bring the pipe ends against the facer. Adjust the gap between the facer bushings and pipe clamps so the pipe will be squared when that length of pipe is faced. The minimum is one complete cut from both pipe ends. Snug the pipe clamps only finger tight to hold the pipe while facing.
   h. Crank the facer handle and apply force on the lever handle, maintaining pipe-to-facer contact.
   i. Continue facing until the facer bushings have bottomed out against the clamps and the crank handle operates freely. Remove the facer from the guide rods.

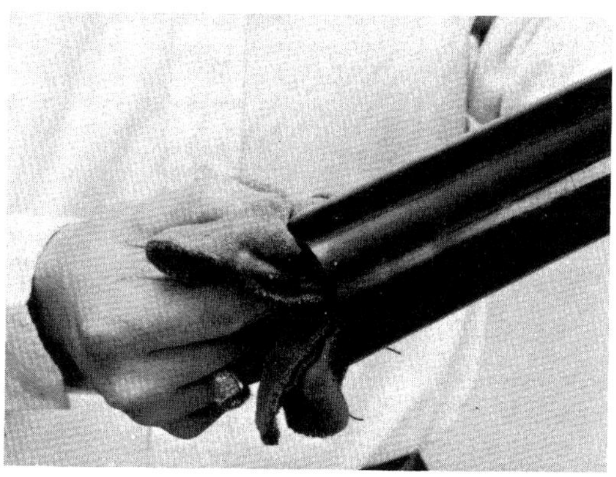

Clean and Dry Pipe

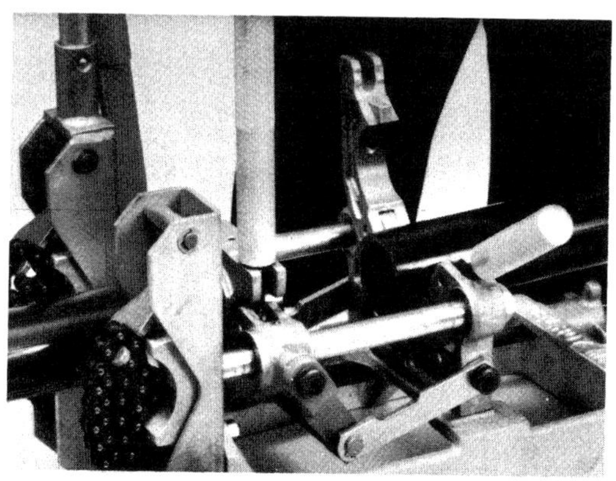

Install Pipe in Machine

5

j. Carefully remove all shavings from the pipe and the machine.

k. *DO NOT TOUCH THE FRESH-FACED SURFACE OF THE PIPE ENDS WITH YOUR BARE FINGER.*

l. Bring the pipe ends together and check for alignment or "high-low."

m. Adjust by tightening the appropriate clamp.

n. When the high-low has been taken out, tighten both clamps (finger tight). Do not overtighten; overtightening can flare the pipe ends resulting in non-parallel, non-mating surfaces.

o. Apply force to the fusion unit lever handle to ensure that the pipe ends are secure and will not slip.

Position Pipe in Jaws

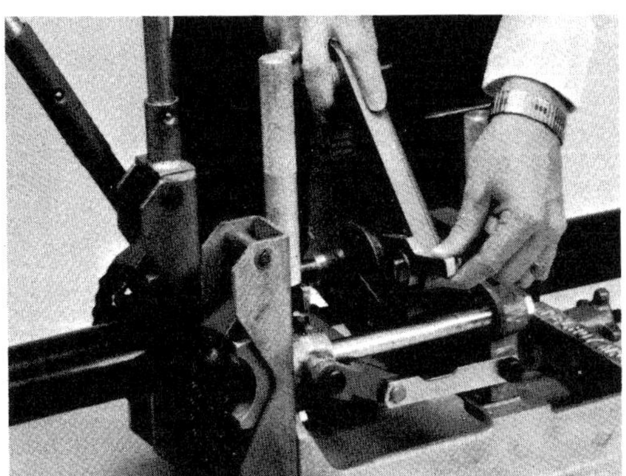

Place Facer

Lock Facer

Adjust Face-Off

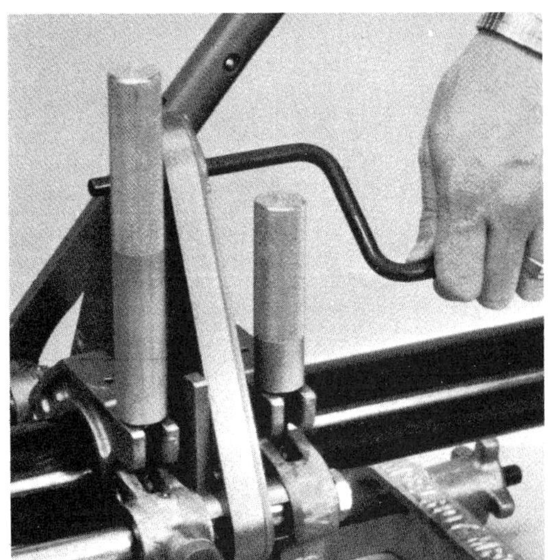

Face Pipe

Bottom-Out Facer Bushings

Remove Shavings

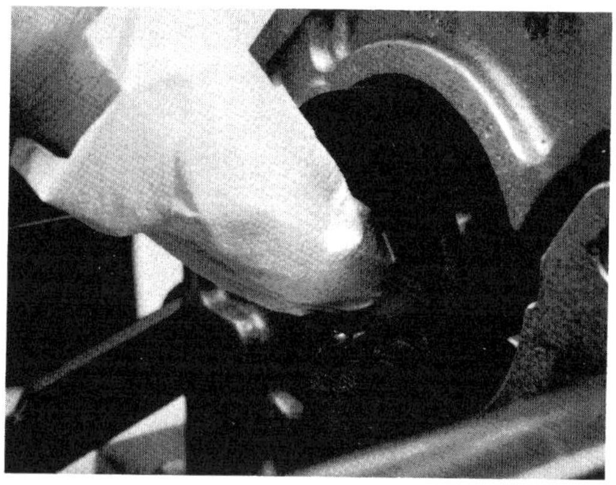

Use Clean Cloth

Check for Alignment

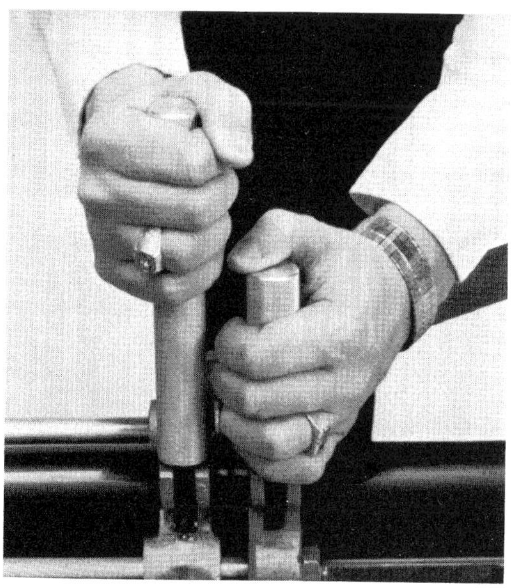

Adjust for Alignment

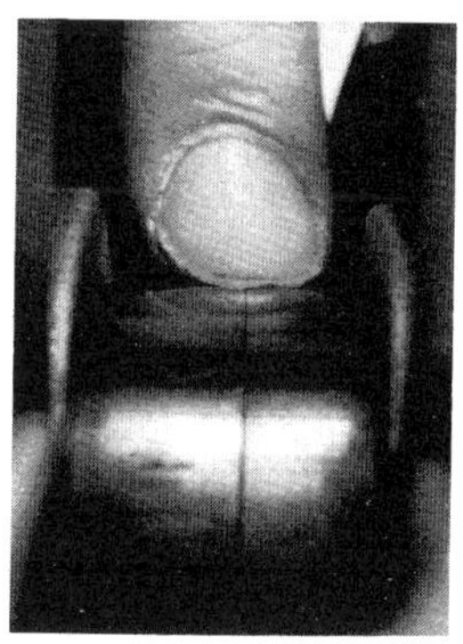

Check Alignment

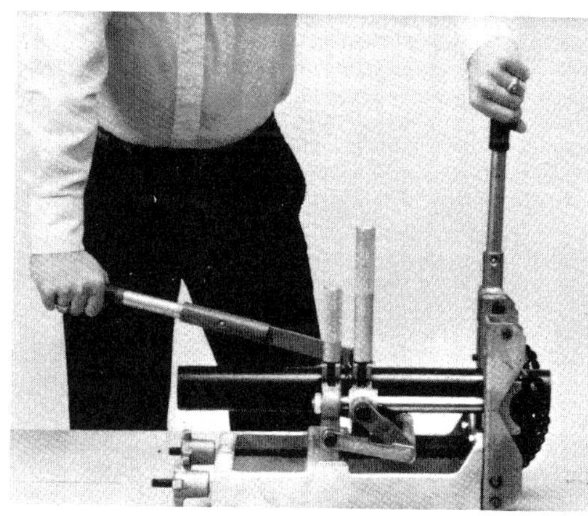

Check for Slippage

5

5. HEAT THE PIPE ENDS.
   a. Place the heater between the two pipe ends. Snap the pipe ends against the heater plate and maintain contact with the heater until the fusion bead begins to appear.
   b. When the melt pattern approximates the size of a round, wooden toothpick, remove the heater plate.
   c. Inspect for even melt around the pipe ends and immediately bring the melted ends together.

   If excessive force is applied during the butting process, the melt will be displaced from the fusion zone and could cause a cold joint.

Melt Pipe

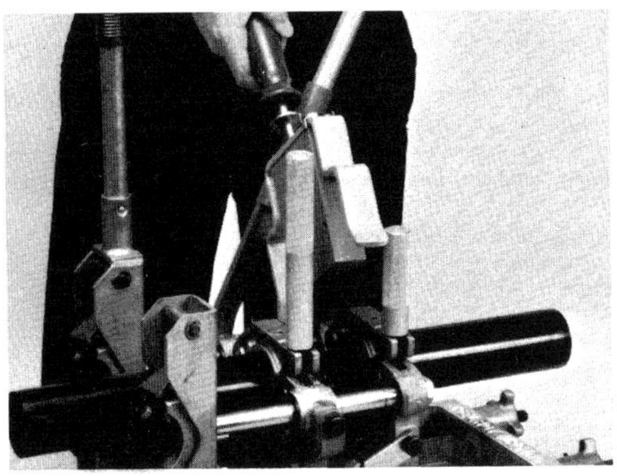

Insert Heater

Inspect Melt

---

6. FUSE THE PIPE PROPERLY.
   a. Do not slap the pipe ends together but apply enough pressure to form a double rollback bead. The bead will be rolled back to the body of the pipe on each side of the fusion joint.
   b. Place the heater in the storage box or container after the fusion is made.
   c. Allow the fusion joint to cool in the machine until a finger can remain comfortably on the bead.
   d. Visually inspect every fusion joint for uniformity and absence of gaps after the specified cooling period.

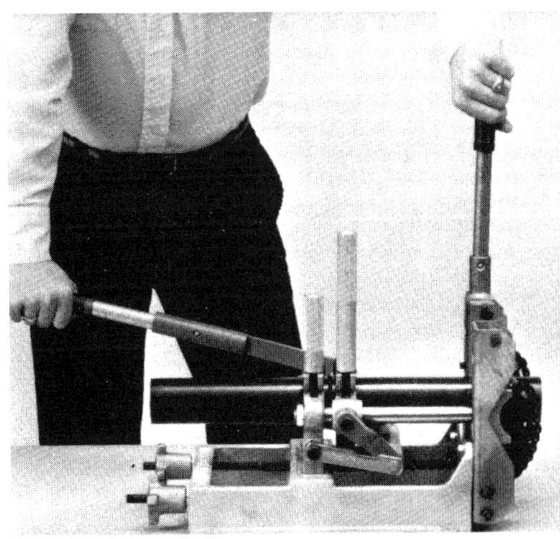

Form Double Rollback Bead

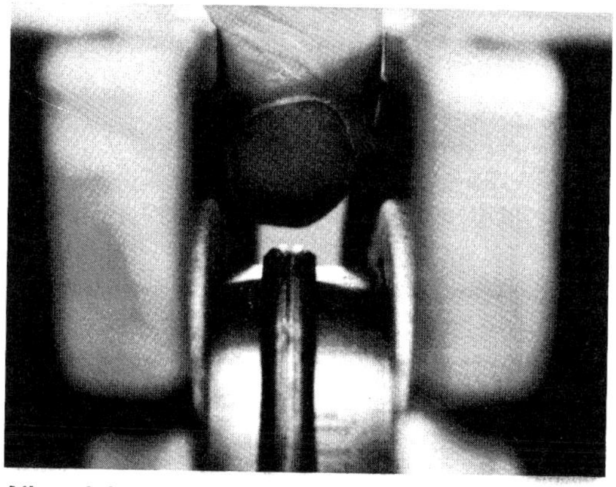

Allow Joint to Cool

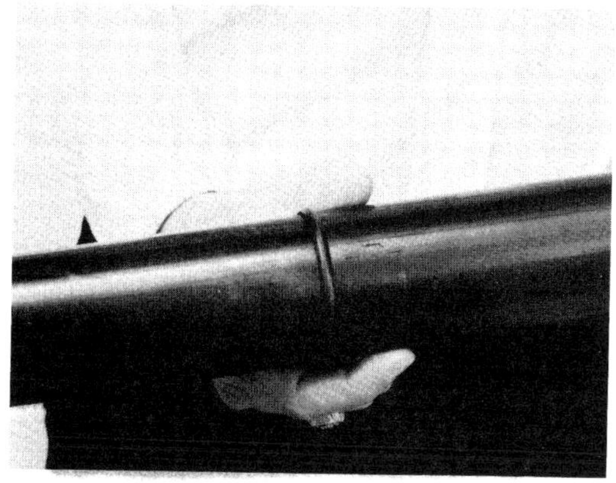

Inspect Fusion Joint

Sidewall Fusion Joining

Sidewall fusion is used to attach parallel loops or pipes to a header pipe. An adapter fitting is heat fused to the side of the pipe using convex and concave heater adapter fittings attached to the straight-faced heater plate. The procedure is as follows:

1. PREPARE THE FUSION MACHINE AND HEATER CORRECTLY.
   a. Install the heater adapters on the heater plate. Install the convex heater on the top face and the concave heater on the bottom face.
   b. Do not overtighten, but ensure that the mating of the surfaces of the heater and the adapters is flush.
   c. Install the proper fitting inserts in the upper and lower movable jaws. These inserts hold the shank of the adapter fitting.
   d. Install the appropriate sidewall clamping inserts into the tailstock for header sizes of 2 inches and smaller. These inserts must be used to ensure precise positioning and alignment.

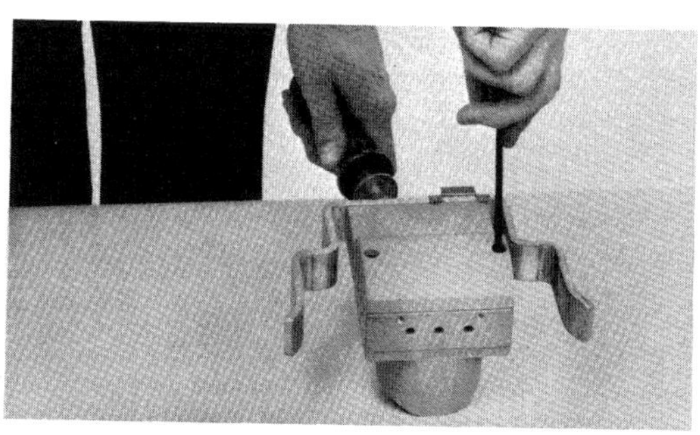

Install Adapters

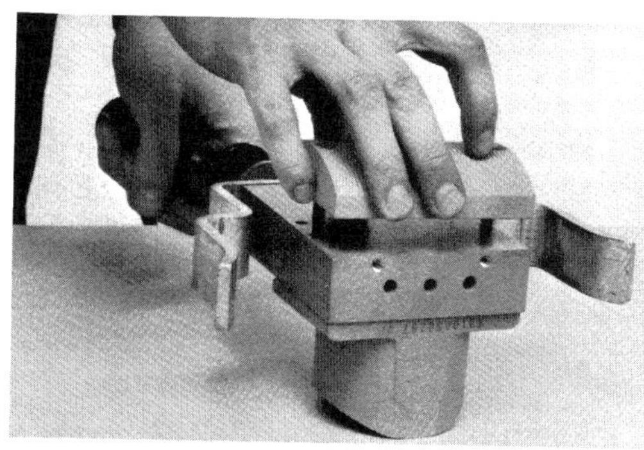

Tighten Adapters

Install Insert

5

2. PROPERLY POSITION PIPE AND FITTING IN MACHINE AND PREPARE FOR FUSION.
   a. Place the machine on the header. For header sizes of 1¼ inch up to 4 inches, install a line or pipe bolster to provide strength to the header pipe. Check to ensure that the line bolster is centered on the pipe header and opposite the machine so the heater adapters will not come in contact with the curved metal bolster plate or backup support.
   b. Tighten the machine onto the header. Do not overtighten, as flattening of the header may occur making it difficult to establish a melt pattern.
   c. Using 50- or 60-grit utility cloth, clean and rough the header pipe at the point of the fusion joint to expose fresh material.
   d. Clean the base of the fitting in a similar way.
   e. Position the fitting on the header and roll back into the insert in the movable jaw.
   f. Apply slight downward force on the fitting, close the jaw, and tighten the clamp knob, while maintaining the fitting-to-pipe alignment.

Clamp Pipe

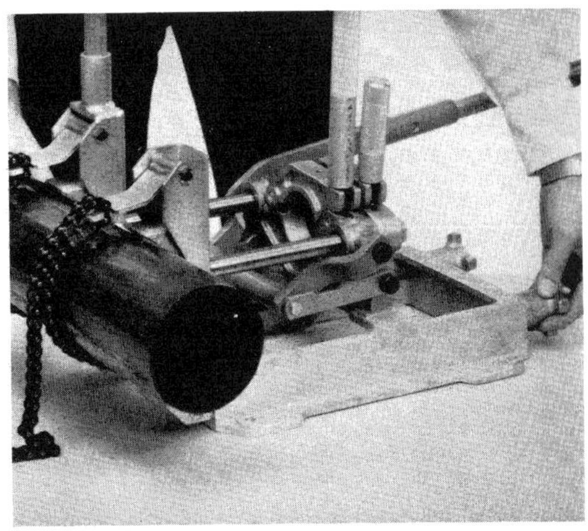

Tighten Pipe Clamps

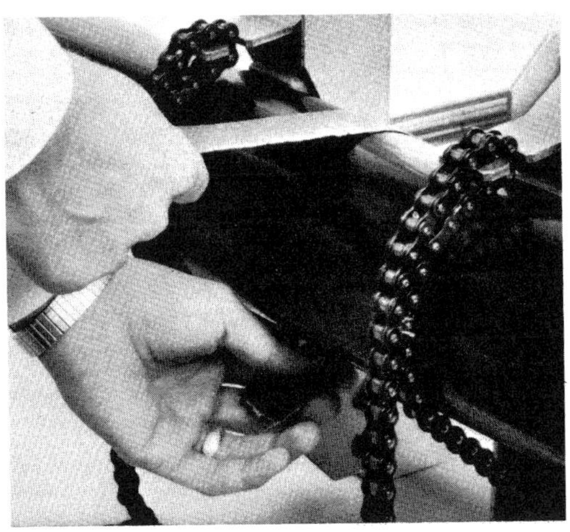

Expose Fresh Material

Clean Base

5

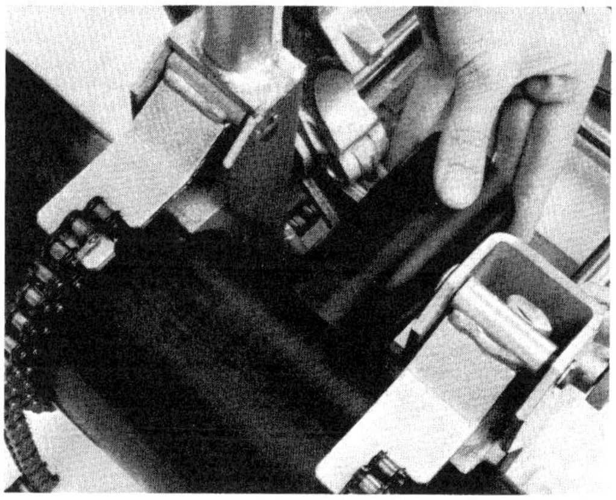

Position Fitting

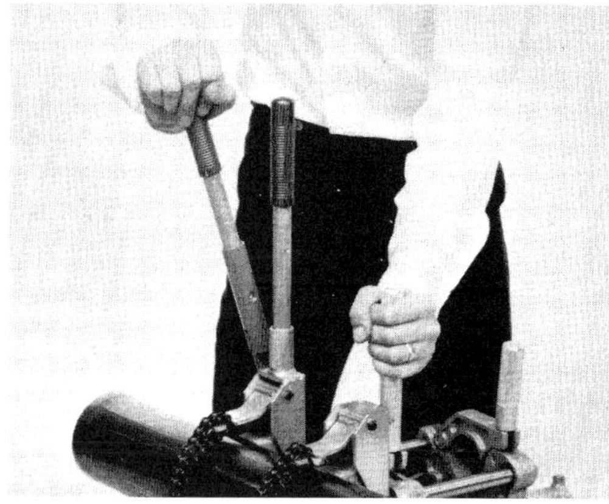

Check for Tightness

3. HEAT PIPE AND FITTING.
   a. Check heater temperature.
   b. Place heater between the fitting and header. The safety guard centers the heater and ensures perfect alignment.

> **NOTE: If the heater plate does not appear to align properly, check the manufacturer's specific recommendations for realignment.**

   c. Apply firm, continuous pressure on the fitting and header until there is an indication a complete melt bead is started on the header. Release the pressure to "very light" pressure and continue the heat-soak cycle on the fitting and header. Watch the base of the fitting for a bead size approximately 1/16 inch for a 1 1/4-inch header and 1/8 inch for a 2-inch header.

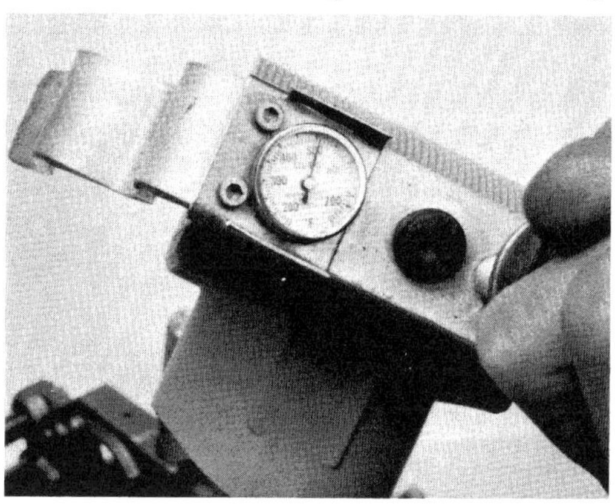

Check Heater Temperature

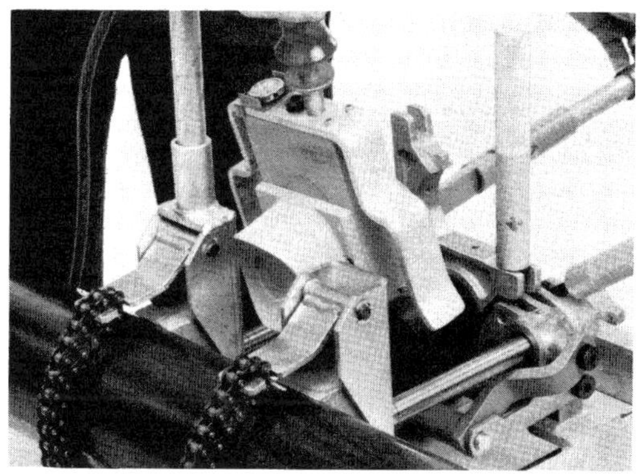

Place Heater Plate

Apply Firm Force

5

4. PROPERLY FUSE PIPE AND FITTING.
   a. Remove the heater with a snap action, quickly inspect the melt pattern, and firmly join the fitting to the header.
   b. Apply continuous pressure until the melt pattern on the fitting rolls back against the fitting and a triple bead is formed. Maintain pressure until the joint has cooled and is comfortable to the touch.
   c. Inspect the finished joint.

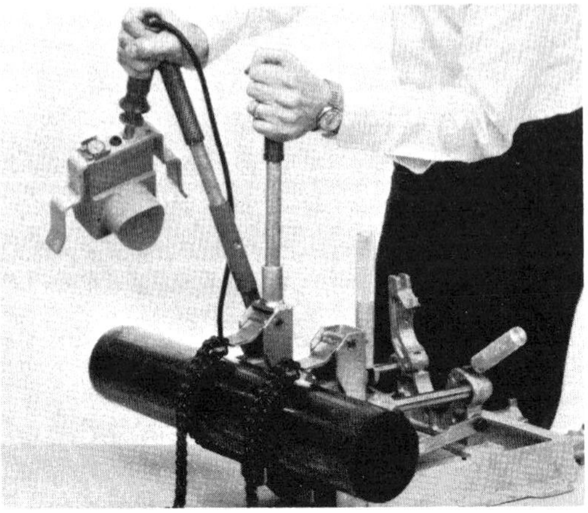

Inspect Melt

Inspect Joint

Form Fusion Joint

5. CUT HOLE IN HEADER.
   a. After the sidewall fusion joint is cooled, drill out the plastic material on the header or main pipe with either a twist or paddle drill bit.
   b. Remove all loose plastic. Be sure that all material has been collected and removed from the header.

Socket Fusion Procedure*

1. ASSEMBLE THE REQUIRED EQUIPMENT AND TOOLS:
   a. appropriate power supply
   b. heater plate with correct socket faces
   c. depth gauge
   d. cold ring
   e. tubing cutters
   f. temperature-sensing equipment
   g. clean, dry cotton rags

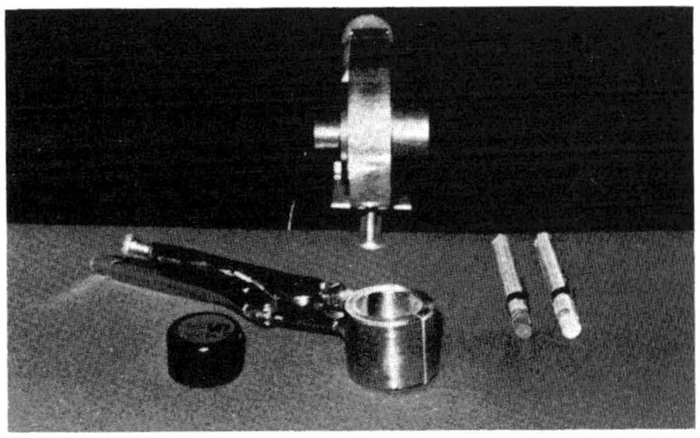

Required Tools

*As adapted from Vanguard Plastics' *Socket Fusion Techniques.*[18]

2. CORRECTLY PREPARE THE SOCKET FUSION MACHINE.

Do's

a. Securely attach the correct socket faces to the heating body of the tool. Do not over tighten.
b. Make sure the socket faces are in good condition.
c. Use an adequate power supply and an adequate power supply cord.
d. Use a power supply not in use with other tools.

Dont's

a. Keep socket faces from becoming loose, which will impair heat transfer.
b. Prevent an inadequate power supply and inadequate power cord from causing loss of amperage and resulting heat loss.
c. Do not use other tools on the same power source which can cause loss of amperage and heat.

---

3. CLEAN AND HEAT THE SOCKET FACES TO THE PROPER TEMPERATURE.

Do's

a. Wipe socket face surfaces without fail with a clean, DRY, cotton rag immediately after each fusion.
b. Maintain critical socket face temperatures throughout the fusion process to assure an adequate bond, especially in cold weather.
c. Remember that the heating tool thermometer indicates internal tool head temperatures, NOT the socket face surface temperatures.
d. Maintain socket face temperatures between 500° and 525°F. This is especially critical in cold weather.
e. Check the socket face surface temperatures OFTEN with a pyrometer or temperature-indicating crayons. Use only temperature-indicating crayons which have the correct temperature label.

Dont's

a. Don't clean the socket face surfaces with a synthetic fabric (polyesters).
b. Don't rely on the heating tool temperature gauge to indicate face surface temperatures.
c. Don't use temperature-indicating crayons on the surface of the non-stick faces where the pipe and fitting come in contact.
d. Don't clean the socket face surfaces with a wet or damp rag, chemicals or solvents, nails, knives, or other sharp, hard objects.
e. Don't place the socket tool on the ground, in grass, or in a location where socket face surfaces may be damaged.

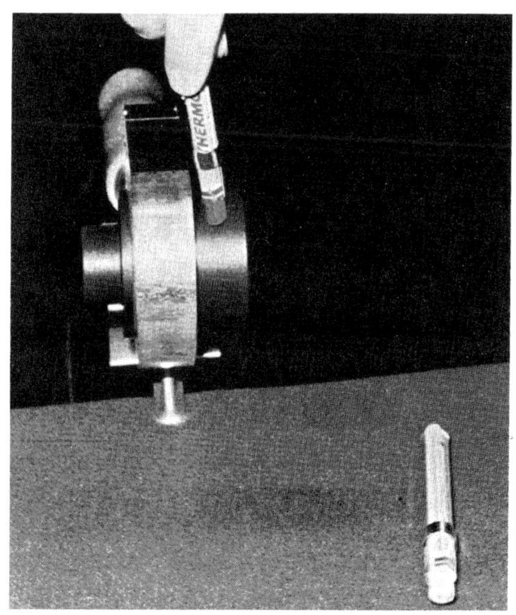

Check Face Temperature

5

4. PROPERLY PREPARE PIPE ENDS AND FITTINGS FOR FUSION.

Do's

a. Cut pipe SQUARE with a tubing cutter designed for plastic pipe.
b. Thoroughly wipe and remove all foreign material at the joining surfaces with a clean, SLIGHTLY damp rag.
c. Inspect each and every fitting for out-of-roundness or damage, and wipe clean. Check inside pipe for contaminants or obstructions.

Dont's

a. Don't cut pipe at an angle; this will produce inadequate joints.
b. Do not use flattened, damaged, or out-of-round pipe.
c. Do not use wet cleaning rags, only damp ones.
d. Do not use gasoline, solvents, or primers to clean the pipe or fittings. Isopropyl alcohol may be used to remove stains or adhesives.

5. USE A PROPER DEPTH GAUGE AND COLD RING.

a. After the pipe end and fitting socket have been cleaned and inspected, place the depth gauge on the pipe end.
b. Use the depth gauge as a guide for positioning the cold ring.
c. After placing the depth gauge on the pipe, hold the gauge firmly in place and clamp the cold ring around the pipe, flush against the depth gauge. Using the depth gauge/cold ring combination assures that the heated pipe and fitting will join at the proper depth, prohibiting the pipe from either penetrating too far into the fitting, creating a flow restriction, or failing to penetrate the fitting deeply enough for adequate joint strength. The cold ring also helps keep the pipe round while it is being heated.

THE PIPE END AND FITTING ARE NOW READY FOR HEATING.

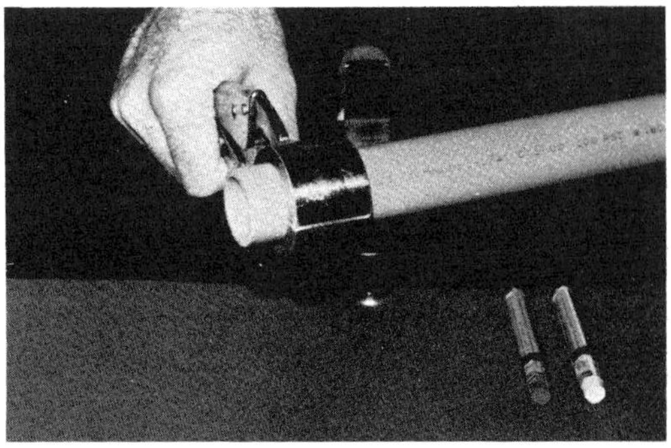

Attach Cold Ring

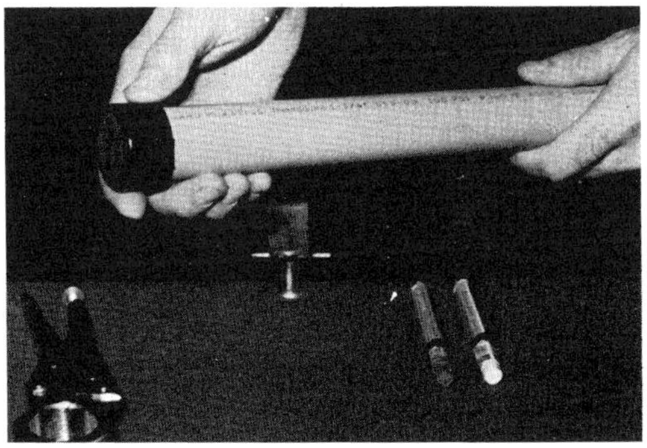

Place Depth Gauge

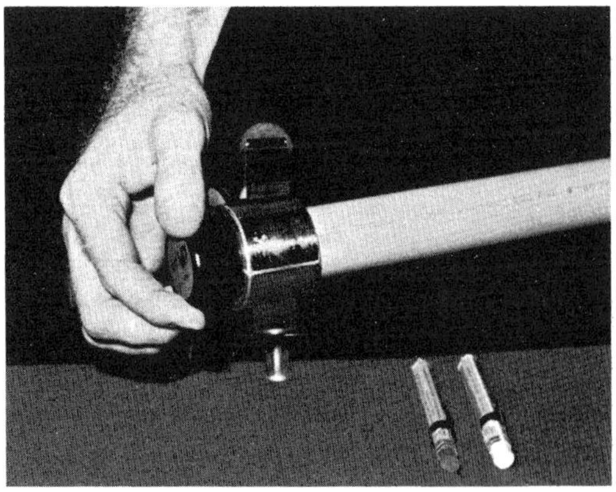

Check Cold Ring Placement

6. SIMULTANEOUSLY HEAT PIPE AND FITTING.
   Do's
   a. Grip the pipe behind the cold ring.
   b. Push the pipe end (with the cold ring attached) onto its socket face while simultaneously pushing the fitting over the opposite socket face.
   c. Push the pipe and fitting squarely together and, at the same rate of pressure, onto their respective socket faces until the cold ring BARELY TOUCHES its socket face.
   d. Align the fitting socket just flush with the socket face. Don't force the fitting against the heated plate.
   e. Begin the timing cycle as soon as the cold ring and fitting edge contact the socket faces or are fully inserted.

   Dont's
   a. Only SLIGHTLY touch the heated socket faces with the pipe and fitting.
   b. Do not use excessive force. Too much pressure will cause a rollback of the fitting edge, shortening the socket depth and creating a gap when the pipe and fitting are joined.

---

7. PROPERLY FUSE PIPE AND FITTING.
   Check the socket face temperature before each fusion in cold weather (32°F). Be sure that wind currents do not affect face temperature.
   Do's
   a. When the heat cycle is complete, remove the fitting and pipe from the socket faces with a quick, snap action.
   b. Quickly (within three seconds), and WITHOUT TWISTING, push the pipe and fitting together squarely until the cold ring is in contact with the fitting edge.
   c. Hold the joined pipe and fitting firmly together for 30 seconds. Releasing pressure before the required holding time has elapsed will cause the pipe to "back out," resulting in a faulty fusion joint.
   d. Move the fused joint only when the holding time has elapsed and it is set enough to allow CAREFUL MOVEMENT. However, the joint is still fresh and MUST NOT be twisted, torqued, bent, or otherwise handled roughly for 20 minutes ($\frac{3}{4}$-inch to $1\frac{1}{2}$-inch) or 30 minutes (2-inch) after the fusing process (contact pipe manufacturer for heat cycle table).

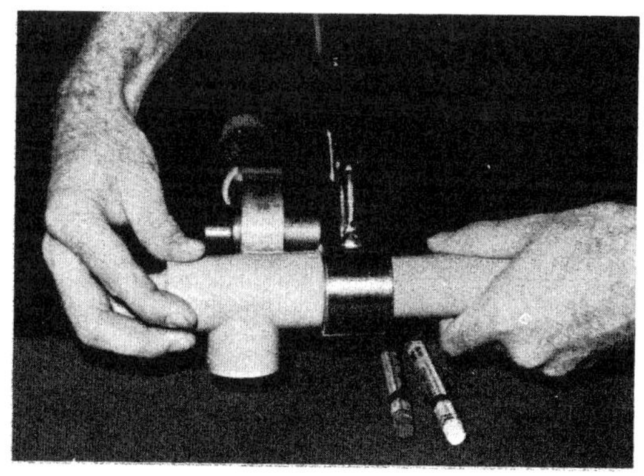

Heat Socket and Pipe

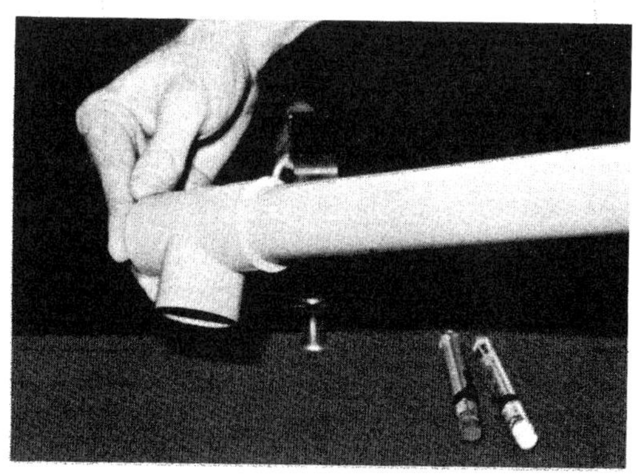

Hold for Cooling

5

WIPE OFF FUSION SOCKET FACES IMMEDIATELY AFTER EVERY FUSION. REMOVE ALL VISIBLE POLYBUTYLENE MATERIAL.

Dont's

a. Do not leave material on the socket faces; it will carbonize and disrupt good heat transfer. It will also form a barrier over the non-stick coating causing poor release. Without a smooth release, dirty socket faces will strip off the molten surface plastic, resulting in a poor joint, and possibly complete joint failure.
b. Do not use a synthetic cloth, wet cloth, or hard or metal objects which could damage the non-stick surface of the socket face. Use only a clean, non-abrasive cotton rag to clean the socket faces. Wet or damp rags could significantly lower the temperature of the faces.

USE CAUTION WHEN HANDLING OR CLEANING HOT FUSION TOOLS.

---

8. TEST FOR LEAKS.

After assembly, the system should be tested for leaks. All systems should be tested hydrostatically.

Do's

a. Test at a maximum test pressure of 225 psi (175 + 50).
b. Ensure that the pressure decreases due to pipe deformation and expansion does not exceed 10% during the initial two hours of the test period, after which the expansion should not cause the pressure to decrease more than 3% during the next two-hour period. During the initial pressurization period, the pressure indicated on the gauges will decrease due to the initial deformation of the pipe, followed by a slow expansion. The pressure drop is dependent on ambient temperature, system capacity, and test pressure. If additional make-up water is introduced into the system, the pressure decrease during the next two hours would be less than the loss in the first 1½ hours.
c. Inspect the system visually for leaks and good workmanship.
d. Examine, repair, and retest every joint as required.
e. Field test a given number of joints and record information as to the job site, operator, date, etc.

IF A LEAK IS DETECTED, REMOVE AND REPLACE. DO NOT ATTEMPT TO "PUTTY" LEAKS AT A FUSION JOINT.

6 Installation of the Ground Heat Exchanger

In This Section: Site plan; installation equipment; trenching and drilling costs; pipe installation considerations; headering up

Installation of a ground heat exchanger is a new skill or trade for most people. As the industry develops, it appears that a new trade is being formed, that of a loop installer whose primary purpose is to install, pressure test, and connect heat exchangers. An HVAC contractor takes care of connecting two water lines or taps to the heat pump as specified by the manufacturer, distributor, or dealer. A loop contractor in many cases is responsible for final grading and restoration. It is important to assign overall responsibility for the project to a single individual.

Site Plan

The site plan is a location description of the buried underground utilities that serve a given area. Two things are certain:

1. Memories fail and utility lines seem to move underground with time.
2. Installation costs go up dramatically as services are cut or severed.

Figure 6.1 shows some of the types of services that need to be identified. The location description and sketch should include:

1. The location of all buried utilities. Locator services for electrical, telephone, and gas service are available and are usually listed in the local telephone book. These services are generally provided at no cost to the homeowner or contractor.
2. The location of the ground heat exchanger, which should be established from two permanent points in case of future excavations.

The owner should be consulted to determine:

1. Special areas to be avoided. Trees, shrubs, and gardens that are not to be disturbed should be identified.
2. Acceptable locations for entry and exit of heavy equipment. Caution should be exercised; the limits of a driveway may not be known, and the contractor is generally liable. A lightweight machine with large flotation tires can reduce pavement and yard loads. Machines with four-wheel steer can maneuver in crowded locations.
3. The location of services like underground water sprinklers which may not be known or easily identified by the contractor. Ask the owner to uncover these buried systems or agree in writing in advance that the contractor is not liable.

6

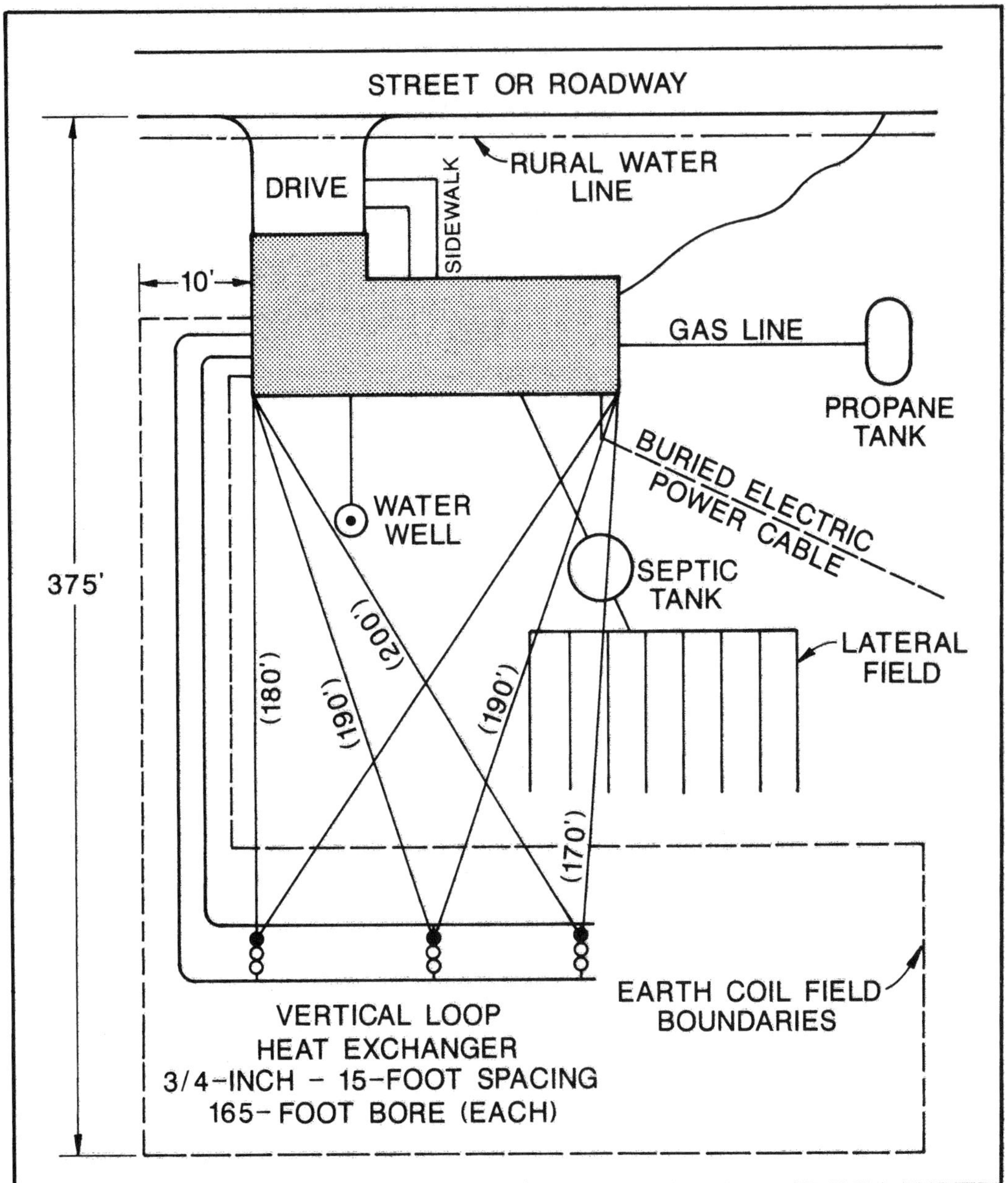

FIGURE 6.1: Location Description

Installation Equipment

Chain trenchers, bulldozers, backhoes, and vibratory plows are used to bury horizontal ground heat exchangers. The choice of machine depends on local site conditions and the cost. Generally, the machine which moves the least amount of soil will be the most cost-effective. Figure 6.2 describes the various machines that are used to bury horizontal ground heat exchangers.

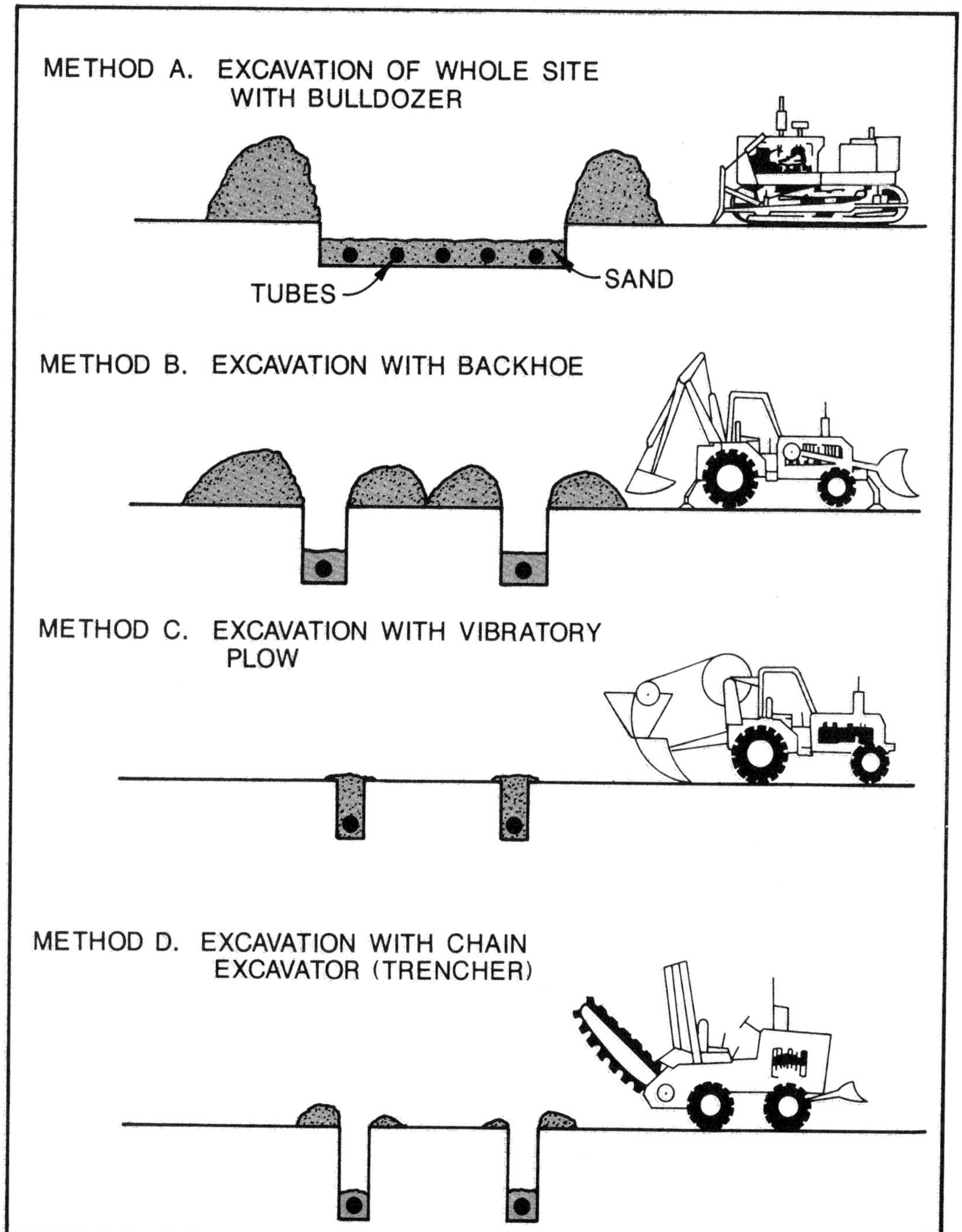

FIGURE 6.2: Horizontal Installation Methods

Chain Trenchers: In many cases, trenchers (Method D) are the most economical choice since the amount of soil or dirt removed is minimal when compared to other methods, and trenching productivity is usually much higher than that of a backhoe. Systems are now being developed which allow the automatic placement and backfilling of multiple pipes in a single trench simultaneously. A trailing vibrator and the addition of water to form a slurry have increased productivity significantly for the multiple-pipe system. Trench widths are about six inches with depths to seven feet for 65 horsepower trenchers. The most common chain trencher has a depth range of three to five feet. A six-foot boom represents about 5% of the cost of a trencher. Multiple pipes in the trench greatly reduce trench length and the area required for a horizontal buried pipe system.

Bulldozers: Method A using a bulldozer would only be appropriate if the soil had been removed for some other purpose or if excavation for a large header system were required.

In larger installations in Europe, track-type machines have been used to plow in horizontal ground heat exchangers and backfill around the pipe in a continuous operation. The backfill process consists of a hopper and chute arrangement which places the material in the immediate pipe area.

Backhoes: Backhoes (Method B) have been used where there are rocks, cobbles, or boulders which are too large or too difficult to remove with a chain trencher. Backhoes have also been used where more than one pipe is placed in a single trench either stacked vertically (such as 1 x 2) or where multiple pipes are placed horizontally in several layers (such as 2 x 3).

NOTE: The nomenclature 1 x 2 defines a pipe system of one level or layer and two pipes. A system of three pipes at two levels or layers would be denoted as 2 x 3.

A backhoe may also be used to open the site from the foundation to the ground heat exchanger. The soil removed by a backhoe tends to be lumpy and can leave air pockets around the pipes, greatly reducing heat transfer capability.

Vibratory Plows: Vibratory plows (Method C) are commonly used at depths not greater than three feet. To facilitate rapid burial, a vibrator or vibratory plow is used to improve the speed of placement and reduce the size of the machine. Loose or unstable soils are also amenable to this type of pipe placement, and these plows function more efficiently than a chain trencher in glacial till. The shallow depth results in larger seasonal soil temperature swings and longer lengths of pipe.

Drilling Machines: Drilling is required to install vertical loops. Two applicable techniques for boring holes are illustrated in Figure 6.3: (a) mud and air rotary drilling, and (b) standard auger or hollow-stem auger drilling. Boring holes for earth loop insertion and boring holes in order to find water and produce wells are two completely different tasks. The ground heat exchanger application is much simpler. The objective is to install a specified length of heat exchanger in a vertical configuration; it is not important to reach a given depth at a particular site. If 600 feet of bore hole are required, two 300-foot vertical bore holes are acceptable and may be more cost-effective. Soil and rock conditions will determine drill rates and consequently an economical bore hole depth. Since ground temperatures fluctuate near the surface, hole depths greater than 50 feet are normally recommended. However, the 50-foot depth is not a minimum standard. There are some general guidelines:

1. Soil moisture and temperature stability generally increase with depth, which favors deeper holes.
2. Fewer holes mean fewer connections at the surface, which can be a time-saver; also, less surface area is required.

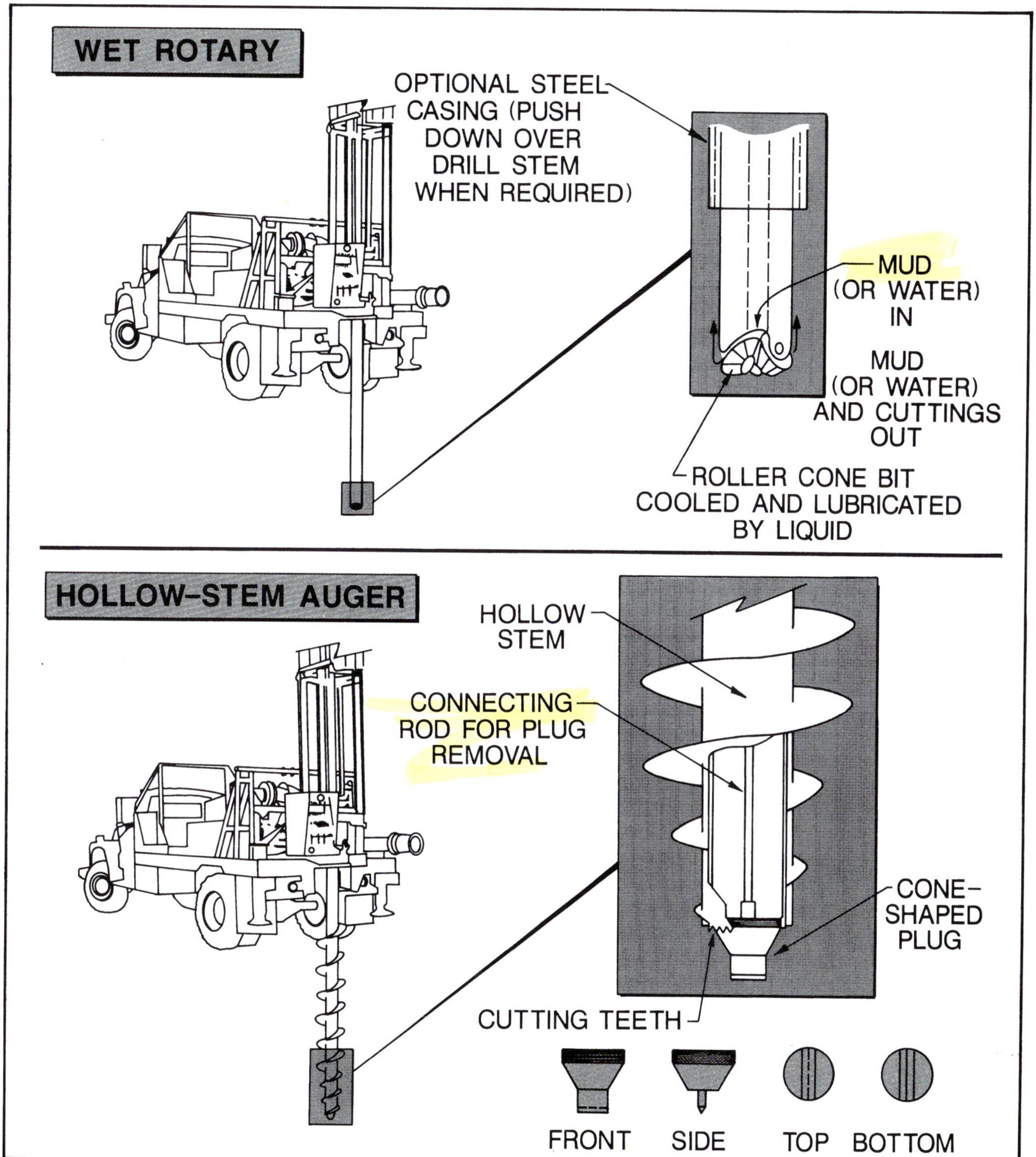

FIGURE 6.3: Vertical Drilling Machines

Source: *Manual of Acceptable Practices for Installation of Residential Earth-Coupled Heat Pump Systems.*[19]

6

3. If many holes are drilled, several bore holes may need to be connected in series before tieing back to the main header. Too many parallel flow paths can result in low flow and reduced performance of the ground heat exchanger. See Section 4 for a procedure to select the number of parallel flow paths.
4. If hard rock layers are encountered, it may be more profitable to move and repeat the process.
5. Extensive hard rock in a drilling service area can easily justify a down-hole hammer or employing a rock quarry drilling contractor with a top-hole hammer.
6. Vertical loops using ¾- or 1-inch pipe and a close U-bend require only a 3- to 4-inch hole. This smaller drilling system uses less mud flow, a smaller and less expensive bit, smaller mud pans, and a smaller mud pump, and results in less wear and tear on the mud pump. In many cases, this will result in lower drilling costs.

The bore hole does not need to be cased for performance reasons. It may need to be cased if the soil is unstable or a formation presents a problem due to lost circulation during drilling. The bore hole must only remain open long enough to insert the ground heat exchanger. Most of these problems can be controlled by proper mud formulation and use. Drilling rates are faster than water well drilling which generally requires sampling and casing. Well completion procedures are not required, which greatly reduces time and cost.

In mud or air rotary drilling, the drill rig rotates the drill pipe and sends high-pressure air, water, or mud down the inside of the drill pipe to lubricate and cool the bit and to bring the cuttings back to the surface along the outside of the drill string. In rotary mud drilling, removed soil or earth is collected in a mud pan to be replaced in the hole or removed from the job site if necessary. Bentonite or fluid additives are added to the drilling mud to control its viscosity and to prevent a number of drilling problems. If the drill cuttings are of a high density, the drilling fluid viscosity must be increased to carry the cuttings to the surface. The mud must also be dense enough both to support the bore hole wall if the surrounding soil is unstable or to be mixed to form a clay deposit on the wall of the hole if the mud begins to leak through the wall of the hole into a porous formation. The proper mud mix can also prevent swelling. A denser mud may be required to keep mud from migrating through surrounding porous formations.

In extreme cases of unstable soils, drill pipes with large inside diameters are used which allow loops with U-bends to be inserted inside the drill string before its removal. The drill bit in this case has a knock-out point which allows the U-bend and loop to remain in the bore hole while the drill string is removed.

In hollow-stem auger drilling, the drill rig rotates a hollow auger with cutting teeth and a temporary point at the bottom. The drilling operation may be completely dry and hence clean at the surface. Most of the cuttings are brought to the surface, but some of them are simply pushed into the walls and compacted in place. The hollow auger essentially serves as a casing in the hole as it is drilled. When a hole is completed, the point at the bottom of the auger is pulled, the ground heat exchanger is inserted, and the hollow auger is then pulled or screwed out. In some designs, a disposable bit point is used which allows a weight on a rope to be used to knock out the disposable point. Auger drilling is applicable in a limited number of different soils and can present some serious problems in clay, adhesive soils, or rock. In moist soils auger drilling is slower than rotary mud or air drilling.

6

Trenching and Drilling Costs

The ownership cost of trenching or drilling equipment is determined mostly by utilization and productivity. A $40,000 trenching machine amortized over five years ($8,000/yr) would cost $40 per installation for 200 installations per year, or $400 per installation for 20 installations per year. A smaller, less-productive machine for $20,000 amortized over five years ($4,000/yr) may only be capable of 100 installations per year. Ownership cost would still be $40 per installation, but 200 installations per year would require two operators, or twice as much labor cost. The larger machine would probably have less maintenance cost if any rock were encountered. Insurance and financing costs are affected in a similar manner.

Pipe Installation Considerations

Consistently successful plastic pipe installations for ground heat exchanger applications are the result of good planning, the use of suitable materials and equipment, and close attention to correct procedures. Time spent in careful planning will substantially reduce the time and cost of the installation.

The following factors should be considered during planning:

1. the depth of the trench, as influenced by climate, soil type, and the size of the trencher available
2. the length of the trench, as influenced by the area available, the heating and cooling load, the number of pipes in the trench, the type of soil, and the moisture content of soil
3. the depth and number of vertical holes, if used, as influenced by the area available, obstructions, the heating and cooling load, and soil and rock types
4. the trench pattern, as influenced by above- and below-ground obstructions, ground slope, trenching turn radius limitations, backfilling, and restoration requirements; be sure all buried conductors are located and marked
5. the type and size of the trenching chain, as influenced by type of soil and rock, depth of the trench, and the size of the trencher
6. the type of plastic pipe: use high-quality polyethylene or polybutylene; PVC is not acceptable in either heat transfer characteristics or strength

Several procedures should be followed in the trenching phase in order to minimize problems, time, and cost:

1. Use a combination of narrow chain, correct teeth, 4-wheel steering for maneuverability, 4-wheel drive for good traction, good backfilling capability with a 6-way blade, and high enough horsepower to finish most jobs in one day, which will minimize expensive interruptions and complications due to weather and variations in soil conditions.
2. Remove rocks left in the bottom of the trench to eliminate damage to the pipe. Use long-handled tongs.
3. Use a backhoe attachment with a 12-inch bucket if needed to dig out large rocks and dig an access hole at a building.
4. Where mucky or sandy conditions cause caving, use a pipe guide mounted on the trenching boom to guide the pipe down to the bottom of the trench as it is dug.
5. To prevent dirt from an intersecting trench from having to be removed from the previous trench, insert and wedge 2 x 6 boards in previously dug trenches. Place a piece of plywood over the near length of the previously dug trench.

The pipe installation phase can be accomplished as the trenching proceeds. Be sure that the following procedures are followed:

1. Inspect the trench to ensure that rocks are removed. Also remove rocks from the edge of the trench and from the top of the spoils pile so they will not fall in with the 4-6" of fines or sand that are first put in the trench to support and cover the pipe for protection.
2. Inspect the pipe as it is laid out beside the trench for cuts, kinks, or other damage.
3. Make all joints while the pipe is laying beside the trench and test with 40 psi air pressure and soap.
4. Ensure that joints are mechanically strong enough to force the pipe to absorb the stress and strain of the expansion and contraction in length caused by the temperature changes in the circulated water. Pipe ends should be extra long and not cut for attaching the end fittings until the rest of the pipe has reached soil temperature in the trench.
5. Carefully lower the pipe into the trench after the bottom has been again checked for rocks and the fines smoothed.
6. If the soil is a heavy clay in very dry climates, it may shrink away from the pipe as the soil dries during the summer cooling cycle operation of the heat pump. Use a sand fill (Figure 6.4) around the pipe or a subsoil drip irrigation line buried a few inches above the pipe. The sand will crumble to maintain contact with the pipe. Add water with a drip line when in–out water temperature indicates reduced heat transfer through the pipe to the soil.

Backfilling is critical. Its successful completion depends on following these procedures:

1. If the rocks have been removed from the edge of the trench and the top of the spoils pile, and the dirt has not formed clods due to rain, an experienced trencher operator can angle-blade the top of the spoils pile into the trench on the first pass. A worker should follow closely and tamp the fines by hand and make sure no rocks fall in.
2. Several more passes with the angle blade should be used to make the backfilling as uniform as possible and to prevent bridging.
3. Several stages of tamping and rolling over with the trencher wheels may be necessary to complete the job.

Additional considerations are required when closed vertical loops are used in limited site areas:

1. Soil conditions will determine whether the auger, recirculating mud, or down-hole hammer method is used. They will also determine the type of drill bit and whether a mud additive is required.
2. The depth and number of holes for vertical loops depend on the heating and cooling load, the drilling rate, the site area, soil and rock types, and moisture level.
3. Each loop should be assembled, laid out straight, taped to reduce springback friction, and carefully tested for leaks and flow before the hole is drilled so that it can be lowered into the hole before it can cave in or the mud can settle to the bottom of the hole. The hole should be 5–10 feet deeper than the length of the loop to accommodate expansion of the loop. Fill the loop with water prior to insertion. If the hole is to be grouted in place, it should be filled with water and pressurized to a level that will prevent the pipe from being crushed by the denser backfill material.

   Taping short lengths of scrap steel rebar on the end of the pipe loop will hold the loop end straight and offset buoyancy of the plastic to make insertion easier and faster.
4. Test the loops after they are filled

6

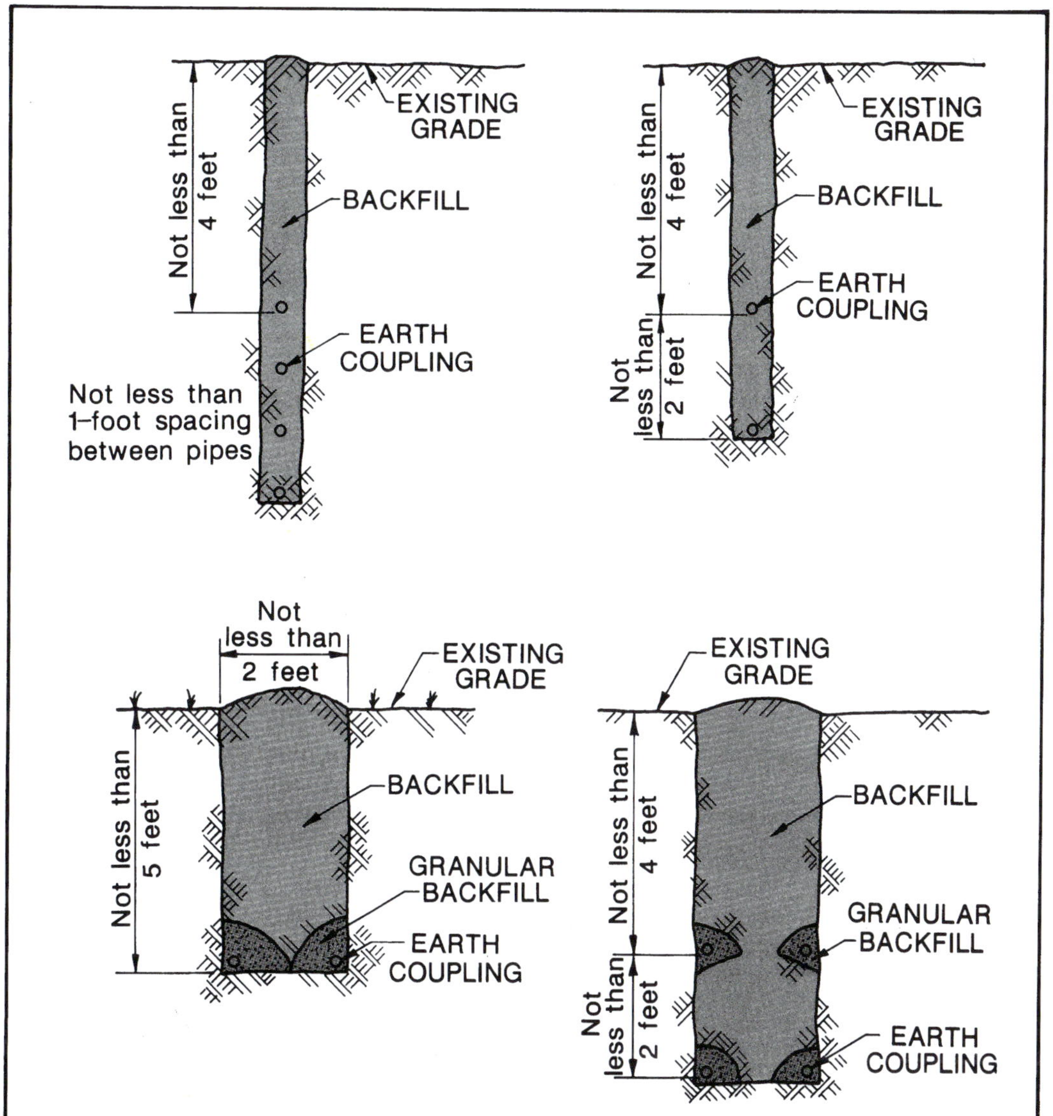

FIGURE 6.4: Typical Pipe Placement: Horizontal Ground Heat Exchangers

with water which is under a pressure of at least 40 psi. This is the typical pressure of a central water supply. Avoid the use of air for testing because a failed joint could result in a flailing pipe.

5. If the water table is low, a high-solids bentonite grout should be used for backfilling so that good thermal contact can be maintained at all times. Refer to *Grouting Procedures for Ground Source Heat Pump Systems,* 1991, IGSHPA.
6. The hole can be filled with grout, if required, using a small plastic discharge pipe inserted with the loop.

   Thermal sand developed by the electric power industry (reported by the Electric Power Research Institute – EPRI) is available but has generally been considered too expensive by the ground-source heat pump industry. Careful backfilling with the excavated material is the norm.

6

Headering Up

The process of connecting loops to the header is called headering up. Consider a system of seven bore holes with drilling placement as shown in Figure 6.5. The contractor has two choices:

1. Run the supply and return headers down the center of the loop field and connect as shown in Figure 6.5.
2. Run the supply and return headers along a single side of the bore holes.

Either method is acceptable. Some contractors prefer to drill bore holes in a single line which reduces the location time between bore holes.

The trenching pattern or sequence connecting the loops to the bell hole is shown in Figure 6.6. By placing the main header as shown in Figure 6.7 and using 90° elbows, pipe loops can be connected without placing undue stress on the pipe. The bell hole must be large enough for two people to work in comfortably.

For bore holes in a straight line, the axis of the service saddle can be laid horizontal and avoid the use of the 90° elbow. Header placement in this case is shown in Figure 6.7.

NOTE: Always hand backfill around pipe bends to avoid pipe crimping. Follow the manufacturer's recommendation for minimum pipe radius.

1. For PE pipe butt fused, the minimum bending radius is 25 times the diameter of the pipe (25 diameters).
2. For PB pipe socket fused, the minimum bending radius is 50 times the diameter of the pipe (50 diameters).

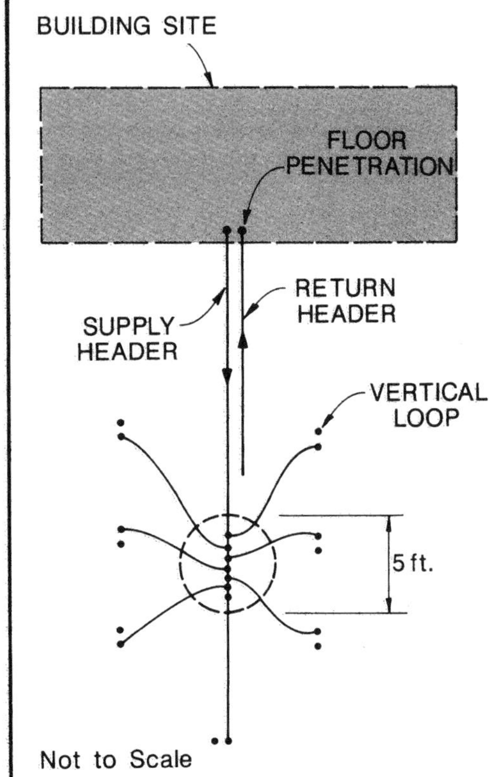

FIGURE 6.5: Header Connections

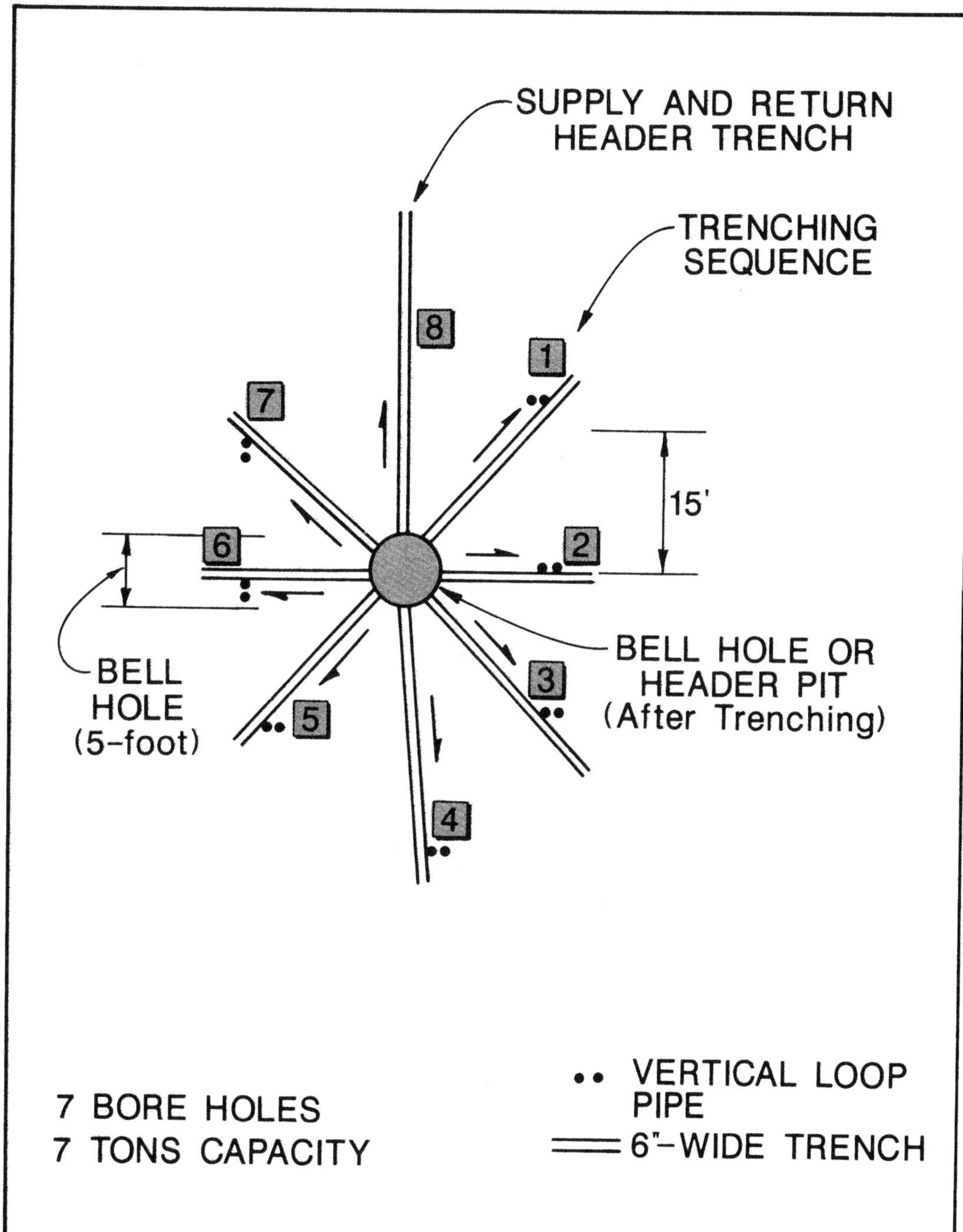

FIGURE 6.6: Trenching Configuration

6

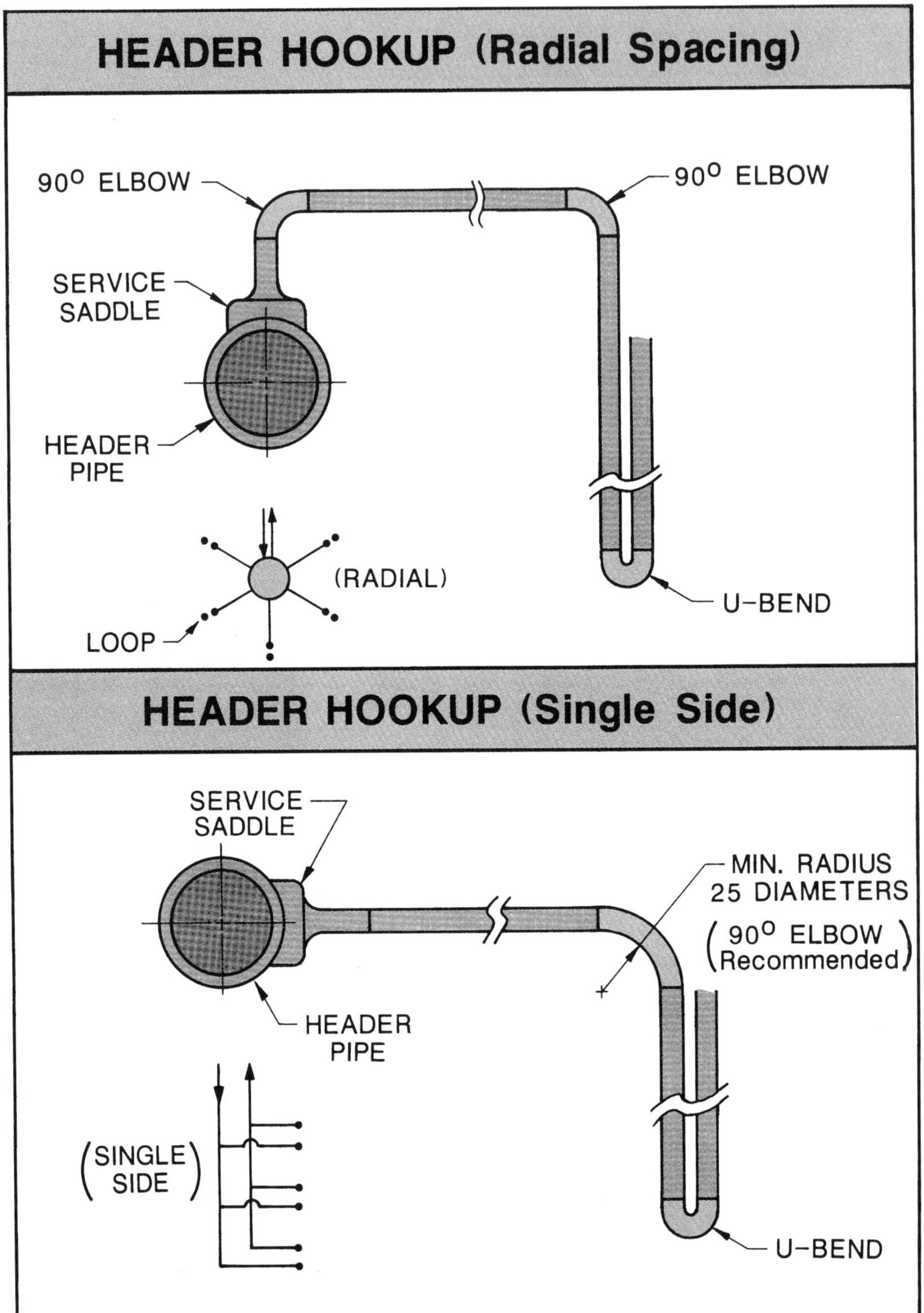

FIGURE 6.7: Header Hookup

7 Flushing and Purging the System

In This Section: Flushing debris; air purging; system flushing; verifying earth coil pressure/flow design; error detection techniques; installation notes; charging the system with an antifreeze solution; pressurizing the closed-loop system

The water-source heat pump and pumping module unit is connected to the ground heat exchanger as shown in Figure 7.1. Before operating the system, the following procedural and verification checks must be made to ensure the system will perform properly and as designed:

1. flushing debris from the ground heat exchanger
2. purging air from the ground heat exchanger
3. verifying ground heat exchanger design (pressure/flow)
4. checking for possible flow blockage if design verification does not check
5. charging ground heat exchanger with antifreeze
6. pressurizing the ground heat exchanger

These procedures and verification checks are detailed below.

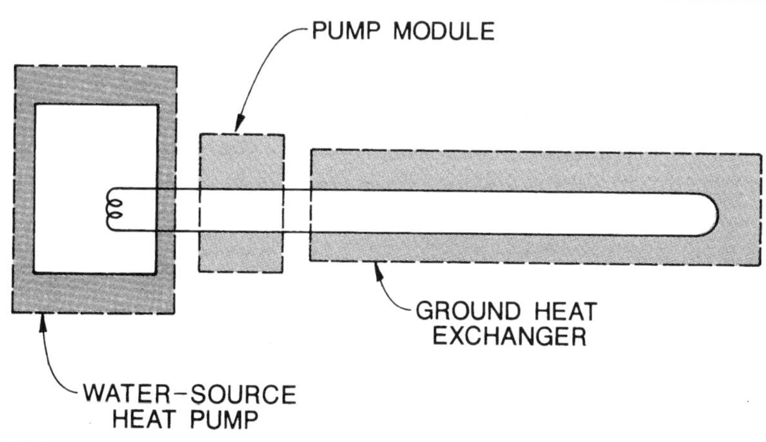

FIGURE 7.1: CL/GS System Configuration

7

Flushing Debris

Before operating the system, it should be flushed with water to remove any debris possibly trapped inside during fabrication. This debris may include dirt, pipe shavings, or other foreign objects. While a small amount of dirt may not be harmful to the piping system, bearings in circulating pumps can be damaged by particulates.

Depending on water quality, it may be necessary to filter the system to remove any particles or foreign materials in the water or water and antifreeze. Some brands of pelletized calcium chloride have harmful particles.

Taping or fusing on end caps to the pipe's ends before it is placed in the trench will help minimize contamination. If individual loops have been flushed and pressure tested, it is recommended that a permanent seal (such as a fused end cap) be used to ensure nothing can inadvertently enter the pipe.

Air Purging

In addition to flushing any debris from the system, it also is necessary to purge the ground heat exchanger of any air trapped in the closed-loop pipe system. Failure to do this can corrode the metallic components in the circulating system, causing their eventual failure. Under certain circumstances, excessive air in the system may block water flow in some branches of a parallel ground heat exchanger system. For example, air trapped at the top of one of several vertical ground heat exchanger loops may cause an air column that will prevent flow through the loop when low-power flushing is used. In this case, the fluid in the loop will not circulate, and the flow will be restricted to the remaining vertical loops which are air-free. A similar flow blockage due to air also could exist if a horizontal installation is laid at a constant depth in sloping terrain. In this case, elevation changes may result in greater pressure heads than can be overcome by the circulating pump. If sufficient flushing power is used, the air will be removed in all cases.

A flow velocity of 2 ft/sec in a piping system will completely remove any trapped air in the system. In parallel systems, the required flow rate for air purging in each parallel loop and each section of the header system should be checked to satisfy the flow velocity condition. Recall that the header flow rate decreases as the flow is branched down each parallel loop. For illustrative purposes, Figure 7.2 details three commonly used header configurations for parallel systems with three branches. These types of header configurations are typical for three-ton heat pump systems. The flushing flow rate required for air removal in each of these header systems varies with the particular design option.

During normal heat pump operating conditions, the header flow (Figure 7.3) at the first loop is 9 GPM (assuming 3 GPM/ton); between the first and second loop, 6 GPM; and between second and third loop, 3 GPM. The typical circulating pump designed for normal heat pump operation is not capable of removing all trapped air. To remove the air, a fluid velocity of 2 ft/sec in each branch must be maintained during air purging.

Table 7.1 shows the required flow rate (GPM) to give a velocity of 2 ft/sec for a number of typical Schedule 40 pipe sizes (inches) used in ground heat exchangers.

$$\text{VEL (FT/SEC)} = \frac{0.4085\ (Q)}{Di^2}$$

$$Q\ (\text{GPM}) = \frac{(2\ \text{FT/SEC})\ (Di^2)}{0.4085}$$

Using the data from Table 7.1 and the header configuration options shown in Figure 7.2, the required flow rates and the resulting pressure heads for air removal for each of these systems can be deter-

7

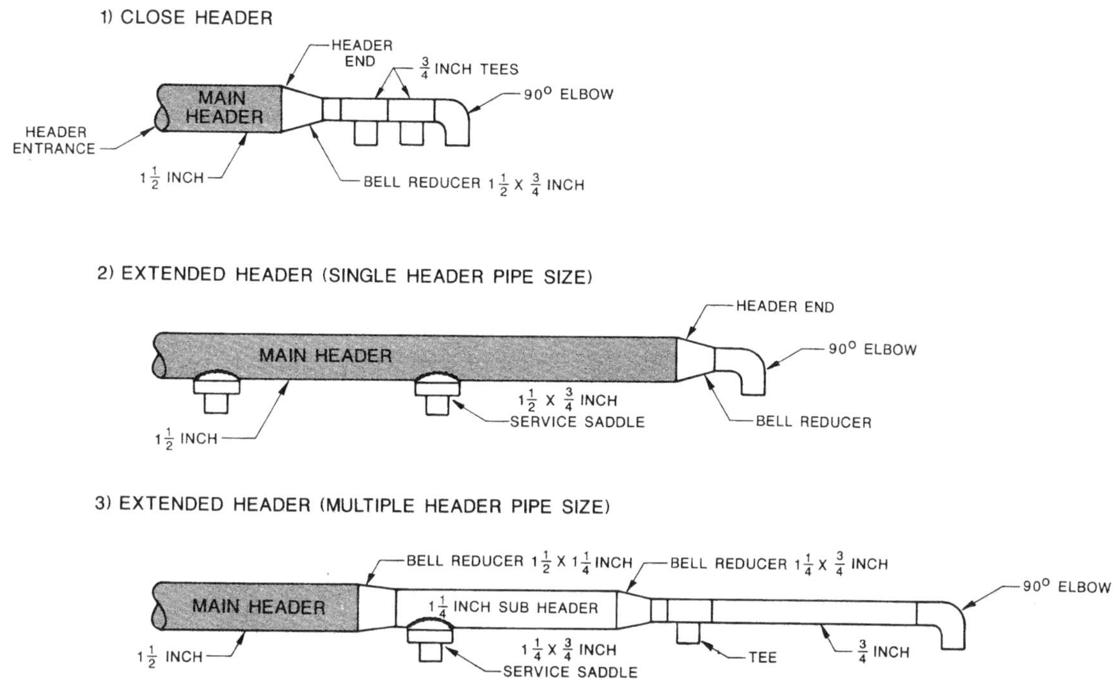

FIGURE 7.2: Header Configuration Options

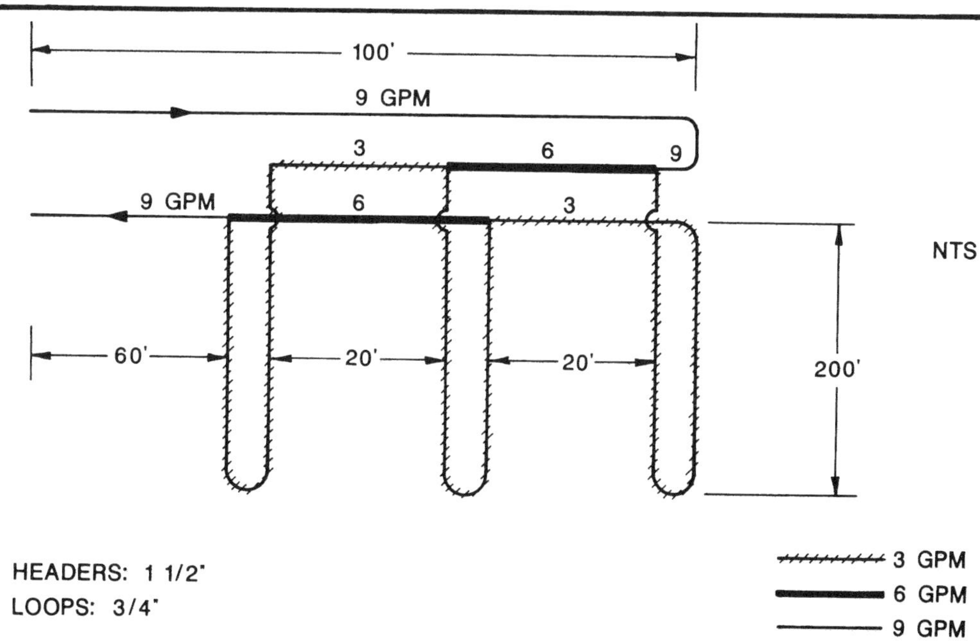

FIGURE 7.3: Spaced Header with Reverse Return

7

TABLE 7.1: Flow Rate (GPM) Required for Debris Flushing and Air Purging to Give an Average Velocity of 2 FPS in Pipe

	NOMINAL PIPE SIZE (inches)	PIPE ID (inches)	FLOW RATE (GPM)
	POLYETHYLENE, SDR-11		
1	¾	0.860	4.3
2	1	1.077	5.7
3	1¼	1.358	9.0
4	1½	1.554	11.8
5	2	1.943	18.4
	POLYETHYLENE, SCH 40		
6	¾	0.824	3.3
7	1	1.049	5.4
8	1¼	1.380	9.3
9	1½	1.610	12.7
10	2	2.067	20.9
	POLYBUTYLENE, SDR-17, IPS		
11	1½	1.676	13.8
12	2	2.095	21.5
	POLYBUTYLENE, SDR-13.5, CTS		
13	1	0.957	4.5
14	1¼	1.171	6.7
15	1½	1.385	9.4
16	2	1.811	16.1

mined. Table 7.2 gives these results for the three types of headers shown.

The required flushing flow rate is determined by the number of loops beyond the header end. If a single loop is attached at the header end, the flow rate at the header end must flow through the end loop. Therefore, for the extended single pipe, the flushing flow rate becomes three times the flow rate of the end loop or 3 x 12.7 or 38.1 GPM.

The flushing flow rates and pressure heads given in Table 7.2 can be determined as follows:

1. flushing flow rate (FFR) for three-ton heat exchangers:

$$= \frac{3 \text{ x Header FFR}}{\text{number of loops beyond the header end}}$$

2. individual loop flow:

$$= \frac{\text{FFR}}{\text{number of loops beyond the header end}}$$

3. flushing pressure head:

DP = header pressure head + loop pressure head

$$DP = \frac{0.0004\, Q_H^{1.85}\, L_H}{DI_H^{4.86}} + \frac{0.0004\, Q_L^{1.85}\, L_L}{DI_L^{4.86}}$$

where DP = pressure drop (psi)

Q_H = the header flushing flow rate, FFR

Q_L = loop flow rate

7

TABLE 7.2: Flow and Pressure Heads for Air Purging

	Flow (GPM)			Pressure Head (feet)	
Option Description	Main Header	Sub Header	Loop	Headers main/sub (ft/100 ft)	Loop 400 ft of pipe
1. Close	12.7	—	4.23	1.0	11.1
2. Extended (single pipe)	38.1	—	12.70	7.7	84.9
3. Extended (multiple pipe)	13.95	9.3	4.65	1.2/1.2	13.2

Example: Assume a ground heat exchanger with 100 feet of headers and three individual loops of 400 feet of pipe each. The header pipe diameter is 1½ inch (1.554 inches) and the loop pipe diameter is ¾ inch (0.86 inches). The header configuration chosen is Option 1 (close header) described in Figure 7.2.

1. flushing flow rate and header flow:
   = (3 x 12.7) / 3 = 12.7 GPM
2. loop flow rate:
   = 12.7/3 = 4.25
3. flushing pressure head:

$$\frac{0.0004 \times 12.7^{1.85} \times 100}{1.61^{4.86}} + \frac{0.0004 \times 4.23^{1.85} \times 400}{0.86^{4.86}}$$

= 0.43 + 4.80 + 5.24 psi or 12.1 feet of head

OPTION 1:

Option 1 brings all the loop heat exchangers to a common point as shown in Figure 7.2. This reduces the distance between attachment points of the loops to the header to a minimum. The advantage here is the flushing flow rate will be lower and the necessity of providing a reverse return is eliminated.

One disadvantage of Option 1 is the loop lengths are not equal. For small residential systems (three tons and less), this does not present a serious problem. Consider the system shown in Figure 7.4 where three vertical loops of unequal length are spaced 20 feet apart. In this case, two loops are constructed from 440 feet of pipe each and the third from 400 feet of pipe. An analysis would show that the flow differences in each loop are slight (2.95, 3.08, and 2.95 GPM).

OPTION 2:

The flushing flow rate required at the header end of the extended single-pipe header system is 12.7 GPM. This will require loop 3 as well as loops 1 and 2 to have the same flow rate. Therefore, a flow of 3 x 12.7 or 38.1 GPM is required at the header inlet. This would require a considerably larger purging unit (38.1 GPM and a pressure head of 84.9 feet) than that required of Option 1.

OPTION 3:

Option 3 uses a reducing header diameter along the flow path as shown in Figure 7.2. In this example, the main header is 1 ½-inch pipe, the subheader between the first and second branch uses 1¼-inch pipe, and third branch uses ¾-inch pipe. When Options 1 and 3 are compared, this configuration reduces the flushing flow requirement from 38 to 14 GPM and the pressure head from 84.9 to 11.1 feet.

7

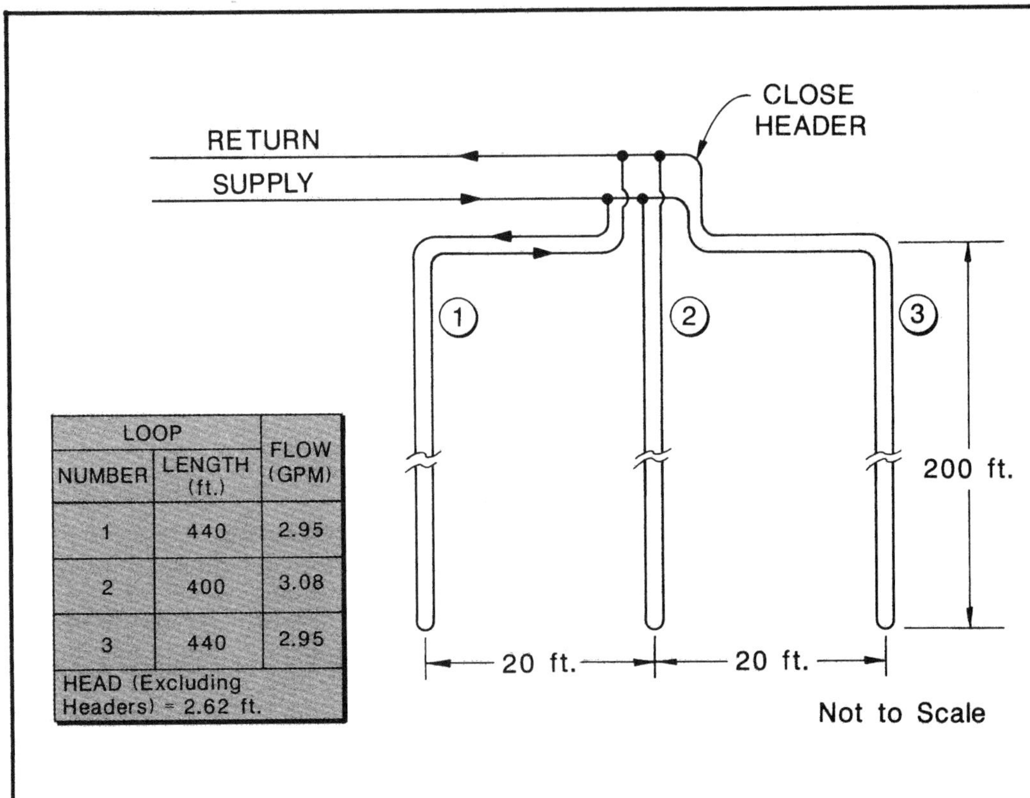

FIGURE 7.4: Unequal Loop Lengths

DISCUSSION OF HEADER OPTIONS

The three options discussed above clearly demonstrate the need for careful planning when flushing procedures and power flushing levels are considered. The advantage of Options 1 and 2 is that only a single header pipe size is required. Option 3 requires two pipe sizes for the header and a third pipe size for the loops. However, Option 2 has a considerably higher flushing power level that will require a high pump power level. The advantages of Options 1 and 3 from a flushing standpoint far outweigh the conveniences of Option 2.

Option 2 has the lowest header pressure drop of all three designs. It also has the advantage that each loop will have the same pressure difference, resulting in equal flow in all loops. Measured results have shown that the difference in pressure requirements of all three designs at normal heat pump flow is less than three feet of head.

System Flushing

The pump modules shown in Figures 7.5 and 7.6 allow for easy flushing and purging of the heat pump and ground heat exchanger system. The heat pump circulating pump used in most residential systems will not normally provide enough pumping power to reach the minimum flow velocities of 2 ft/sec necessary to purge the system of trapped air.

The pressure head and necessary flow capability to purge the system of air is provided by a temporary external pump-

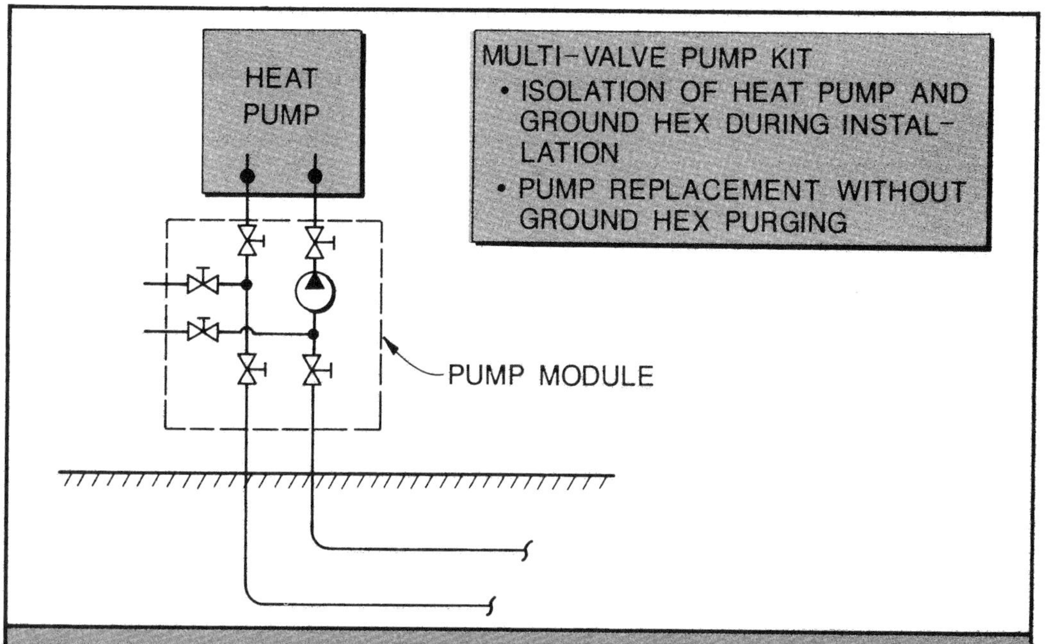

FIGURE 7.5: Pump Module and Hookup — Six Valves
Source: *WaterFurnace Liquid Source Heat Pumps*[2]

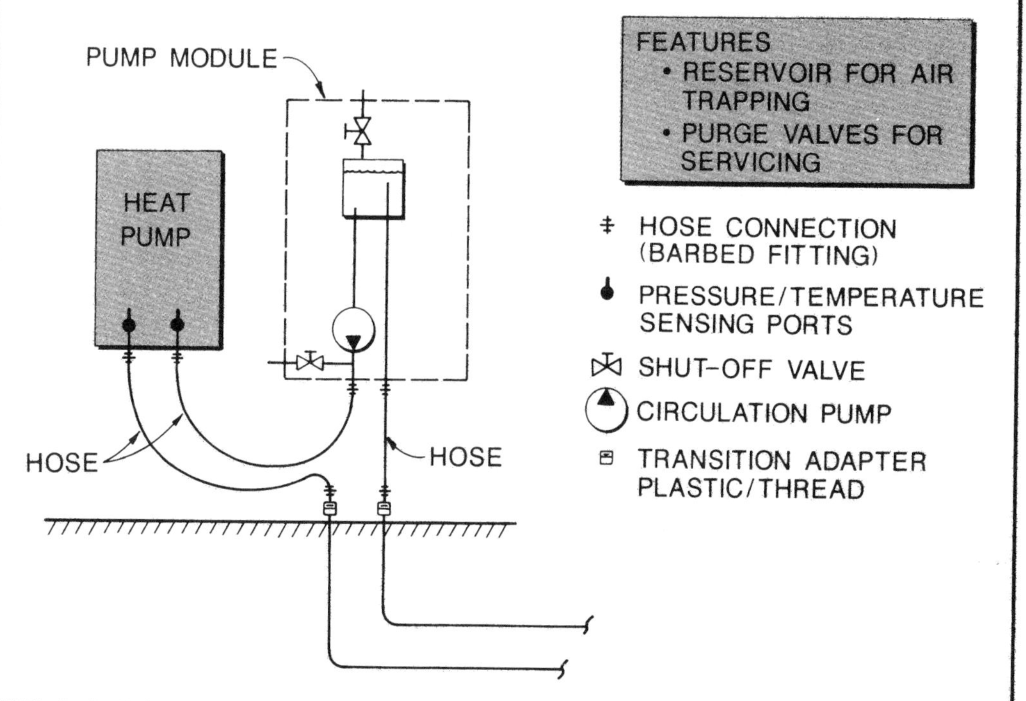

FIGURE 7.6: Pump Module and Hookup — Air Tank
Source: *Closed Loop Earth Coupled Applications, Specifications/Installation Instructions*[20]

7

ing power source called a flushing or purging unit. The purging unit (Figure 7.7) is connected to the loop using flexible rubber hoses and band clamps. The purging unit is connected to the closed-loop system so the ground heat exchanger can be flushed independently of the water-source heat pump. This procedure minimizes the possibility of bringing debris from the earth coil into the heat pump water heat exchanger. Also, the purging unit's total pumping power can be totally directed to the ground heat exchanger.

This purging and flushing procedure should be followed:

1. Clamp the purging unit pump outlet hose to the barb fitting on valve 2 as shown in Figure 7.8. Valve 2 is located on the inlet side of the water-source heat pump during normal heat pump operation.
2. Clamp the return hose on a barb fitting on valve 1. Valve 1 is located on the water-source heat pump water outlet port.
3. Open valves 1 and 2.
4. To flush the ground heat exchanger without flowing through the heat pump, close valve 3. This may be a separate valve or an integral flange valve mounted on the pump.
5. Start the purging unit pump. The flow through the ground heat exchanger will be in the opposite direction of normal heat pump operation.
6. To check for complete air purging, hold the return hose of the flushing unit below the surface of the water in the purging unit reservoir. If air is still being purged from the system, air bubbles will rise to surface from the end of the hose. The hose must be held deep enough so the normal agitation of the water (due to the flow out of the end of the hose) is not confused with air bubbles coming from the loop.

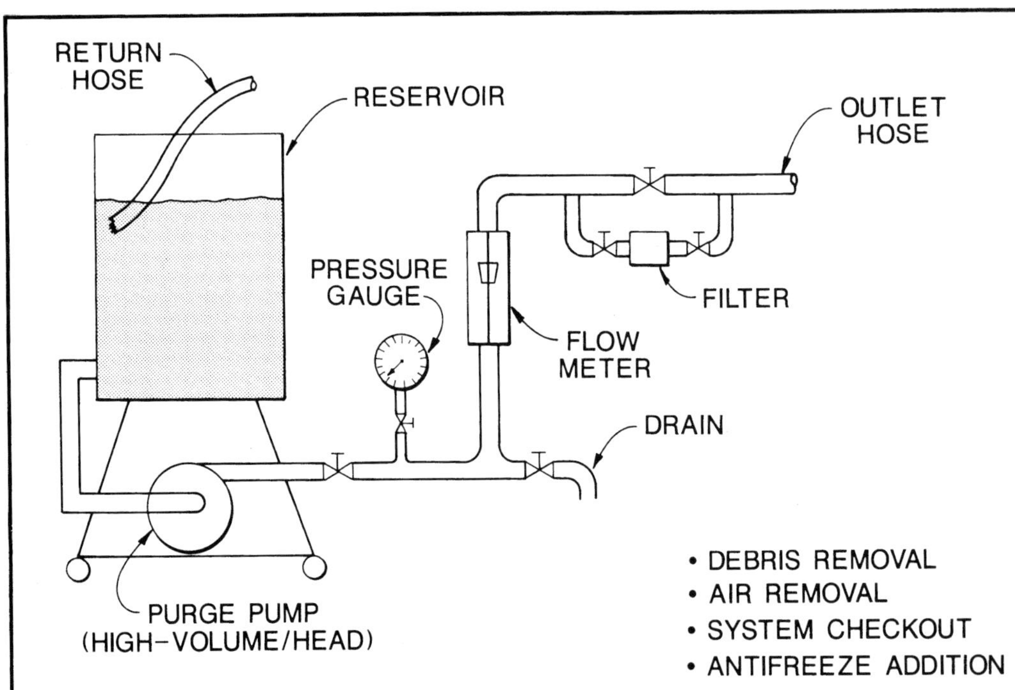

FIGURE 7.7: Purging Unit

7

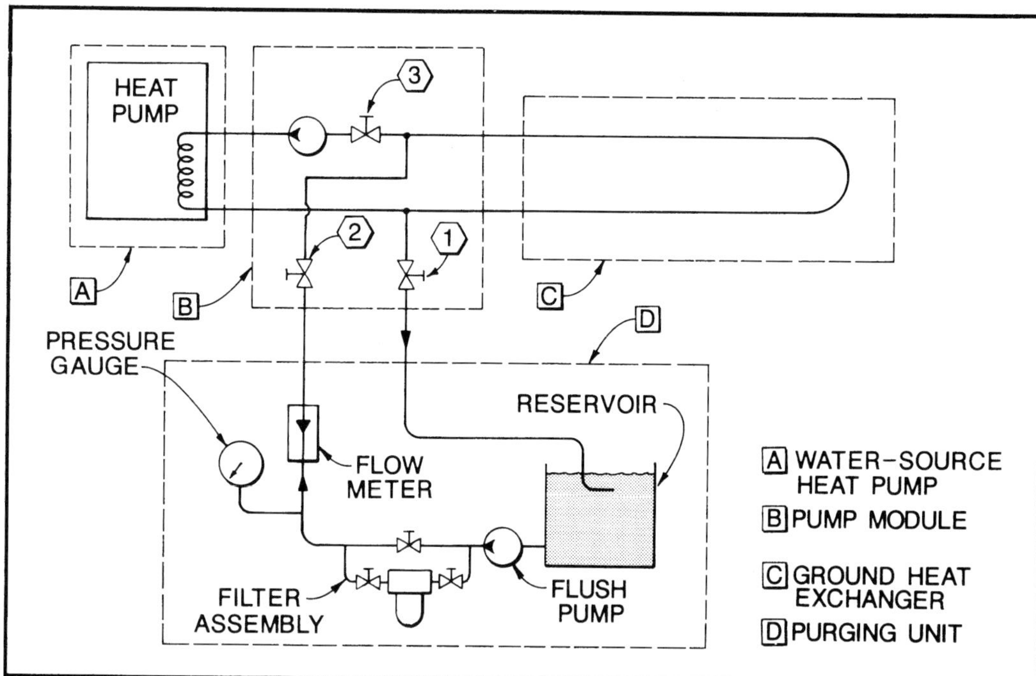

FIGURE 7.8: Servicing Configuration

7. After the ground heat exchanger is purged of air and debris, open valve 3. This allows a portion of the flow to now go through the heat pump system and flush any additional air or debris from the heat pump water heat exchanger circuit.

After completion of steps 1 through 7, the entire heat pump system (liquid loop or closed-loop) will be purged and ready for service.

Verifying Earth Coil Pressure/ Flow Design

At this point in the installation and check-out procedure, the system's closed loop is flushed and purged of air. With header trenches still open, it is suggested that a check be made to determine that no fabrication or installation errors have been made that might later necessitate digging up parts of the system. The most probable fabrication or installation errors made up to this time are:

1. Failure to drill out the header material after making a sidewall fusion joint of a service tee onto the header pipe. This obviously would not allow any flow through the loop heat exchanger attached to the header at this point.
2. A kinked line in a ground heat exchanger loop which would restrict the flow rate through that loop. The most likely place where this might occur would be the end of a horizontal loop where a large radius bend was used to make the return loop. A kink may possibly occur at this bend if, during the back-filling process, this radius was mistakenly reduced (by, *e.g.*, a large boulder or a tractor tire accidentally driven into in the trench). Kinked lines can be avoided

7

by using either two 90° elbows or a close-return U-bend in place of the large-radius turn (Figure 7.9).

The purging unit (Figure 7.7) can be used to verify the ground hex loops will operate as designed and to rule out the possibility that any of the errors mentioned above exists.

For purposes of illustration, assume the ground heat exchanger is installed as shown in Figure 7.10. The ground heat exchanger design parameters are as follows:

Ground heat exchanger flow rate (GPM)---- 9
Headers: Pipe inside diameter (inches)------- 1.554
Total length (feet) --------------------- 200
Loops: Number ------------------------------------- 3
Pipe inside diameter (inches) ---------- 0.86
Pipe length (feet) ------------------------ 400

The pressure drop can be estimated using the Hazen-Williams equation:

HAZEN–WILLIAMS EQUATION

$$DP = \frac{452\ Q^{1.85}}{C^{1.85}\ Di^{4.86}}$$

DP = Pressure drop per 100 feet of pipe (psi)
Q = Flow rate (GPM)
C = Hazen-Williams roughness coefficient
= 155 for smooth pipe (PE and PB)
Di = Pipe inside diameter (inches)
452 = A constant

NOTE: Pressure drop is the difference between the inlet and outlet pressures (P1-P2) of the ground heat exchanger. See Figure 7.10 for definition and location of P1 and P2.

Using this equation, the pressure drop for the header and three loops at 9 GPM is calculated as follows:

HEADER PRESSURE LOSS:

$$DP = \frac{452\ (9)^{1.85}}{(155)^{1.85}\ (1.554)^{4.86}} = 0.27 \text{ psi/100 feet}$$

LOOP PRESSURE LOSS:

$$DP = \frac{452\ (3)^{1.85}}{(155)^{1.85}\ (0.86)^{4.86}} = 0.63 \text{ psi/100 feet}$$

TOTAL PRESSURE LOSS:

400 x 0.63/100 + 200 x 0.27/100 = 3.1 psi

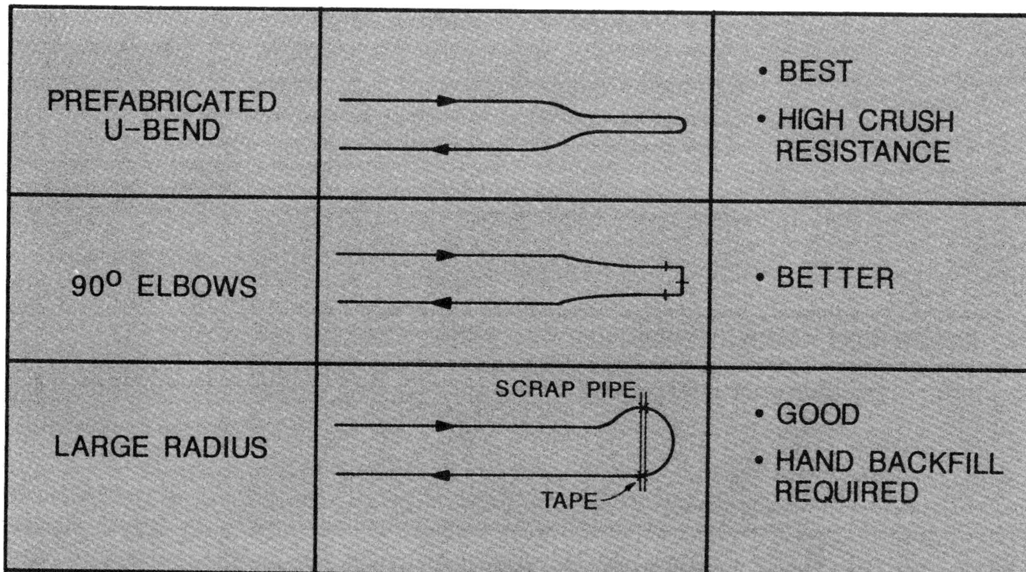

FIGURE 7.9: Horizontal Return Loop Options

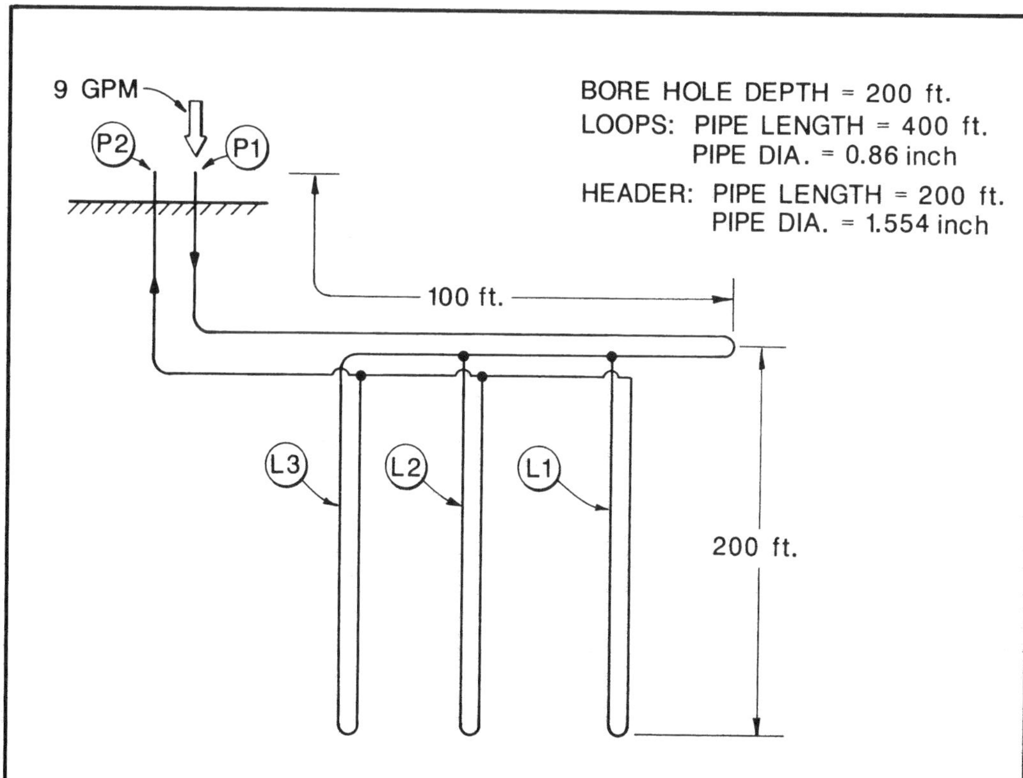

FIGURE 7.10: Ground Heat Exchanger Example

7

If the measured pressure increase (for a flow of 9 GPM through the ground heat exchanger) is approximately 3.1 psi, then it can be assumed the system is operating as designed and all flow paths are clear.

> **NOTE: When verifying the system pressure drop through the heat pump or ground heat exchanger, do not include the pressure drop due to the valves and hoses connecting the flushing unit to the pump module. These values must be subtracted from the total system pressure drop.**

By varying the flow in the ground heat exchanger, a pressure/flow characteristic curve (Figure 7.11) can be constructed for the three-loop ground heat exchanger defined above. Any measured variation from this curve (curve 1) will indicate that something is wrong and further testing is required. If further purging and reverse flow does not solve the problem, then it is possible one of the above installation errors has occurred.

The first step in determining that a problem exists is to understand the pressure flow characteristics of the ground heat exchanger. The Hazen-Williams equation given above can be used to construct the following table (Table 7.3).

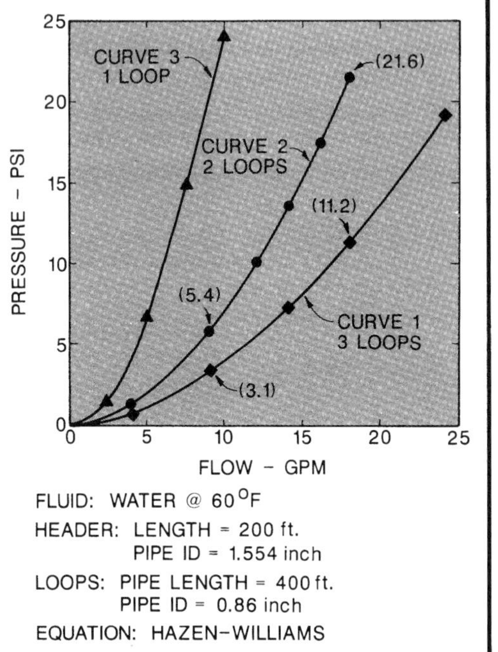

FIGURE 7.11: Pressure/Flow Characteristic Curve

Assume that one of the three service saddle tees had not been drilled out. If this were the case, the 9 GPM flow would be divided between the two (N = 2) open ¾-inch ground heat exchanger loops. The resulting two-loop characteristic curve (curve 2) is shown in Figure 7.11 along with the previously defined three-loop sys-

TABLE 7.3: Flow Versus Pressure Drop

	PRESSURE DROP (PSI)		
FLOW (GPM)	**N = 3**	**N = 2**	**N = 1**
2.5	0.28	0.6	1.9
5.0	1.04	2.0	6.7
10.0	3.70	7.2	24.3
15.0	7.90	15.3	51.4
20.0	13.60	26.0	87.6
25.0	20.40	39.3	132.3

N = number of loops

tem (curve 1). It is obvious the two-loop heat exchanger has a higher pressure difference than the three-loop system at a system flow rate of 9 GPM. This higher pressure difference may indicate one of the loops is not flowing properly. The calculation procedure for the two-loop system at 9 GPM is as follows:

1. Calculate the pressure drop per 100 feet for each of the two ¾-inch (0.86-inch) loops at 4.5 GPM as:

$$DP = \frac{452\,(4.5)^{1.85}}{(155)^{1.85}\,(.86)^{4.86}} = 1.3 \text{ psi/100 feet}$$

2. Sum the header and loop pressure drop to obtain the total ground heat exchanger pressure drop as:

$$DP = 0.27/100 \times 200 + 1.3/100 \times 400 = 5.74 \text{ psi}$$

With one horizontal ¾-inch loop blocked off, the measured pressure drop across the ground heat exchanger would be higher than the sample's design value by about 5.74 – 3.1 or 2.64 psi. Since the difference between these values is relatively small, experimental error in taking the readings may not allow verification of the system's performance. One way to check the system would be to increase the flow rate and observe the difference in heat exchanger pressure at the higher flow rate.

The higher powered flushing unit and the characteristic pressure/flow curves of Figure 7.11 can be used to determine if an installation error has been made. If the pump flow rate is increased to 18 GPM and the system is operating normally with all three ¾-inch loops open, the flow rate in each of the ¾-inch loops will be 6 GPM and the heat exchanger pressure drop can be read as 11.2 psi from curve 1 of Figure 7.11.

Now, assuming as before that one of the ¾-inch loops was blocked, the 18 GPM pump flow rate would again be divided equally between the two open ¾-inch loops resulting in 9 GPM in each loop.

For this condition, the pressure drop for each of these two ¾-loops would be:

$$DP = \frac{452\,(9)^{1.85}}{(155)^{1.85}\,(.86)^{4.86}} = 4.9 \text{ psi/100 feet}$$

The total system pressure drop with the one blocked ¾-inch loop at 18 GPM would then be:

$$DP = 200 \times 0.99/100 + 400 \times 4.9/100 = 21.6 \text{ psi}$$

This drop could have been read from curve 2 of Figure 7.11.

At this increased flow rate (18 GPM), the pressure difference of 21.6 – 11.3 = 10.3 psi is sufficiently large to determine that a possible blocked loop exists.

If an increased pressure drop of the magnitude indicated above is observed, the conclusion to be reached is that one of the loops is blocked. The next step in the troubleshooting procedure would be to identify which of the three loops is blocked. This can be done by squeezing off (one at a time) each of the ¾-inch loops with a pinch-off tool (Figure 7.12) and noting the pressure drop. When the blocked ¾-inch loop is squeezed off there will be no pressure drop change since there already is no flow in this loop. When either of the other open ¾-inch loops is squeezed off, the pressure drop across the entire system will increase since all of the flow now will be forced through the one remaining open loop.

Error Detection Techniques

1. Kinked lines generally occur at points where large-radius loops are used to make sharp turns such as those found in horizontal parallel loops. Avoid the large-radius turn by substituting a fabricated close-return bend or two 90° elbows closely fused together.
2. Use some identification procedure (special colored tape, spray paint,

7

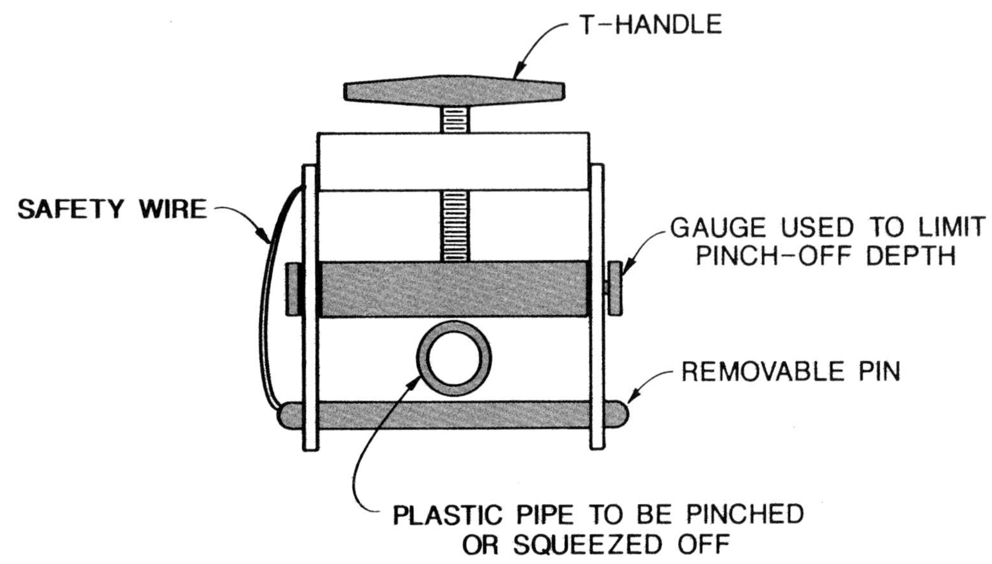

FIGURE 7.12: Pinch-Off Tool

etc.) to note that a service saddle tee has been drilled. If two installers are doing the work, the most common error is, "I thought you drilled that one out!" Careful attention should be paid to ensure the drill coupon or loosely drilled materials are removed. They commonly lodge in the circulating pump or heat pump liquid exchanger either stopping or degrading the system flow rate.

POSSIBLE KINKED LINE IN GROUND HEAT EXCHANGER

A kinked line in one of the ¾-inch horizontal loops would produce a pressure drop somewhere between the 11.2 and 21.6 psi observed in the previous example where the flow rate was 18 GPM. The pressure drop observed for the kinked line would be dependent on the severity of the kink. If there is only a slight kink resulting in only a small flow restriction, the observed pressure drop would only be slightly higher than the 11.2 psi value expected in the example. For such a case, the inaccuracy of the measuring equipment in the field probably would not allow determination of a slight kink. A slight flow restriction in one loop would not significantly deteriorate the performance of that single loop.

The checkout procedure for determining if there is a possible blocked line is the same as that described above for a kinked line. It may be necessary to squeeze off both remaining loops and compare this pressure drop against all loops checked under a full system flow condition.

Once identified, the end of the kinked loop on a horizontal system should be dug up to find the kink. If possible, the pipe should be repositioned to form a larger radius bend. In some cases the kink should be cut out and a U-bend fused onto the two ends of pipe.

7

Installation Notes

1. Select only the better grades of pipe so that kinks can be worked out instead of having to be cut out. Kinks in lower grades of pipe will have to be cut out and the pipes rejoined by heat fusion, using up valuable time.
2. When possible, use close return U-bends on horizontal systems. This prevents kinks resulting from improper backfilling or soil settling. It is almost impossible to crush a U-bend assembly with a vehicle tire.
3. Choose a high-power purging unit with high pressure and flow characteristics to be able to check systems out at the higher flow rates.
4. Many of the problems noted do not occur if the proper pipe materials, return bends, etc., are selected prior to installation.
5. Proper crew training is essential. Only professionals make it look simple and easy.

Charging the System with an Antifreeze Solution

The flushing unit is used to charge the system with an antifreeze solution. The procedure is as follows:

1. Isolate the ground heat exchanger from the heat pump water heat exchanger by closing valve 3 (Figure 7.8).
2. Add the precalculated amount of antifreeze to the flushing unit reservoir. As the antifreeze is added to the reservoir, the return hose from the ground heat exchanger is removed from the reservoir and the initial flow of water returning from the ground heat exchanger is discarded. The amount of fresh water discarded should be the same as the amount of antifreeze solution which will be added.
3. After all the antifreeze has been added, place the return hose in the flushing unit reservoir and continue until the antifreeze is uniformly mixed throughout the ground heat exchanger.
4. Open valve 3 to circulate the antifreeze into the water heat exchanger of the heat pump loop.

Pressurizing the Closed-Loop System

The final step prior to operating the system is to pressurize the system for closed-loop operation. The system must be pressurized to a level that will prevent cavitation (formation of air bubbles in solution) of the circulating pump and maintain closed-loop pressure during the heating and cooling temperature cycles. Cavitation in the centrifugal circulating pump can be prevented by maintaining a minimum of 5 psi on the suction side of the pump. Changes in system pressure due to temperature changes can be managed with initial pressures of 30 psi.

The flushing unit is used to pressurize the system. The pressurization procedure is as follows:

1. Check that valves 1, 2, and 3 (Figure 7.8) are all open.
2. Start the purging unit pump.
3. Close valve 1. This prevents return flow to the flushing unit reservoir and allows the flushing unit pump to develop increased pressure in the ground heat exchanger loop.
4. When the pressure peaks on the purging unit pump outlet pressure gauge, close valve 2, trapping this pressure in the closed loop.
5. Shut the flushing unit pump off.

The flexible hoses used with the flushing unit can now be disconnected from the ends of valves 1 and 2. The barb fittings on valves 1 and 2 should be removed and a service plug installed. Recommended practice is to remove the handles on valves 1 and 2 to prevent accidental opening of these valves and loss of pressurization.

System pressure in this charged state is usually 20 to 30 psi. This is for two reasons:

1. to prevent cavitation in the circulating pump
2. to keep the system under pressure when loop temperatures increase. Experience has shown that the system's pressure decreases when loop temperatures increase. This is a result of the pipes' internal volume expanding at a greater rate than the liquid on the inside.

COMMENTS ABOUT PERFORMANCE CHECKING

1. Performance checking can only be as good as the accuracy of the measurements. Temperature measurements must be made to within 0.1°F on the water side to give reasonable results. Electronic measuring equipment is available at reasonable prices to obtain this accuracy.
2. Accurate air-side measurements are difficult to make in the field. Airflow is especially important since it will give some indication of whether the duct work is adequate to handle the design flow rate. A simple hand-held manometer and heat pump blower performance data will produce a lot of important data rapidly. Table 3.12 gives blower performance data for a manufacturer's line of water-source equipment.
3. Performance of the heat pump will depend on the sizing of the unit to the load. Common complaints are:
   a. The unit runs all the time! If the unit is sized properly, then, during extreme heating and cooling, the unit will run long hours, thus giving higher performance than units with large numbers of on-off cycles.
   b. I'm using supplemental heat and I wanted to avoid this. In areas where cooling loads are used to size the heat pump, supplemental heat will be required in extreme heating periods. It can be more economical for the homeowner to use supplemental heat during these short periods than to go to the expense of installing oversized ground heat exchangers and heat pumps. Sizing the heat pump for the cooling load is often the most economical choice for the homeowner.

8 Heat Pump System Start-Up and Checkout

In This Section: Heating mode operation; cooling mode operation; heat pump performance checkout

Once the system has been flushed, antifreeze added, and the system pressurized, the heat pump system can be started and checked out. Depending on the season, the performance of the heat pump should be checked against the manufacturer's published data.

Heating Mode Operation

During the heating cycle, the performance can be checked by measuring the heat absorbed from the earth and adding this to the amount of heat from the compressor, fan, and other electrical inputs, or by following the equation below:

Equation 8.1

Heat pump capacity = Heat absorbed from the earth + Electrical power input

The amount of heat absorbed from the ground heat exchanger can be calculated by measuring the flow rate and temperature change at the inlet and outlet of the water-source heat pump. The formula, Equation 8.2, is shown below:

Equation 8.2

Heat absorbed (Btu/hr) = (Q) x (DT) x (K)

Where Q = Flow rate in gallons per minute (GPM)

DT = Temperature change between inlet and outlet of the heat pump (°F)

K = Flow constant

= 500 for water

= 8.34 lb/gal x 60 min/hr x 1 Btu/lb – °F

8

Flow rate through the heat pump can be measured directly if a flow meter is installed in the circulation loop. Since this is usually not the case, it is necessary to measure the pressure drop across the heat pump's water heat exchanger and then compare the manufacturer's water flow rate data with this pressure drop.

Equation 8.3 can be used to convert the manufacturer's data listing pressure drop in feet of water to psi:

Equation 8.3

Feet of water (wpd) = PSI/0.433
= PSI x 2.31

The temperature difference (DT) is determined by measuring the heat pump inlet and outlet water temperatures. To make a reasonable check on the heat pump's performance, it is necessary to measure these temperatures to the closest 0.1°F. It is suggested the same temperature probe be used for measuring both inlet and outlet temperatures. This reduces the possibility of introducing error in DT should different probes have varying bias errors.

The electrical power input to the heat pump can be measured directly using a wattmeter, or estimated by Equation 8.4:

Equation 8.4

Power (watts) = 0.85 (voltage) (amperage)
where 0.85 = an assumed power factor

The conversion from watts to Btu/hr is given by Equation 8.5:

Equation 8.5

Power (Btu/hr) = 3.412 x Watts
= 3,412 x kilowatts (kWh)

Having determined the water flow rate through the heat pump, the temperature difference across the heat pump, and the electrical power input to the heat pump, the terms on the right-hand side of Equation 8.1 can be calculated. The sum of these terms should equal the heat pump heating capacity shown in graphical or tabular form on the manufacturer's performance data sheets for the operating conditions under which the above measurements were made. If the system is operating normally, the measured data should agree closely with the manufacturer's data.

Cooling Mode Operation

Equation 8.6 gives the heat balance relationship for the cooling mode:

Equation 8.6

HP cooling capacity = Ground heat exchanger load – Electrical power input

The ground heat exchanger load is calculated using measured data across the water side of the heat pump. Using the same type of measured information completed for the heating mode, the terms on the right side of Equation 8.5 are calculated. Notice that, for the cooling mode, the electrical power input is subtracted from the heat rejected to the ground heat exchanger. This estimated cooling value should agree closely with the manufacturer's published value of heat pump cooling capacity for the specified operating conditions. A nominal value for a three-ton heat pump operating in a cooling mode would be approximately 50,000 Btu/hr with the building at normal comfort conditions.

Heat Pump Performance Checkout

Performance data on water-source heat pumps is available from the manufacturer. The data will be presented for a number of entering water and air temperatures. Table 8.1 is typical manufacturer's data that can be used in evaluating a particular product.

HEATING PERFORMANCE CHECK

During heating operations, the following readings are obtained from measurements taken on the water side and on the air side of the unit.

1. Water pressure drop across the coil------ 5.75
   PSI Feet of head = 5.75 psi x 2.31 ---------- 11.3
   From manufacturer's data @ 11.3 ft hd---- 2.5 GPM
2. Temperature of water into heat pump ----- 50.2°F
3. Temperature of water out of heat pump --- 42.2°F
4. Electrical power input to heat pump ----- 1,150 Watts
5. Inlet air conditions -------- 70°F
6. Airflow (Standard Cubic Feet per Minute) ---------- 375 SCFM

The accuracy of these measurements will determine how well the unit's performance can be checked. On the water side, given the flow rate and the power input, the output of the heat pump can be estimated as follows:

1. Heat absorbed from the earth:
   2.5 GPM x (50.2–40.2) x 500 = 10,000 Btu/hr
2. Heat from electrical power input:
   1,150 Watts x 3.412 Btu/hr/Watt = 3,924 Btu/hr
3. Unit capacity at the above conditions:
   Heat pump capacity = 10,000 + 3,924 = 13,924 Btu/hr

The manufacturer's performance data (Table 8.1), gives a value of 14,200 Btu/hr at the above conditions. This is close to the calculated value of 13,924 Btu/hr, and it can be assumed the heat pump is operating normally. It must be remembered that ordinary field measurements give approximate values and they should be treated as such. The measurement accuracy of the inlet air temperature and airflow are critical when comparing performance data to manufacturer's data. Generally stated, it is easier and more accurate to check the heat pump on the water side than on the air side. Air-side measurements taken in the field are difficult, and great care must be taken in determining air properties under operating conditions.

COOLING PERFORMANCE CHECK

During the cooling cycle, the following measured data is obtained:

1. Water pressure drop --------- 5.0 psi
2. Temperature of water in ------ 90°F
3. Temperature of water out ------ 102°F
4. Electrical power input -- 1,200 Watts
5. Inlet air conditions:
   Dry bulb temperature ---------- 75°F
   Wet bulb temperature --------- 63°F

As before, the flow rate at 5.0 psi water pressure (5.0 x 2.31 = 11.7 wpd) is 2.5 GPM. Therefore, the amount of heat rejected to the ground heat exchanger will be:

2.5 GPM x (102°F – 90°F) x 500 = 15,000 Btu/hr

8

TABLE 8.1: Heat Pump Performance Data

			COOLING						HEATING				
EWT	GPM	WPD	EA	TC	SC	KW	HR	EER	EA	HC	KW	HE	COP
30	1.5	6.9	75/63	13.9	9.8	0.82	16.7	17.0	60	10.0	0.92	6.8	3.17
			80/67	15.1	10.2	0.83	17.9	18.1	70	9.6	0.97	6.3	2.91
			85/71	16.3	10.6	0.85	19.2	19.2	80	9.3	1.02	5.8	2.67
	2.5	15.8	75/63	14.2	9.9	0.78	16.8	18.2	60	10.5	0.95	7.3	3.24
			80/67	15.4	10.3	0.79	18.1	19.4	70	10.1	1.00	6.7	2.97
			85/71	16.6	10.7	0.81	19.4	20.5	80	9.8	1.05	6.2	2.73
	3.5	27.0	75/63	14.4	10.0	0.75	17.0	19.1	60	10.9	0.97	7.6	3.29
			80/67	15.7	10.4	0.77	18.3	20.4	70	10.5	1.02	7.0	3.01
			85/71	17.0	10.8	0.79	19.6	21.6	80	10.1	1.07	6.4	2.76
50	1.5	5.8	75/63	12.8	9.2	0.99	16.2	13.0	60	13.5	1.10	9.8	3.60
			80/67	13.9	9.6	1.01	17.4	13.8	70	13.1	1.16	9.1	3.30
			85/71	15.0	10.0	1.03	18.5	14.7	80	12.6	1.22	8.5	3.04
	2.5	13.3	75/63	13.1	9.3	0.94	16.3	13.9	60	14.2	1.14	10.3	3.65
			80/67	14.2	9.7	0.96	17.5	14.8	70	13.7	1.20	9.6	3.35
			85/71	15.3	10.1	0.98	18.7	15.7	80	13.2	1.26	8.9	3.08
	3.5	22.8	75/63	13.3	9.4	0.91	16.5	14.6	60	14.5	1.16	10.6	3.68
			80/67	14.5	9.8	0.93	17.7	15.6	70	14.1	1.22	9.9	3.38
			85/71	15.7	10.2	0.95	18.9	16.5	80	13.6	1.28	9.2	3.11
70	1.5	5.4	75/63	11.8	8.8	1.16	15.7	10.2	60	16.4	1.24	12.2	3.89
			80/67	12.8	9.1	1.18	16.8	10.8	70	16.0	1.30	11.5	3.59
			85/71	13.8	9.5	1.20	17.9	11.5	80	15.5	1.37	10.8	3.32
	2.5	12.4	75/63	12.0	8.8	1.10	15.7	10.9	60	17.2	1.28	12.9	3.95
			80/67	13.0	9.2	1.12	16.9	11.6	70	16.7	1.34	12.1	3.65
			85/71	14.1	9.6	1.15	18.0	12.3	80	16.2	1.41	11.4	3.37
	3.5	21.2	75/63	12.2	8.9	1.07	15.9	11.5	60	17.7	1.30	13.3	3.98
			80/67	13.3	9.3	1.09	17.0	12.2	70	17.2	1.37	12.5	3.67
			85/71	14.4	9.7	1.11	18.2	12.9	80	16.7	1.44	11.8	3.39
90	1.5	5.1	75/63	10.7	8.3	1.32	15.2	8.1	60	19.5	1.41	14.7	4.06
			80/67	11.6	8.6	1.35	16.2	8.6	70	19.1	1.48	14.0	3.77
			85/71	12.6	9.0	1.38	17.3	9.1	80	18.6	1.56	13.3	3.51
	2.5	11.7	75/63	10.9	8.4	1.26	15.2	8.6	60	20.3	1.45	15.3	4.09
			80/67	11.9	8.7	1.29	16.3	9.2	70	19.8	1.53	14.6	3.80
			85/71	12.8	9.1	1.31	17.3	9.8	80	19.4	1.61	13.9	3.53
	3.5	20.0	75/63	11.1	8.4	1.23	15.3	9.1	60	20.7	1.48	15.6	4.09
			80/67	12.1	8.8	1.25	16.4	9.7	70	20.2	1.56	14.9	3.80
			85/71	13.1	9.2	1.28	17.4	10.2	80	19.7	1.64	14.2	3.53
110	1.5	4.8	75/63	9.4	7.5	1.49	14.5	6.3					
			80/67	10.2	7.8	1.52	15.4	6.7					
			85/71	11.0	8.2	1.56	16.3	7.1					
	2.5	11.1	75/63	9.6	7.6	1.42	14.4	6.7					
			80/67	10.4	7.9	1.45	15.3	7.2					
			85/71	11.2	8.2	1.48	16.3	7.6					
	3.5	19.0	75/63	9.8	7.7	1.38	14.5	7.1					
			80/67	10.6	8.0	1.41	15.4	7.5					
			85/71	11.4	8.3	1.44	16.4	8.0					

1. Performance data at EWT below 35°F are based upon 15% methanol antifreeze solution.

Source: *WaterFurnace Liquid Source Heat Pump.*[8]

- EWT = entering water temperature, °F
- GPM = gallons per minute water flow
- WPD = water pressure drop, feet of water
- EA = entering air temperature, °F (dry bulb/wet bulb)
- HC = total air heating capacity, Btu/hr (1000)
- TC = total cooling capacity, Btu/hr (1000)
- SC = sensible cooling capacity, Btu/hr (1000)
- KW = total unit power input, kilowatts
- HR = total heat rejection, Btu/hr (1000)
- HE = total heat absorption, Btu/hr (1000)
- DC = desuperheater capacity, Btu/hr (1000)
- EER = Energy Efficiency Ratio = Btu output/Watt input
- COP = Coefficient of Performance = Btu output/Btu input

The electrical power input is 1,200 Watts or 1,200 Watts x 3.412 Btu/Watt = 4,094 Btu/hr

The cooling capacity can then be estimated to be:

15,000 − 4,094 = 10,906 Btu/hr

The manufacturer's performance data for cooling at these conditions (Table 8.1) is 10,900 Btu/hr. It can therefore be concluded that the heat pump is performing normally in the cooling mode.

APPENDIX A

Physical Properties and Selection of Ground Heat Exchanger Circulating Fluids

In This Appendix: Antifreeze solutes; circulating fluids' solute requirements; circulating fluids and antifreeze solutions; materials for salt solutions

The following possible ground heat exchanger circulating fluids will be considered here:

1. water
2. sodium chloride in water
3. calcium chloride in water
4. ethylene glycol in water
5. propylene glycol in water
6. methyl alcohol in water
7. ethyl alcohol in water

Table A.1 gives selected physical properties of some of the pure substances listed above.

The selection of an earth coil circulating fluid is based upon:

1. safety (toxicity and flammability)
2. freezing point
3. heat transfer characteristics (primarily thermal conductivity and viscosity)
4. frictional pressure drop and pumping requirements (primarily viscosity)
5. cost (initial charge, makeup, and inhibitor costs)
6. corrosiveness
7. physical and chemical compatibility with all system components
8. availability of product in local market area
9. ease of transporting and storing

The earth coil circulating fluid freezing point is particularly important and determines the solute concentration in the circulating fluid that is required to protect the evaporator of the heat pump against freezing. The coldest expected evaporator exit temperature of the earth coil circulating fluid during heating operation determines the required freezing point of the circulating fluid.

The refrigerant temperature in the evaporator runs approximately 10°F lower than the exit water temperature. Therefore, it is necessary to provide freeze protection for the minimum operating evaporator refrigerant temperature.

Antifreeze Solutes

Figure A.1 gives freezing point versus weight percent solute in water for calcium chloride and sodium chloride. Freezing points of methanol, ethylene glycol, and propylene glycol versus weight percent solute are given in Figure A.2.

These systems actually do not have a fixed freezing point. The temperatures

A

TABLE A.1: Earth Coil Fluids/Pure Substances' Physical Properties

	WATER	ETHYLENE GLYCOL	PROPYLENE GLYCOL	METHANOL
Molecular Weight	18.01	62.07	76.10	32.04
Specific Gravity 20/20	1.000	1.116	1.038	0.7917
Density @ 68°F (lb/gal)	8.33	9.28	8.65	6.60
Freezing Point, °F	32	8.6		-143.8
Normal Boiling Point, °F	212	387	370	148
Specific Heat @ 60°F (Btu/lbm – °F)	1.000	0.522	0.598	0.590
Viscosity @ 32°F, CP	1.79	57.4	243	
Viscosity @ 68°F, CP	1.01	20.9	60.5	0.6
Viscosity @ 104°F, CP	0.655	9.5	18.0	
Thermal Conductivity (Btu/hr – ft – °F)	0.338	0.166		0.124
Flash Point, °F		240	225	58

Source: *Chemical Engineers' Handbook*[21]

given are the temperatures at which freezing begins, creating a slush.

An earth coil circulating system should be designed to function at a mean temperature 10°F above the circulating fluid's freezing point. The mean temperature is the average between the inlet and outlet of the heat pump.

Circulating Fluids' Solute Requirements

The desired freezing point yields the needed solution concentration of solute and water. Figures A.1 and A.2 yield weight percent solute versus freezing point. A 10°F interval is usually desirable or attempted between the mean fluid operating temperature and the actual freezing point.

The pounds mass of solute needed per 100 gallons of 60°F water is given by:

Equation A.1

$$M_A = \left[\frac{X_{MA}}{100 - X_{MA}} \right] \times 833.8$$

Where: M_A = pounds mass of solute per 100 gallons of water at 60°F

X_{MA} = the weight percent

A

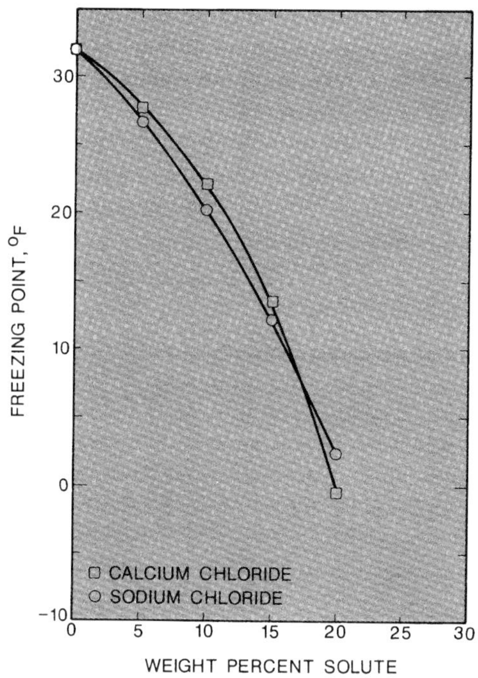

Source: *Chemical Engineers' Handbook*[21]

FIGURE A.1: Freezing Point vs. Weight Percent Solute (Calcium Chloride and Sodium Chloride)

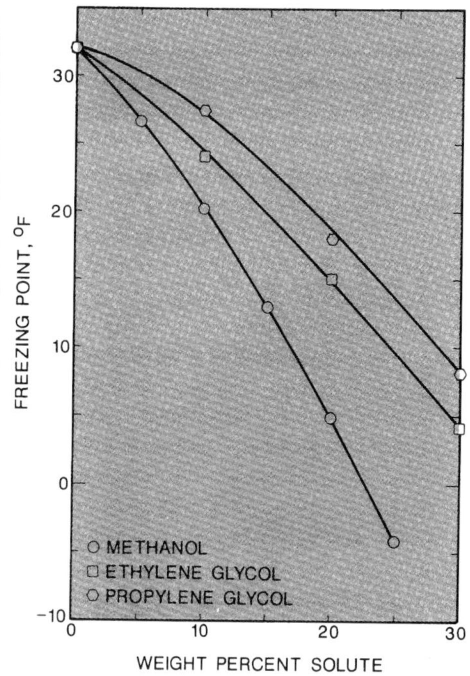

Source: *Chemical Engineers' Handbook*[21]

FIGURE A.2: Freezing Point vs. Weight Percent Solute (Methanol, Ethylene Glycol, and Propylene Glycol)

solute to achieve the correct freezing point
Figure A.1: calcium chloride
sodium chloride
Figure A.2: methanol
ethylene glycol
propylene glycol

The gallons of pure solute (Q_{FA}) needed per 100 gallons of 60°F water is given by:

Equation A.2

$$Q_{FA} = \left[\frac{X^{MA}}{100 - X_{MA}}\right] \times \left[\frac{100}{SG_A}\right]$$

Where: Q_{FA} = gallons of solute per 100 gallons of water

S_{GA} = specific gravity of solute:
Figure A.3: methanol
Figure A.7: ethanol
Figure A.11: ethylene glycol
Figure A.15: propylene glycol
Figure A.19: sodium chloride
Figure A.23: calcium chloride

If the total volume of the fluid in the system is to be 100 gallons, then:

Equation A.3

$$M_A = \frac{X_{MA}}{100 - X_{MA}} \times 8.33.8 \times \frac{Q_w}{100}$$

A

and

Equation A.4

$$Q_{FA} = \frac{X_{MA}}{100 - X_{MA}} \times \frac{100}{SG_A} \times \frac{Q_w}{100}$$

Where: Q_W = 100 gallons – gallons of methanol

Table A.2 gives physical properties for pure water at various temperatures. Additional tables have been calculated for other antifreeze solutions:

Table A.3: methanol
Table A.4: propylene glycol
Table A.5: ethylene glycol
Table A.6: sodium chloride
Table A.7: calcium chloride

Values are given for common freeze protection points. A sample calculation follows.

EXAMPLE – ANTIFREEZE MIXTURE CALCULATIONS

Assume that 100 gallons of a 20°F antifreeze solution of methanol are desired. The calculation procedure to determine the number of gallons of water and methanol required is as follows:

1. From Figure A.2 determine the weight percent solute (X_{MA}) for 20°F methanol as:

$$X_{MA} = 10\%$$

2. Using Equation A.1 calculate the pounds mass of methanol/100 gallons:

$$M_A = \frac{10}{100 - 10} \times 833.8 = 92.64 \frac{\text{lbs mass of methanol}}{\text{100 gallons of water}}$$

TABLE A.2: Pure Water Physical Properties

Temperature (°F)	Density (lb/gal)	Viscosity (centipoise)	Thermal Conductivity (Btu/hr – ft – °F)
32	8.344	1.789	0.327
40	8.344	1.550	0.332
50	8.339	1.310	0.338
60	8.334	1.200	0.344
70	8.338	0.979	0.349
80	8.311	0.860	0.355
90	8.303	0.764	0.360
100	8.287	0.682	0.364
110	8.267	0.616	0.368

NOTE: Specific heat ≈ 1.000 Btu/lbm – °F

Source: *Chemical Engineers' Handbook*[21]

A

3. Using Equation A.2 calculate the gallons of methanol/100 gallons of water:

$$Q_{FA} = \frac{10}{100-10} \times \frac{100}{0.815}$$

$$= 13.63 \frac{\text{gallons of methanol}}{\text{100 gallons of water}}$$

4. If the total volume of the fluid in the system is to be 100 gallons, then from Equation A.4:

$$Q_{FA} = \frac{10}{100-10} \times \frac{100}{0.815} \times \frac{100-13.63}{100} = 11.78 \text{ gallons}$$

The 100-gallon mixture would consist of:

Methanol	= 11.78 gallons
Water = 100 − 11.78	= 88.22
	100.00 gallons

Circulating Fluids and Antifreeze Solutions

Water. Water is the least expensive and most readily available circulating fluid. Water has a relatively low viscosity and high thermal conductivity which can yield low frictional pressure drops and high heat transfer coefficients.

The purity and softness of the water are important. The presence of impurities, as ions or dissolved solids, can play a significant role in scale build-up, heat exchanger fouling, and pump maintenance and life in open systems. This would not be a major concern in a closed-loop system.

Water would normally be the chosen earth coil circulating fluid if it were not for its following two disadvantages:

1. Water has a relatively high freezing point of 32°F.
2. Water expands upon freezing.

These two disadvantages can eliminate water from being the chosen earth coil circulating fluid in northern climates. A fluid with a lower freezing point must be chosen if evaporator refrigerant temperatures approach 32°F.

Table A.2 states pure water density, viscosity, and thermal conductivity at various temperatures. Water's specific heat can be taken to be 1.000 Btu/(lbm–°F), for all practical purposes.

Alcohols

Methyl Alcohol (Methanol). Methyl alcohol, sometimes referred to as methanol, wood alcohol, or carbinol, has been widely used as an antifreeze. Methanol in water offers low cost, low corrosivity, low viscosity, and good thermal conductivity. Methanol water offers relatively low frictional pressure drops and relatively high heat transfer coefficients. Methanol, however, presents the disadvantages of high volatility, high flammability, and high toxicity. Pure methanol has a flash point of 54°F to 60°F, while a 30% methanol-in-water solution has a flash point of 75°F. These flash points are poor.

Figures A.3, A.4, A.5, and A.6 give specific gravities, specific heats, viscosities, and thermal conductivities of methanol-in-water solutions. Table A.3 gives pertinent data for methanol waters at 25°F, 30°F, 35°F, and 40°F.

Ethyl Alcohol in Water. Ethyl alcohol in water, sometimes referred to as ethanol water, has found some applications in refrigeration. Ethanol waters are less toxic and less flammable than methanol waters. But ethanol waters are still highly flammable and require careful handling and storage. Ethyl alcohol solutions are more viscous than methyl alcohol solutions resulting in larger frictional pressure drops and lower heat transfer coefficients when compared to methyl alcohol mixtures. Corrosion inhibitors can be added to make ethanol waters noncorrosive. Ethanol waters are more expensive than methanol waters but less expensive than the ethylene glycol solutions. Ethanol-in-

A

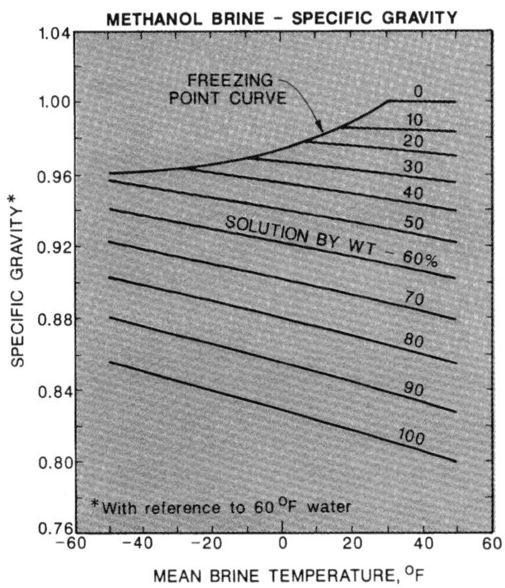

FIGURE A.3: Methanol Brine — Specific Gravity

Source: Chemical Engineers' Handbook[21]

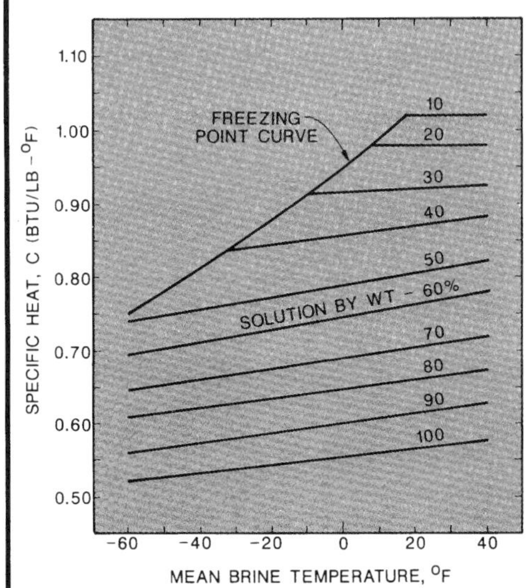

FIGURE A.4: Methanol Brine — Specific Heat

Source: Chemical Engineers' Handbook[21]

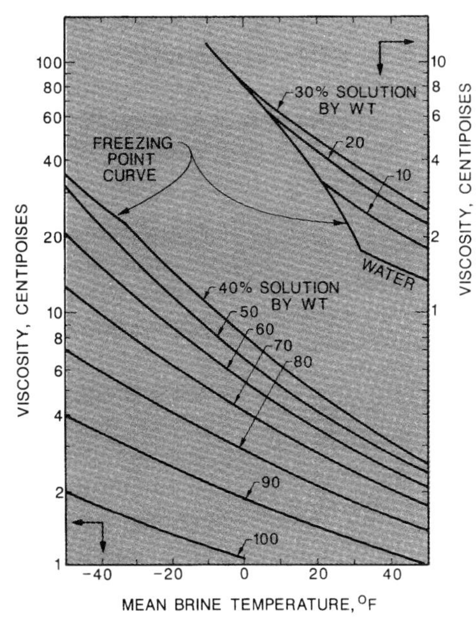

FIGURE A.5: Methanol Brine — Viscosity

Source: Chemical Engineers' Handbook[21]

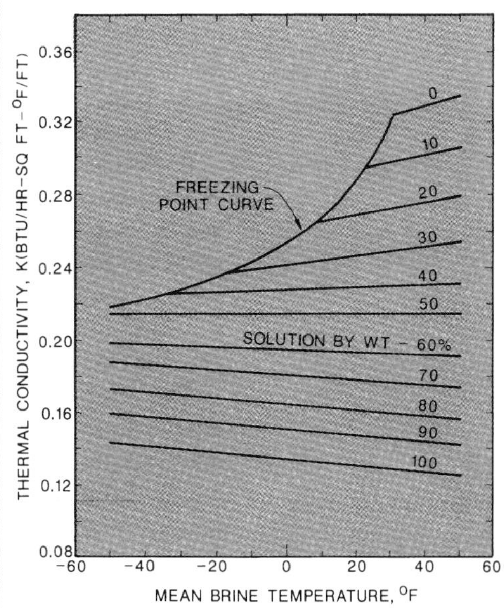

FIGURE A.6: Methanol Brine — Thermal Conductivity

Source: Chemical Engineers' Handbook[21]

TABLE A.3: Methanol Solution Physical Properties

Mean Temperature, °F	25	30	35	40
Freezing Point, °F	15	20	25	30
Weight % Methanol	13.6	10.0	6.3	2.0
Methanol Mass(1), lbm	131.6	92.7	56.1	17.0
Methanol Volume(2), gal	19.41	13.63	8.29	2.53
Specific Gravity, $SG^T/_{59}$	0.813	0.815	0.811	0.807
Density, lb/gal	8.180	7.546	8.254	8.257
Specific Heat Btu/lbm – °F	1.01	1.02	1.025	1.02
Viscosity, Centipoise	3.30	2.70	2.15	1.60
Thermal Conductivity Btu/hr – ft – °F	0.286	0.296	0.310	0.324

NOTE: (1) Pounds mass of pure methanol needed per 100 gallons of pure water.
(2) Gallons of pure methanol needed per 100 gallons of pure water.

Source: *Chemical Engineers' Handbook*[21]

water solutions are normally denatured to give them some toxicity.

Figures A.7, A.8, A.9, and A.10 give specific gravities, specific heats, viscosities, and thermal conductivities of ethanol water mixtures.

GLYCOLS

Ethylene Glycol in Water. Ethylene glycol in water is a common and often favorable alternative to salt brines. Ethylene glycol is colorless, practically odorless, and completely miscible with water. Properly prepared and properly inhibited aqueous ethylene glycol solutions are relatively noncorrosive. Aqueous ethylene glycol solutions have relatively low volatilities and present a low flammability hazard.

Ethylene glycol solutions have two disadvantages:

1. They are somewhat toxic and should be stored and handled with care.
2. The viscosity may become too high at low temperatures.

Figures A.11, A.12, A.13, and A.14 give specific gravities, specific heats, viscosities, and thermal conductivities of ethylene glycol-in-water solutions. Ethylene glycol solutions have been used at temperatures as low as – 30°F but find best application at temperatures above 15°F (primarily due to high viscosity limitations).

Table A.5 gives pertinent data for ethylene glycol-in-water solutions at 25°F, 30°F, 35°F, and 40°F.

The quality of the water used to prepare ethylene glycol solutions is important. The water should be "soft" and have a low concentration of chloride and sulfate ions. Ethylene glycol inhibitor is commonly added by the manufacturer. Ethylene

A

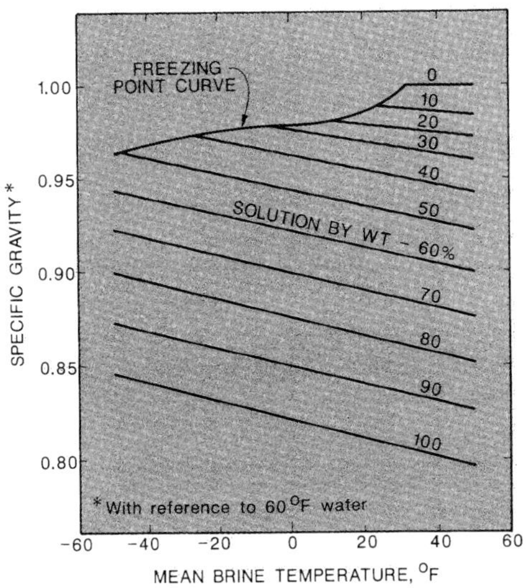

FIGURE A.7: Ethanol Brine — Specific Gravity

Source: *Chemical Engineers' Handbook*[21]

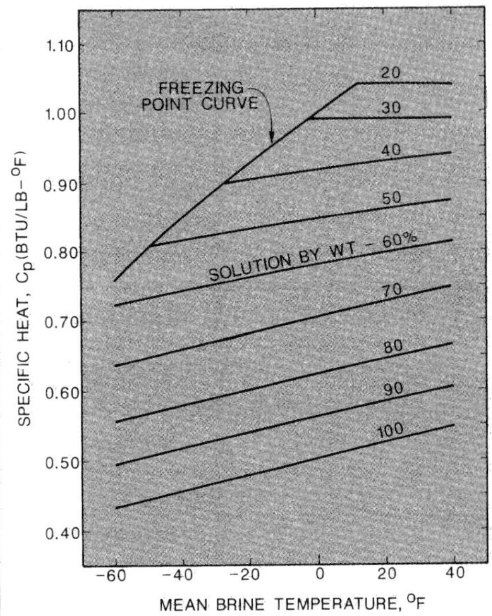

FIGURE A.8: Ethanol Brine — Specific Heat

Source: *Chemical Engineers' Handbook*[21]

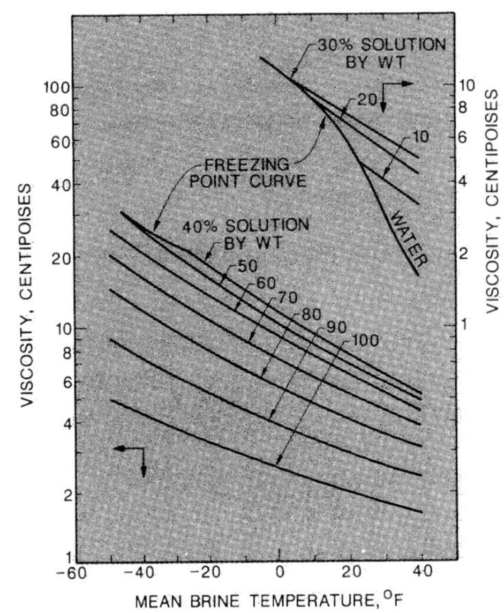

FIGURE A.9: Ethanol Brine — Viscosity

Source: Chemical Engineers' Handbook[21]

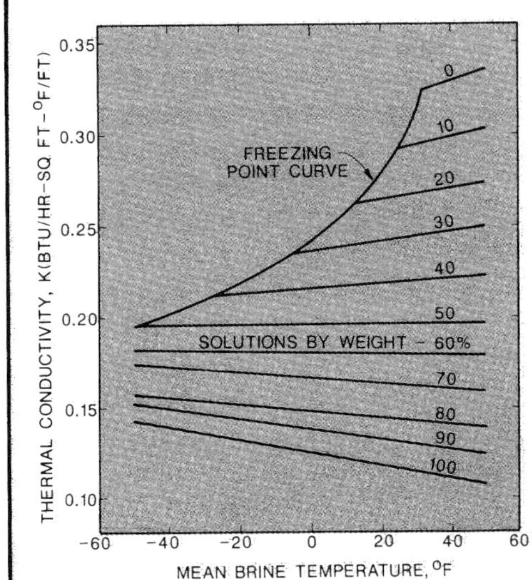

FIGURE A.10: Ethanol Brine — Thermal Conductivity

Source: Chemical Engineers' Handbook[21]

A

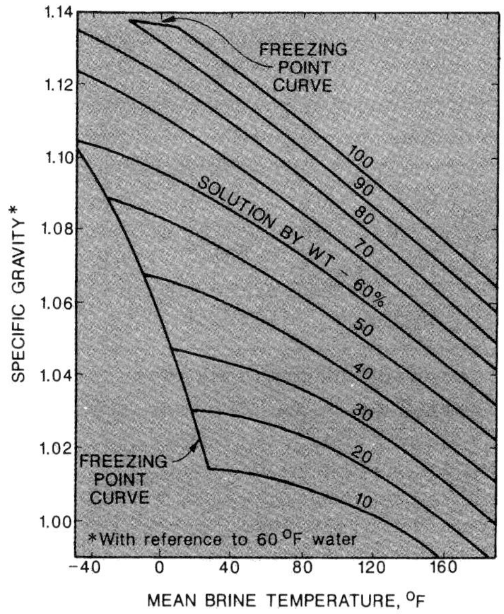

FIGURE A.11: Ethylene Glycol — Specific Gravity

Source: Chemical Engineers' Handbook[21]

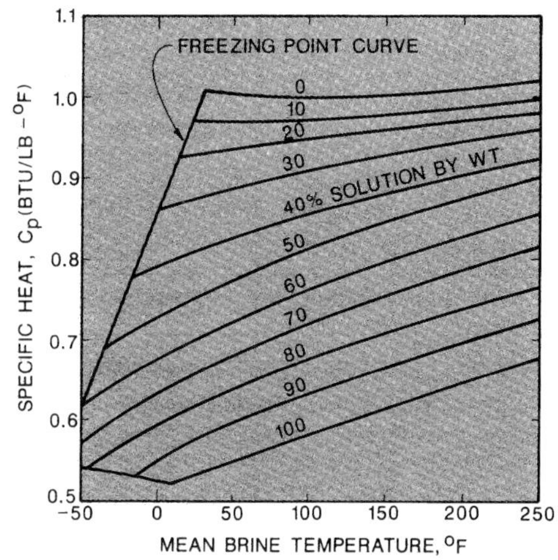

FIGURE A.12: Ethylene Glycol — Specific Heat

Source: Chemical Engineers' Handbook[21]

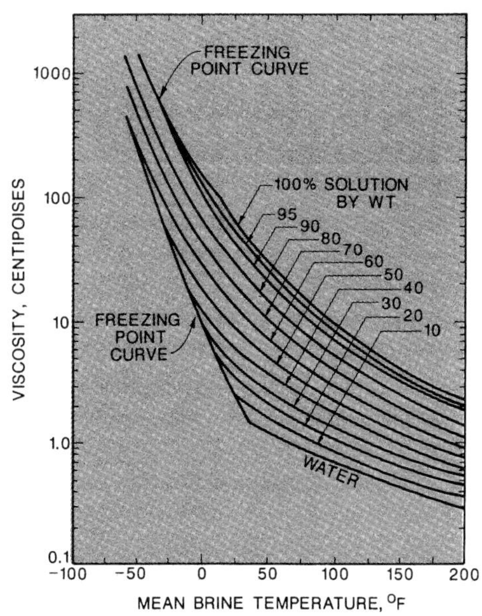

FIGURE A.13: Ethylene Glycol — Viscosity

Source: Chemical Engineers' Handbook[21]

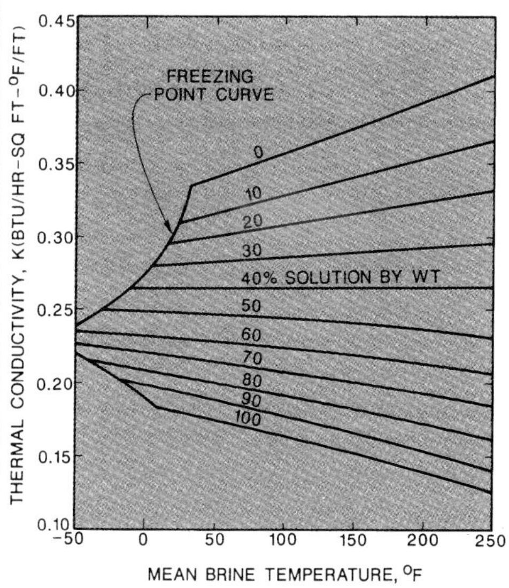

FIGURE A.14: Ethylene Glycol — Thermal Conductivity

Source: Chemical Engineers' Handbook[21]

A

TABLE A.4: Propylene Glycol Solution Physical Properties

Mean Temperature, °F	25	30	35	40
Freezing Point, °F	15	20	25	30
Weight % Glycol	23.5	18.3	12.9	5.9
Glycol Mass(1), lbm	256	187	124	52
Glycol Volume(2), gal	29.9	22.0	14.7	6.2
Specific Gravity, $SG^{T}/_{59}$	1.025	1.018	1.013	1.006
Density, lb/gal	8.55	8.49	8.45	8.39
Specific Heat Btu/lbm – °F	0.96	0.97	0.98	0.99
Viscosity, Centipoise	6.3	3.0	1.9	1.5
Thermal Conductivity Btu/hr – ft – °F	0.225	0.236	0.25	0.275

NOTE: (1) Pounds mass of pure propylene glycol needed per 100 gallons of pure water.
(2) Gallons of pure propylene glycol needed per 100 gallons of pure water.

Source: *Chemical Engineers' Handbook*[21]

glycol solutions normally have a pH of 8.8 to 9.2. Ethylene glycol solutions having a pH that falls below 7.5 should be replaced. Indiscriminate mixing of differently inhibited solutions should be avoided unless the solutions are known to be compatible.

Chromite treatment should never be used because it will result in rapid buildup of sludge. Ethylene glycol solutions, even though inhibited, do not have an indefinite service life and should be checked yearly. Spent glycol solution should be removed from the system and replaced with fresh glycol solution.

Propylene Glycol in Water. Propylene glycol solution may be considered as an alternative to ethylene glycol. Lower toxicity is the only true advantage of propylene glycol over ethylene glycol. Propylene glycol, like ethylene glycol, can offer low corrosivity and low volatility, and presents a low flammability hazard. However, propylene glycol is more costly and yields more viscous solutions than ethylene glycol. Propylene glycol solutions may be considered for use down to about 15°F. Use of propylene glycol solutions below 15°F presents excess viscosity which, in turn, yields excessively high frictional pressure drops and excessively low heat transfer coefficients.

The 15°F temperature is definitely the maximum lower limit for closed-loop systems. A more reasonable lower limit threshold would be a maximum 25% mixture by volume and operating above 25°F. Lower operating temperatures and/or higher concentrations of propylene glycol are not economical when the energy required to pump the fluid and maintain turbulent flow under those conditions is considered.

In order to obtain good heat transfer

within the buried pipe system, the calculated Reynolds number should not fall below 3000 for the coldest exit fluid temperature from the heat pump.

Figures A.15, A.16, A.17, and A.18 give specific gravities, specific heats, viscosities, and thermal conductivities of propylene glycol-in-water solutions. Table A.4 gives pertinent propylene glycol properties at 25°F, 30°F, 35°F, and 40°F.

The comments made earlier for inhibiting ethylene glycol solutions also apply to propylene glycol solutions.

SALTS

Sodium Chloride in Water. Sodium chloride in water has the lowest brine cost per gallon. Sodium chloride brines have relatively low viscosity and relatively high thermal conductivity. Sodium chloride brines have low toxicity, low volatility, and are not flammable.

There are two serious drawbacks for sodium chloride brines:

1. Freezing point depression achieved with sodium chloride brines is moderate and, thus, sodium chloride brines are limited to 15°F and above.
2. Sodium chloride brines are very corrosive and require inhibitors that must be regularly checked and replenished to maintain pH greater than 7.0.

Figures A.19, A.20, A.21, and A.22 give specific gravities, specific heats, viscosities, and thermal conductivities of sodium chloride-in-water solutions. Table A.6 gives selected sodium chloride brine physical properties at 25°F, 30°F, and 35°F.

Calcium Chloride in Water. Calcium chloride in water offers a greater freezing point depression than sodium chloride and has wide use as a circulating brine. Calcium chloride brines are the second lowest cost brines next to sodium chloride brines. Calcium chloride brines have been used to as low as –30°F; however, thermal and

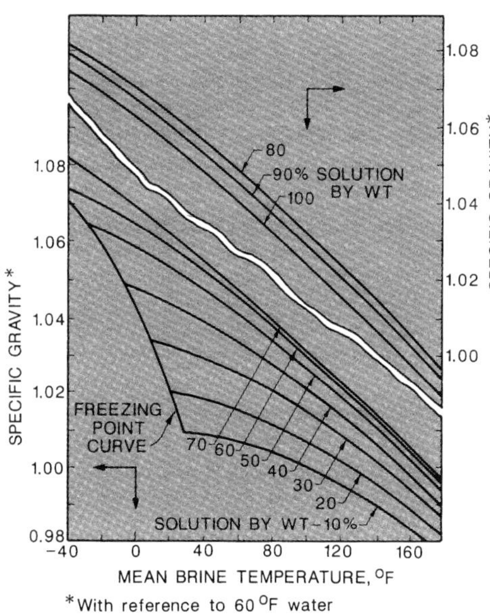

FIGURE A.15: Propylene Glycol — Specific Gravity

Source: Chemical Engineers' Handbook[21]

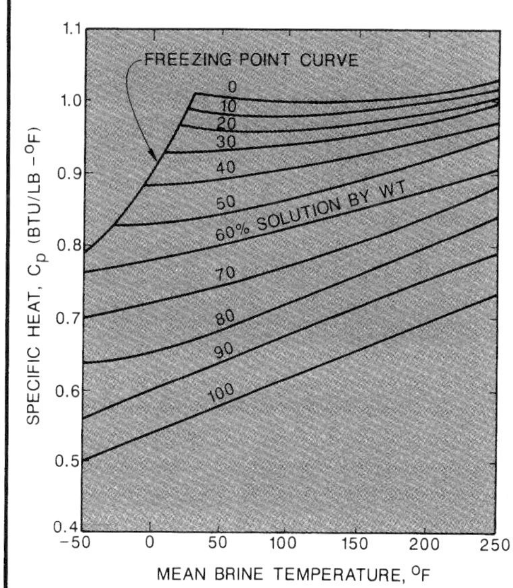

FIGURE A.16: Propylene Glycol — Specific Heat

Source: Chemical Engineers' Handbook[21]

A

TABLE A.5: Ethylene Glycol Solution Physical Properties				
Mean Temperature, °F	25	30	35	40
Freezing Point, °F	15	20	25	30
Weight % Glycol	20.0	14.6	8.8	2.5
Glycol Mass[(1)], lbm	209	143	81	23
Glycol Volume[(2)], gal	24.2	16.7	9.5	2.6
Specific Gravity, $SG^T/_{59}$	1.035	1.025	1.018	1.004
Density, lb/gal	8.63	8.55	8.49	8.37
Specific Heat Btu/lbm – °F	0.92	0.95	0.98	1.00
Viscosity, Centipoise	2.7	1.9	1.5	1.35
Thermal Conductivity Btu/hr – ft – °F	0.24	0.26	0.27	0.28

NOTE: (1) Pounds mass of pure ethylene glycol needed per 100 gallons of pure water.
(2) Gallons of pure ethylene glycol needed per 100 gallons of pure water.

Source: *Chemical Engineers' Handbook*[21]

TABLE A.6: Sodium Chloride Solution Physical Properties			
Mean Temperature, °F	25	30	35
Freezing Point, °F	15	20	25
Weight % NaCl	13.3	10.1	6.4
Mass[(1)], lbm	127.9	93.7	57.0
Specific Gravity, $SG^T/_{59}$	1.104	1.078	1.049
Density, lb/gal	9.21	8.99	8.75
Specific Heat, Btu/lbm – °F	0.851	0.880	0.918
Viscosity, Centipoise	2.6	1.8	1.7
Thermal Conductivity Btu/hr – ft – °F	0.276	0.280	0.300

NOTE: (1) Pounds mass pure sodium chloride needed per 100 gallons of pure water.

Source: *Chemical Engineers' Handbook*[21]

A

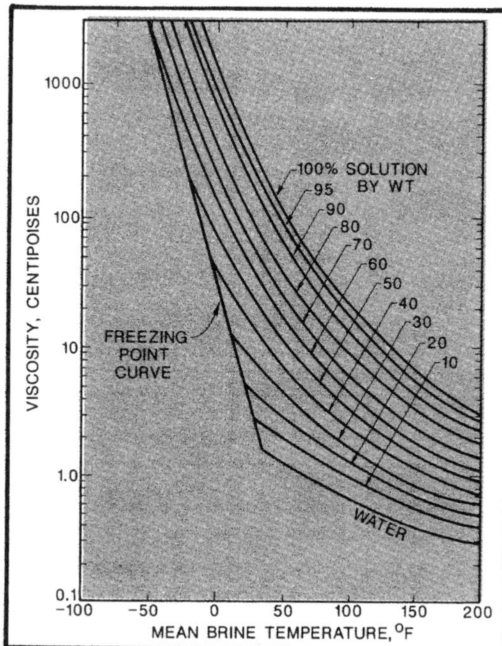

FIGURE A.17: Propylene Glycol — Viscosity

Source: Chemical Engineers' Handbook[21]

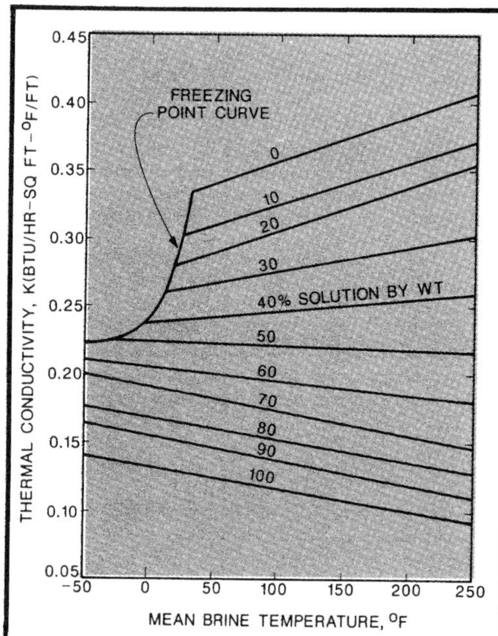

FIGURE A.18: Propylene Glycol — Thermal Conductivity

Source: Chemical Engineers' Handbook[21]

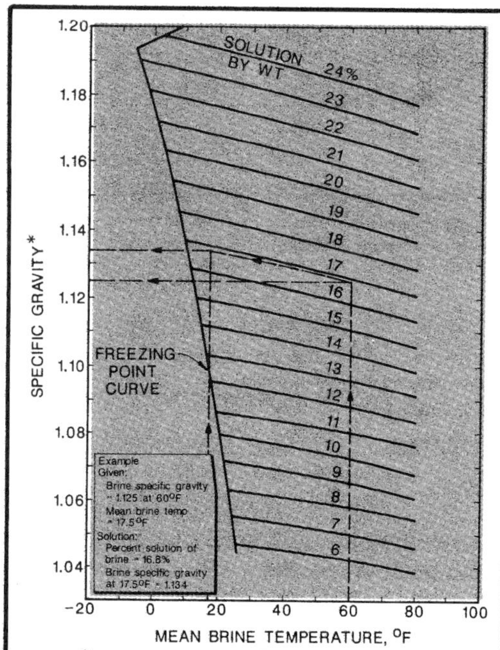

*With reference to 60 °F water

FIGURE A.19: Sodium Chloride — Specific Gravity

Source: Chemical Engineers' Handbook[21]

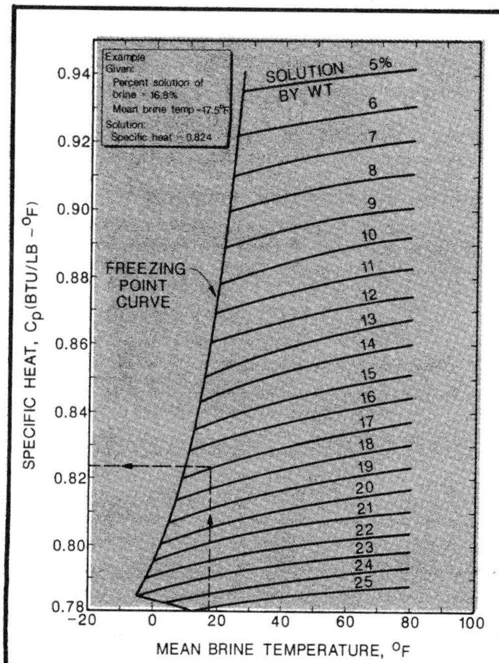

FIGURE A.20: Sodium Chloride — Specific Heat

Source: Chemical Engineers' Handbook[21]

A

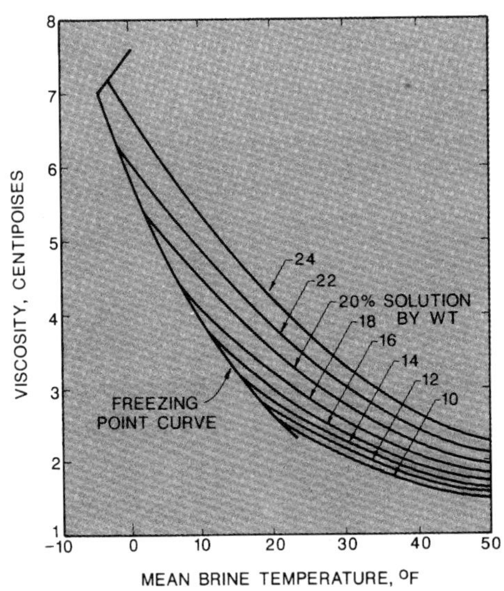

FIGURE A.21: Sodium Chloride — Viscosity

Source: Chemical Engineers' Handbook[21]

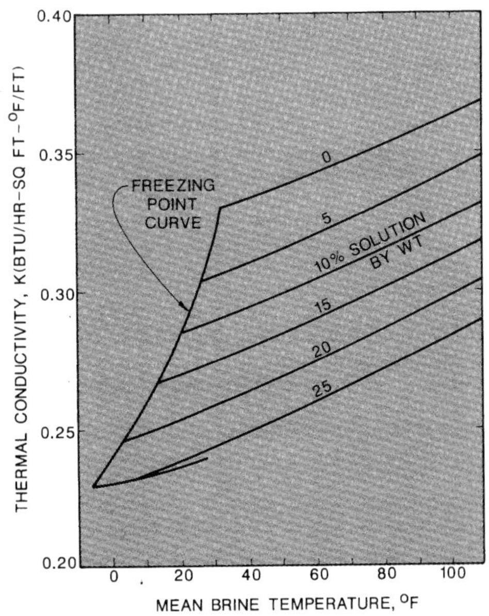

FIGURE A.22: Sodium Chloride — Thermal Conductivity

Source: Chemical Engineers' Handbook[21]

flow characteristics deteriorate rapidly below –5°F.

Figures A.23, A.24, A.25, and A.26 give specific gravities, specific heats, viscosities, and thermal conductivities of calcium chloride brines. Table A.7 gives calcium chloride solution pertinent physical properties for mean temperatures of 25°F, 30°F, 35°F, and 40°F.

Calcium chloride is marketed in flake, solid, and solution forms. The flake form is used most extensively and is available as Type 1, which is 77 minimum weight percent calcium chloride, and Type 2, which is 94 minimum weight percent calcium chloride. Flaked lime should be checked to ensure that the remaining 23% is not lime or calcium hydroxide since this could cause sludge.

Both of the salt brines previously discussed can cause considerable corrosion problems. Regular monitoring and maintenance may be necessary. Seclusion of the brines from oxygen and carbon dioxide helps prevent corrosion; however, closed systems such as the earth coil still tend to have some corrosion. Strangely enough, dilute solutions tend to be more corrosive than concentrated solutions. Corrosion is best controlled by minimizing contact with air and maintaining pH between 7.5 and 8.0. Acidic, neutral, or strongly alkaline brines tend to be most corrosive.

Sodium dichromate inhibitor is generally considered to be the most economical and most effective means of combating corrosion with salt brines. Sodium hydroxide, which is used for pH control in many inhibitors, cannot be used because calcium hydroxide would be formed which is insoluble and would precipitate out. Sodium dichromate comes as orange crystals that are readily dissolved in warm water. **Sodium dichromate dissolves very slowly in cold water and should never be put into a brine tank in crystal form.** Dissolve the dichromate crystals in warm water and add the resulting solution to the brine. The dichromate quantities rec-

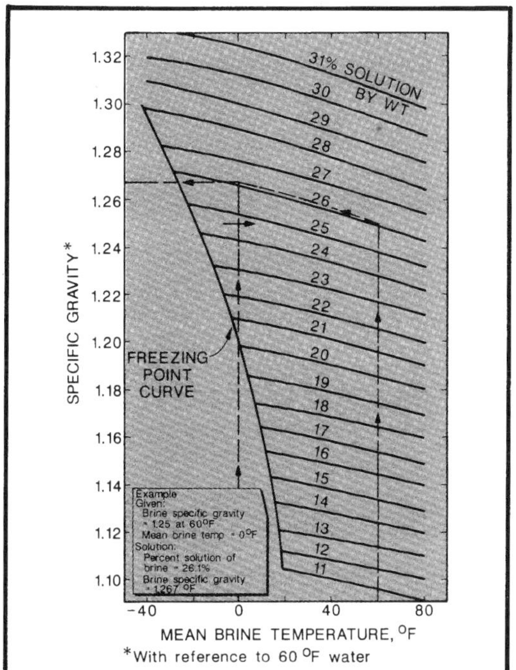

FIGURE A.23: Calcium Chloride — Specific Gravity

Source: Chemical Engineers' Handbook[21]

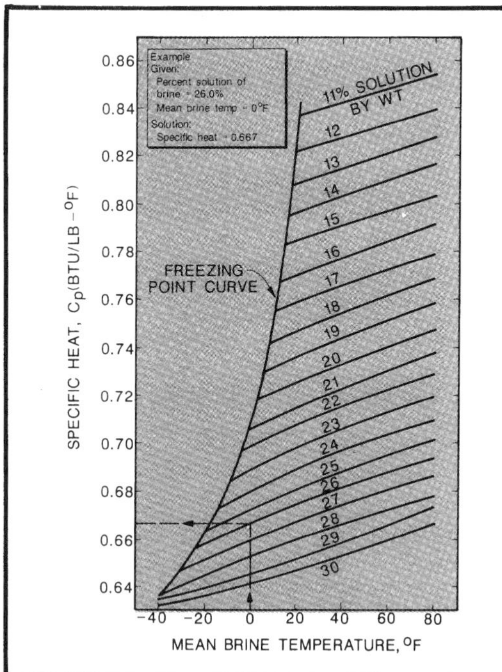

FIGURE A.24: Calcium Chloride — Specific Heat

Source: Chemical Engineers' Handbook[21]

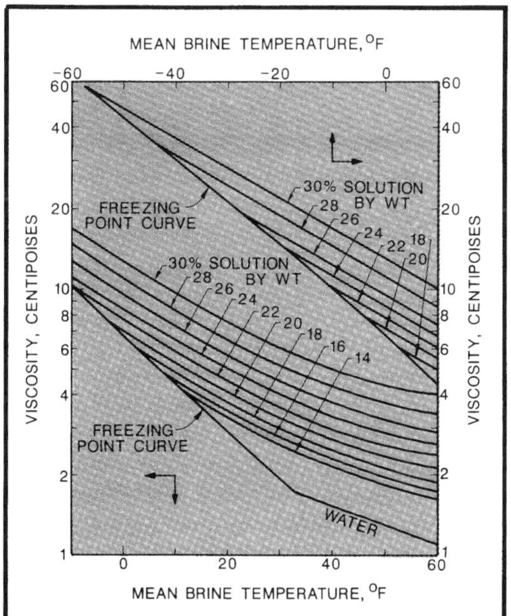

FIGURE A.25: Calcium Chloride — Viscosity

Source: Chemical Engineers' Handbook[21]

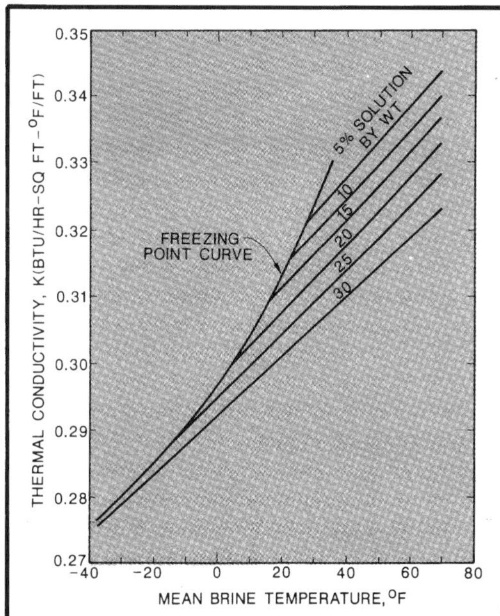

FIGURE A.26: Calcium Chloride — Thermal Conductivity

Source: Chemical Engineers' Handbook[21]

A

TABLE A.7: Calcium Chloride Solution Physical Properties

Mean Temperature, °F	25	30	35	40
Freezing Point, °F	15	20	25	30
Weight % $CaCl_2$	14.3	11.4	7.5	2.6
$CaCl_2$ Mass,$^{(1)}$, lbm	139.2	107.3	67.6	22.3
Specific Gravity, $SG^T/_{59}$	1.135	1.104		
Density, lb/gal	9.47	9.21		
Specific Heat, Btu/lbm – °F	0.795	0.835		
Viscosity, Centipoise	3.0	2.4	1.9	
Thermal Conductivity Btu/hr – ft – °F	0.315	0.318	0.323	

NOTE: (1) Pounds mass of pure calcium chloride needed per 100 gallons of pure water.

Source: *Chemical Engineers' Handbook*[21]

ommended by ASHRAE for long-term protection are as follows:

1. 1.25 pounds of sodium dichromate per 10 cubic feet (74.8 gallons) of calcium chloride brine
2. two pounds of sodium dichromate per 10 cubic feet (74.8 gallons) of sodium chloride brine

Addition of the sodium dichromate to the brine makes the resulting solution acidic whereas a slightly basic solution is desired. Before the salt brine is treated with sodium dichromate, determine the pH of the salt brine. If untreated salt brine is acidic or neutral, add sufficient caustic soda to bring the pH up to 8.0. If the untreated salt brine has a proper pH (7.5 to 8.0), add .27 pounds of 76% commercial flake caustic soda for each pound of sodium dichromate added. The caustic soda must be thoroughly dissolved in warm water before adding it to the brine.

Materials for Salt Solutions

If salt brine solutions are selected, the following points have been observed in systems that have successfully operated in this environment for long periods of time with little or no difficulty:

1. During system startup, all air must be purged or removed. Fluid velocities greater than 2 feet per second must be achieved during the charging, purging, and startup operation in order to accomplish this objective.
2. Air traps at the end of headers and manifolds must be eliminated. Many of these air traps used in residential systems are made from high zinc content materials. They eventually leak, causing a spill, and corrosion develops. Circulation systems that are properly designed do not need air vents.
3. Metal components with a high zinc content must be eliminated. These

include circulating pumps, fittings, and valves. Red brass (15% zinc or less) is satisfactory. **Yellow brass must not be used because dezincfication occurs.** This can be partially accomplished in the design process by reducing the number of metal components in the system.

4. Alloys that are high in zinc content (for about 15%) should be avoided and substituted by 300 and 400 series stainless steel, copper-nickel, or other more noble (cathodic) materials.
5. Copper-nickel heat exchangers are more likely to survive the salt brine environment than standard copper heat exchangers. However, there is no strong evidence that copper is attacked.
6. Dissimilar metal connections should be avoided, such as cast iron to copper. An electrical insulator such as plastic should be used. Care should be taken in selecting pump flanges since there are some on the market that have a high zinc content. If a cast iron pump is used, it is best to use cast iron pump flange brackets.
7. If the system is drained, flush thoroughly at once. Corrosion pits will develop in just a few hours.

There have been many failures with salt circulation systems. Extreme care must be taken in the design and selection of metal components. However, there are a large number of such systems operating satisfactorily using both cast iron and stainless steel pumps.

Copper Materials. The following discussion on copper and other alloys is for the designer and installer of systems that will operate in a corrosive environment.

Copper Alloys. There are well over 900 commercial copper alloys. This fact, coupled with the loose handling of words like "brass" and "bronze," presents a real problem to the end user of these products. For purposes of discussion, the copper alloys will be divided into the following groups:

1. copper-nickel
2. bronzes (copper-tin, copper-aluminum, etc.)
3. brasses (copper-zinc)

Copper-Nickel. Copper-nickel or cupronickel, as it is often called, is the most corrosion-resistant of all the copper alloys. The cupro-nickel alloys contain from 10 to 30% nickel. The three used alloys are:

C70600 (90 Cu, 10 Ni)
C71000 (80 Cu, 20 Ni)
C71500 (80 Cu, 30 Ni)

Cupro-nickel heat exchangers are commonly used in the water-source heat pump industry. The additional cost of the copper-nickel liquid refrigerant heat exchanger can easily be justified in a salt brine system.

Bronzes. As with the stainless steels, all bronzes are not the same and do not respond the same way in a corrosive environment. A true bronze is a copper-tin alloy. Most often it is referred to as phosphor bronze, from the use of phosphorus to deoxidize the material. A common wrought phosphorus bronze is C51000 (94.8 Cu, 5 Sn, 0.2 P). A cast bronze used for pump bodies is C90500 (gun metal) (88 Cu, 10 Sn, 2 Zn). Tin in the range of 1 to 10% is added to the copper for increased strength and corrosion resistance.

Other bronzes such as aluminum bronze and silicone bronze are available. Aluminum bronze would perform better than the silicone bronze in the operating temperature range of the ground-coupled heat pump.

If bronzes are to be used, the system must be closed, pressurized, and free of air and other contaminants. The degree of alkalinity or acidity (pH) must be monitored and controlled through the life of the system. A brine pH of 7.5 for sodium or calcium chloride is considered ideal. As was mentioned earlier, a pH of 7.5 to 8 should

A

be maintained. A brine pH of 7.5 for sodium or calcium chloride is considered ideal.

Brasses. Brass is the result of adding zinc to copper in amounts ranging in content from 5 to 45%. As the zinc content increases, corrosion resistance decreases. Brasses with a zinc content greater than 15% should be avoided in salt brine systems. This alloy is referred to as admiralty brass. Admiralty brasses are modified yellow brasses (71 Cu, 28 Zn, 1 Sn).

The main problem with high zinc content alloys is dezincification. The accepted theory is that the copper and zinc both dissolve and then the copper plates back. In any event, the dissolved elements form a porous, highly abrasive film or layer on the shaft and bearings of the circulating pump. This re-deposited material eventually destroys the pump bearing and the pump is rendered inoperable.

Stainless Steels. Passive stainless steels, such as 300 and 400 series materials, are slightly more corrosion resistant in salt brine solutions than copper-nickel alloys. This material would be very suitable for fittings and valves.

A word of caution about stainless steel: All stainless steel materials do not have the same resistance to corrosion. Some alloys have a tendency to pitting corrosion.

CONCLUSIONS ON METAL ALLOYS

Economics and availability usually act as the driving forces in the selection and application of materials used in ground-coupled heat pump systems. With all this in mind there are some first, second, and third choices for metal components in salt brine environments. The following list is not intended to be all inclusive, but rather as a guide to decision-making when all factors are nearly equal.

The materials are listed in order of most desirable to least desirable from a corrosion-resistance standpoint:

1. 300 SERIES STAINLESS STEEL (pitting can be a problem in chloride solutions)
2. 400 SERIES STAINLESS STEEL (pitting can be a problem in chloride solutions)
3. COPPER NICKEL ALLOYS
4. BRONZES
5. COPPER
6. BRASSES
7. CAST IRON
8. STEEL
9. ALUMINUM

Many of the above materials will operate successfully with salt solutions. However, in some cases, using them together in the same system with oxygenated solutions will doom them to failure.

APPENDIX B

B Trench and Pipe Spacing

In This Appendix: Horizontal trench and pipe spacing; vertical ground heat exchanger spacing; long-term thermal effects of multiple bore holes

Horizontal Trench and Pipe Spacing

Spacing pipes within trenches and trench spacing should be far enough apart that thermal interference between pipes is minimal. Table B.1 gives soil resistance values as a function of pipe radial distance. Data are presented for a number of different types of soils and materials found in the field. The table is useful in determining soil/field thermal interference when two or more pipes are close to each other. For example, assume two pipes with radii of 0.75 inches are buried six feet apart. The soil/resistance (R_S) for each pipe (R_P) for a heavy damp soil (K_S = 0.75 Btu/hr-ft-°F) and thermal diffusivity (A = 0.025 ft^2/hr) at 1,000 hours is 1.015. For the pipe located at six feet or 72 inches, the soil/field resistance is 0.082. The combined effect for the two pipes is then:

RS = (1.015 + 1.015 + 0.082 + 0.082)/2 = 1.097

A pipe multiplier can then be defined as:

PM = Single-pipe value/Two-pipe value
= 1.097/1.015 = 1.08

Eight percent more pipe is required because of thermal interference between pipes. Recall that no mention was made of buried depth in the above example. This example would apply to vertical heat exchangers since each pipe is surrounded by soil without any surface. To account for surface effects as found in horizontal systems, a different model must be used. The model uses the concept of an image pipe to account for the heat flow to and from the earth's surface.

Ignoring the surface effects found in horizontal systems will result in higher calculated thermal resistance values between pipes than those that occur during operation. For single horizontal pipes spaced four feet apart, the error is about 5%. For two-pipe systems (pipes at four and six feet), the error for trenches spaced five feet is about 10%. To account for the surface effects, an image pipe (Figure B.1) is used.

The procedure (using Table B.1) to consider the image pipe and thus include surface effects with multiple horizontal pipes or trenches is as follows:

1. Select the value of soil resistance (R_S) from Table B.1 for a radial distance equal to the pipe radius. In Table B.2 this will be "Pipe 1 to 1" or "Pipe 2 to 2."
2. Calculate the distance from each buried pipe to all other pipes including the image pipes. The distance can be calculated (Figure B.1) as:

   $$L = [B^2 + (2D)^2]^{1/2}$$

   From Table B.1 read the corresponding soil resistance value for the given radial distance.

B

TABLE B.1: Soil Resistance Values at Selected Radial Distances

Units: K_s(Btu/hr – ft – °F), $\propto$(ft²/hr), ρ(lb/ft³), C_p(Btu/lb – °F)

t(hrs) \ r(in)	0.25	0.5	0.75	1.0	1.25	1.5	12	24	36	48	60	72	84	96	108	120	132	144
DENSE ROCK			$k_s = 2.0$				$\propto = 0.05$					$\rho = 200$				$c_p = 0.20$		
500	.468	.413	.381	.358	.340	.326	.161	.107	.076	.056	.042	.031	.022	.017	.012	.008	.006	.004
1000	.496	.441	.408	.386	.368	.353	.188	.133	.102	.081	.065	.052	.042	.034	.028	.025	.018	.014
1500	.512	.425	.427	.401	.384	.369	.204	.149	.118	.096	.079	.066	.055	.047	.039	.038	.028	.023
2000	.524	.468	.436	.413	.395	.381	.216	.161	.129	.107	.090	.076	.063	.056	.048	.050	.036	.031
AVERAGE ROCK			$k_s = 1.4$				$\propto = 0.04$					$\rho = 175$				$c_p = 0.20$		
500	.656	.578	.531	.499	.473	.453	.217	.140	.098	.070	.050	.036	.025	.018	.012	.008	.005	.004
1000	.696	.617	.570	.538	.513	.492	.256	.178	.134	.104	.081	.064	.051	.040	.031	.024	.019	.015
1500	.719	.640	.593	.561	.536	.515	.279	.201	.156	.125	.102	.083	.069	.057	.047	.038	.032	.026
2000	.735	.656	.610	.578	.522	.531	.295	.217	.172	.140	.116	.098	.082	.070	.059	.050	.042	.036
DENSE CONCRETE			$k_s = 1.0$				$\propto = 0.033$					$\rho = 150$				$c_p = 0.20$		
500	.904	.793	.729	.683	.647	.618	.289	.182	.123	.085	.059	.040	.027	.018	.012	.008	.005	.003
1000	.959	.848	.784	.738	.703	.674	.343	.235	.173	.131	.101	.078	.060	.046	.036	.027	.020	.015
1500	.991	.881	.816	.770	.734	.706	.375	.266	.204	.161	.129	.104	.084	.068	.054	.040	.035	.028
2000	1.01	.904	.839	.793	.758	.729	.398	.288	.226	.182	.149	.123	.102	.085	.071	.059	.049	.040
SOLID MASONRY			$k_s = 0.75$				$\propto = 0.025$					$\rho = 143$				$c_p = 0.21$		
500	1.180	1.028	.942	.881	.833	.795	.356	.215	.139	.091	.059	.038	.024	.015	.009	.005	.003	.002
1000	1.248	1.102	1.015	.954	.907	.869	.428	.284	.204	.150	.111	.082	.061	.045	.032	.023	.017	.011
1500	1.291	1.145	1.059	.998	.950	.912	.471	.326	.244	.187	.146	.114	.090	.070	.055	.042	.033	.024
2000	1.322	1.175	1.089	1.028	.981	.942	.501	.356	.273	.215	.172	.139	.113	.091	.074	.059	.048	.038
HEAVY SOIL DAMP			$k_s = 0.75$				$\propto = 0.025$					$\rho = 131$				$c_p = 0.23$		
t(hrs) \ r(in)	0.25	0.5	0.75	1.0	1.25	1.5	12	24	36	48	60	72	84	96	108	120	132	144
500	1.180	1.028	.942	.881	.833	.795	.356	.215	.139	.091	.059	.038	.024	.015	.009	.005	.003	.002
1000	1.248	1.102	1.015	.954	.907	.869	.428	.284	.204	.150	.111	.082	.061	.045	.032	.023	.017	.011
1500	1.291	1.145	1.059	.998	.950	.912	.471	.326	.244	.187	.146	.114	.090	.070	.055	.042	.033	.024
2000	1.322	1.175	1.089	1.028	.981	.942	.501	.356	.273	.215	.172	.139	.113	.091	.074	.059	.048	.038
HEAVY SOIL DRY			$k_s = 0.50$				$\propto = 0.020$					$\rho = 125$				$c_p = 0.20$		
500	1.727	1.507	1.378	1.286	1.215	1.157	.499	.290	.179	.112	.069	.041	.024	.014	.007	.004	.002	.001
1000	1.837	1.617	1.488	1.396	1.326	1.268	.608	.392	.273	.195	.139	.100	.071	.049	.034	.023	.016	.010
1500	1.902	1.682	1.522	1.461	1.390	1.332	.671	.454	.332	.249	.189	.144	.110	.803	.062	.047	.034	.025
2000	1.948	1.728	1.598	1.507	1.435	1.378	.717	.499	.375	.290	.228	.180	.142	.112	.088	.069	.054	.041
LIGHT SOIL DAMP			$k_s = 0.50$				$\propto = 0.020$					$\rho = 100$				$c_p = 0.25$		
500	1.727	1.507	1.378	1.286	1.215	1.157	.499	.290	.179	.112	.069	.041	.024	.014	.007	.004	.002	.001
1000	1.837	1.617	1.488	1.396	1.326	1.268	.608	.392	.273	.195	.139	.100	.071	.049	.034	.023	.016	.010
1500	1.902	1.682	1.522	1.461	1.390	1.332	.671	.454	.332	.249	.189	.144	.110	.803	.062	.047	.034	.025
2000	1.948	1.728	1.598	1.507	1.435	1.378	.717	.499	.375	.290	.228	.180	.142	.112	.088	.069	.054	.041
LIGHT SOIL DRY			$k_s = 0.20$				$\propto = 0.011$					$\rho = 90$				$c_p = 0.20$		
500	4.081	3.529	3.206	2.978	2.800	2.655	1.018	.518	.273	.141	.070	.033	.014	.005	.002	0	0	0
1000	4.356	3.805	3.483	3.253	3.076	2.931	1.285	.760	.479	.305	.193	.120	.072	.043	.025	.013	.007	.004
1500	4.518	3.966	3.644	3.415	3.237	3.092	1.443	.910	.616	.425	.294	.202	.137	.092	.061	.039	.025	.015
2000	4.633	4.081	3.758	3.529	3.352	3.207	1.556	1.018	.717	.518	.377	.274	.198	.141	.100	.070	.048	.033

B

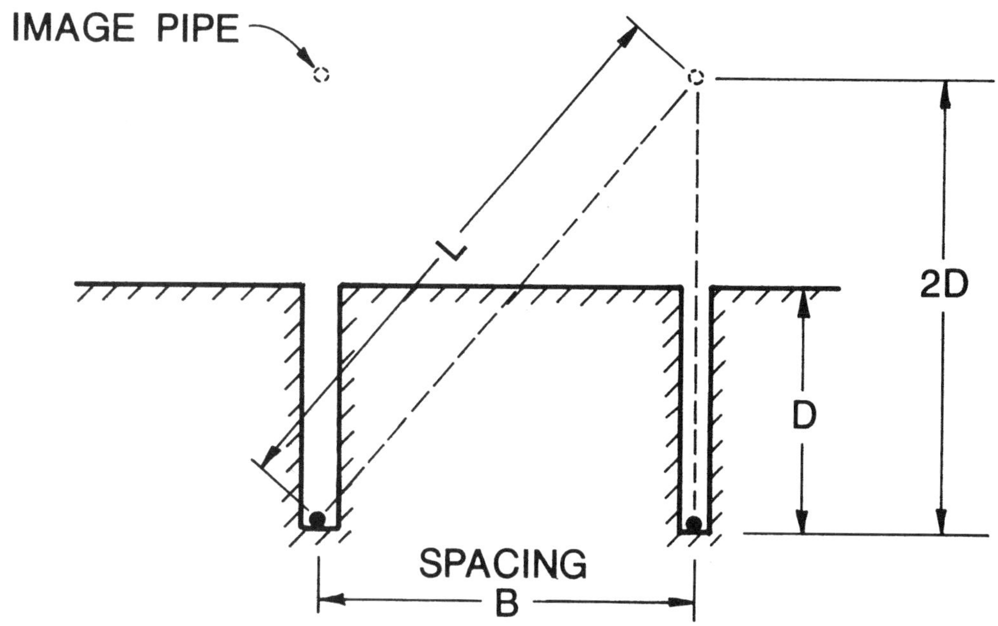

$$L = [B^2 + (2D)^2]^{\frac{1}{2}}$$

FIGURE B.1: Trench Spacing — Two Horizontal Pipes

3. Add the soil resistance of all the buried pipes and subtract the values of the image pipes.
4. The soil/field resistance value will be the sum from Step 3 divided by the number of buried pipes.

Table B.2 details the procedure for three different ground heat exchanger configurations.

For multiple-pipe systems, it is recommended that a computer with available software be used to increase accuracy of the estimates. For example, the number of calculations required to study the effects of soil/field resistance is given as:

Number of calculations = 2 x (number of buried pipes)^2

one trench, 4 pipes $2 \times 4^2 = 32$
two trenches, 4 pipes each $2 \times 8^2 = 128$
three trenches, 4 pipes each $2 \times 12^2 = 288$

Tables B.3 to B.8 give soil/field resistance values for a number of commonly used ground heat exchangers. The given values are for single, two, and three trenches. For a 10% decrease in efficiency, the following spacing guidelines can be determined from Tables B.3 through B.8:

1. For single pipe, Table B.4 —— 4 feet
2. For two pipes, Tables B.5 and B.6 —— 6 feet
3. For four pipes, Tables B.7 and B.8 —— 8 feet

B

TABLE B.2: Sample Calculations of Thermal Interference in Multiple-Pipe Horizontal Ground Heat Exchangers

Example 1:

Horizontal Ground Heat Exchanger: one-pipe system
Buried Depth: 4 feet or 48 inches
Pipe Radius: 0.5 inches
Thermal Diffusivity: 0.025 ft^2/hr
Thermal Conductivity: 0.75 Btu/hr-ft/°F
Operating Hours: 1000

	Radial Dist. (inches)	Soil/Field Res.
Pipe 1 to 1:	0.5	1.10
1 to 2:	96	-0.045 (Img. Pipe)
		1.055

Example 2

Horizontal Ground Heat Exchanger: one-pipe system
Buried Depth: 4 feet or 48 inches
Pipe Radius: 1.25
Thermal Diffusivity: 0.025 ft^2/hr
Thermal Conductivity: 0.75 Btu/hr-ft-°F
Operating Hours: 1000

Pipe 1 to 1:	1.25	0.907
1 to 2:	96	-0.045 (Img. Pipe)
		0.862

Example 3:

Horizontal Ground Heat Exchanger: two-pipe system
Thermal Diffusivity: 0.025 ft^2/hr
Thermal Conductivity: 0.75 Btu/hr-ft-°F
Buried Depth: 4 and 6 feet
Pipe Radius: 0.75 inches

Pipe 1 to 1	0.75 (Pipe Radius)	1.016
Pipe 1 to 2	24	0.284
1 to 3	96	-0.045 (Img. Pipe)
1 to 4	120	-0.023 (Img. Pipe)
2 to 1	24	0.284
2 to 2	0.75 (Pipe Radius)	1.016
2 to 3	120	-0.023 (Img. Pipe)
2 to 4	144	-0.012 (Img. Pipe)
		2.497

Soil Resistance per Pipe = 2.497/2 = 1.24

TABLE B.3: Nomenclature — Horizontal Ground Heat Exchanger Soil/Field Resistance Values (R_S)

HS Damp — Heavy soil damp, K_S = 0.75 A = 0.6
HS Dry — Heavy soil dry, K_S = 0.55 A = 0.48
LS Damp — Light soil damp, K_S = 0.5 A = 0.48
R_S — Soil thermal resistance, hr-ft-°F/Btu
K_S — Soil thermal conductivity, Btu/hr-ft-°F
A — Soil thermal diffusivity, ft^2/day
S — Trench spacing, feet
TR — Trench

4 FT-2TR = 4-foot trench separation, two trenches

(1 x 1) — one pipe layer x one pipe
(2 x 2) — two pipe layers x two pipes per layer

0, 4 — Computer definition; Horizontal distance, vertical distance (ft)

TABLE B.4: Soil/Field Resistance Values (1 x 1)

Pipe Size: Nominal 1½ inches (OD = 1.90 inches)

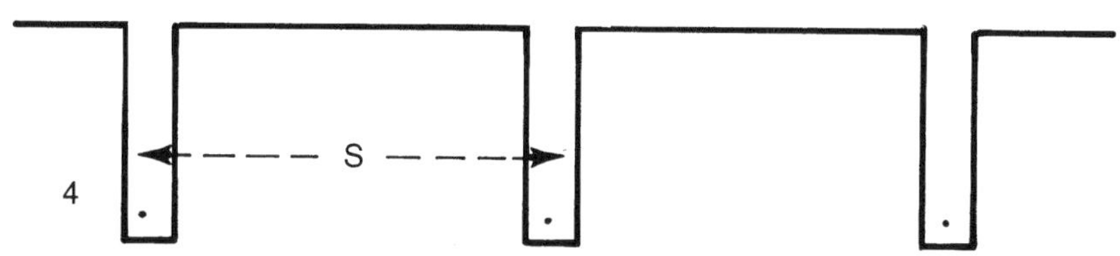

TYPE 1 (1 x 1)	SPACING (S)	HS DAMP	HS DRY	LS DAMP
	One Trench	0.93	1.26	1.39
0, 4	4 FT-2TR	1.07	1.43	1.58
	3TR	1.14	1.52	1.67
	6 FT-2TR	1.01	1.35	1.49
	3TR	1.04	1.39	1.53
	8 FT-2TR	0.97	1.31	1.44
	3TR	0.99	1.33	1.46

B

TABLE B.5: Soil/Field Resistance Values (2 x 1)

Pipe Size: Nominal ¾-inch (OD = 1.05 inches)

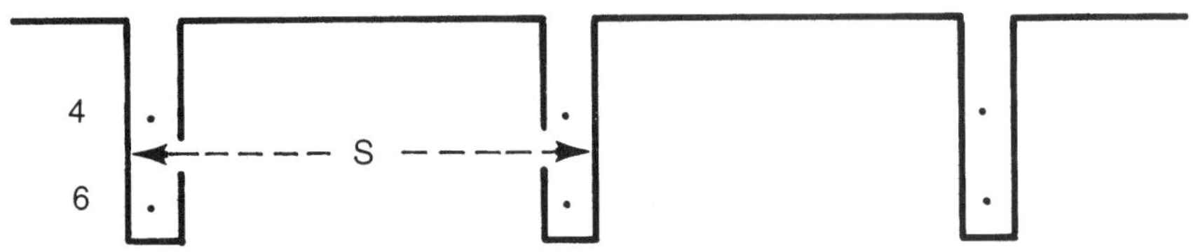

TYPE 2 (2 x 1)	SPACING (S)	HS DAMP	HS DRY	LS DAMP
	One Trench	1.37	1.83	2.02
0, 4	4 FT-2TR	1.65	2.19	2.40
0, 6	3TR	1.80	2.38	2.61
	6 FT-2TR	1.53	2.03	2.23
	3TR	1.60	2.11	2.33
	8 FT-2TR	1.46	1.94	2.13
	3TR	1.50	1.98	2.18

TABLE B.6: Soil/Field Resistance Values (1 x 2)

Pipe Size: Nominal ¾-inch (OD = 1.05 inches)

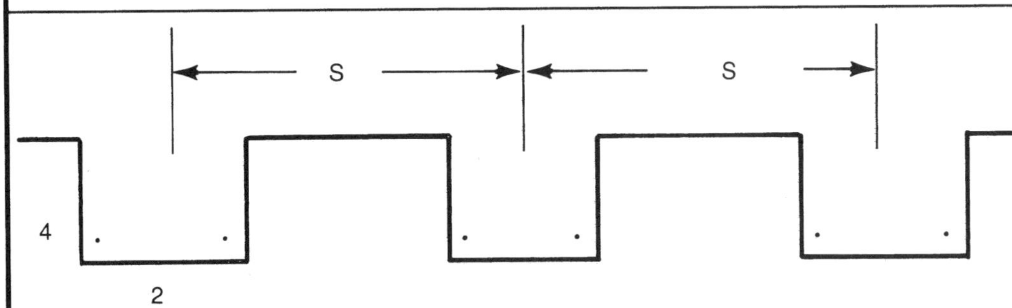

TYPE 3 (1 x 2)	SPACING (S)	HS DAMP	HS DRY	LS DAMP
1, 4	One Trench	1.32	1.78	1.96
1, 4	4 FT-2TR	1.62	2.16	2.38
	3TR	1.78	2.36	2.60
	6 FT-2TR	1.48	1.97	2.17
	3TR	1.55	2.06	2.26
	8 FT-2TR	1.41	1.88	2.07
	3TR	1.44	1.92	2.11

B

TABLE B.7: Soil/Field Resistance Values (4 x 1)

Pipe Size: Nominal ¾-inch (OD = 1.05 inches)

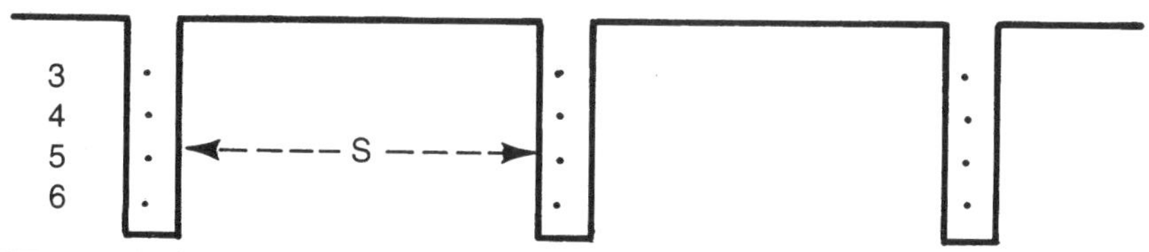

TYPE 4 (4 x 1)	SPACING (S)	HS DAMP	HS DRY	LS DAMP
	One Trench	2.05	2.74	3.02
0, 3	4 FT-2TR	2.56	3.39	3.73
0, 4	3TR	2.85	3.74	4.11
0, 5				
0, 6	6 FT-2TR	2.34	3.10	3.41
	3TR	2.48	3.25	3.58
	8 FT-2TR	2.22	2.94	3.23
	3TR	2.28	3.01	3.31
	10 FT-2TR	2.15	2.84	3.13
	3TR	2.18	2.88	3.17

TABLE B.8: Soil/Field Resistance Values (2 x 2)

Pipe Size: Nominal ¾-inch (OD = 1.05 inches)

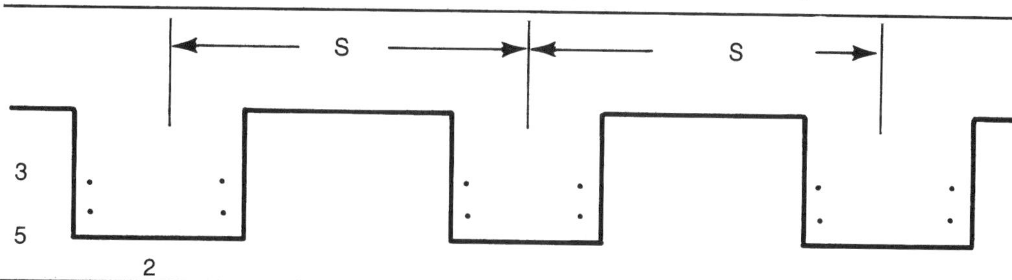

TYPE 5 (2 x 2)	SPACING (S)	HS DAMP	HS DRY	LS DAMP
0, 3	One Trench	1.75	2.34	2.58
0, 5				
2, 3	4 FT-2TR	2.28	3.02	3.32
2, 5	3TR	2.56	3.37	3.71
	6 FT-2TR	2.04	2.70	2.97
	3TR	2.17	2.86	3.14
	8 FT-2TR	1.91	2.53	2.79
	3TR	1.98	2.61	2.87

B

Vertical Ground Heat Exchanger Spacing

Spacing of vertical bore holes must take into account both short- and long-term effects. The short-term or annual heating and cooling effects last only four to six months. In this case, the design method considers only time periods of two to four months since the soil/field thermal resistance from the beginning to the peak season will be of this duration.

Long-term effects must consider the net annual load to the ground heat exchanger. If the amount of energy taken from the earth (heating season) is equal to that rejected, (cooling season), then there are no long-term effects.

Table 3.4 gives annual loads for both heating and cooling. The difference in the summer and winter loads is approximately (75,679,662 – 64,673,044) = 11,006,618 Btu/yr. This small difference would not result in any need to examine long-term effects.

For vertical systems, the image pipe as detailed in Figure B.1 is not applicable. The annual heating and cooling effects resulting from bore hole spacing can be considered by taking the radial distance in Table B.1 equal to the distance between bore holes. Adding the soil/field thermal resistance values of each bore hole gives a good first approximation for short-term effects, *i.e.*, one to two years.

For example:

1. Two bore holes separated 10 feet (120 inches)
2. 1-inch PB SDR-13.5 loop:
   D_{OE} = 1.59 inches (Table 4.1)
   R_{OE} = 0.795; use 0.75 inches
3. Soil thermal properties:
   K_S = 1.0 Btu/hr–ft–F
   A = 0.033 ft^2h/hr.
4. Heat pump run hours: 1,500

From Table B.1:

1. Bore hole 1:

R_{OE} = 0.75 inches, R_S = 0.816
R_{OE} = 120 inches, R_S = 0.040
$R_{S/F}$ = 0.856

2. Bore hole 2: In this case the procedure and values are the same as in Step 1.

The soil/field resistance values for the two bore holes are averaged to get the field value. In this case the field value is the average of the two bore holes:

(0.856 + 0.856)/2 = 0.856 F–ft–hr/Btu

The soil/field resistance value of 0.856 should then be used in the design equations to determine ground heat exchanger length. The same procedure can be followed for multiple bore holes in a ground heat exchanger field.

The above procedure, while approximate, is valid for periods up to the time at which the net heat flow from the ground heat exchanger is in steady state with the earth's surface. This time can be estimated as follows:

$T = H^2/[9 \times A]$

Where:

T = Time in days
H = Bore hole depth (feet)
A = Thermal diffusivity (ft^2/day)

For example, a bore hole 200 feet deep and a thermal diffusivity of 0.033 ft^2/hr (0.79 ft^2/day) would require a characteristic time before surface effects are noticed:

$T = 200^2/[\ 9 \times 0.79\ ]$ = 5,626 days or 15.4 years

For systems that have an annual net excess of heat rejection or heat absorption, the designer is referred to the ASHRAE or Oklahoma State University CL/GC manuals for methods to determine long-term effects.

Note: It would be very difficult to give a general rule for vertical ground heat exchanger spacing. In regions with small net annual energy loads, spacing of 10 feet is common. In northern Swedish climates, spacing of 50 feet is common. An analysis should be performed to determine if a potential problem exists. See the following reference for a more comprehensive treatment of this subject.

B

Eskilson, Per. *Thermal Analysis of Heat Extraction Boreholes*, ISBN 91-7900-298-6.
Distributed by:

Lund Institute of Technology
Department of Mathematical Physics
P.O. Box 118
221 00 Lund, Sweden

Long-Term Thermal Effects of Multiple Bore Holes

The long-term thermal effects of placing many bore holes in a grid configuration can be analyzed using the methods given in P. Eskilson's *Thermal Analysis of Heat Extraction Boreholes*. The equation to determine the increase in temperature from a continuous annual net heat flow is as follows:

$DT = [Q/(2 \times 6.28 \times K_S \times H)]\ g(t/t_1, R_O/H)$

- Q = Hourly annual net heat flow, Btu/hr
- K_S = Soil thermal conductivity, Btu/hr-ft-F
- H = Bore hold depth, feet
- R_O = Bore hole radius, feet
- t = Time, years
- t_1 = Characteristic time = $H^2/9A$
- A = Soil thermal diffusivity, ft^2/year

The function $g(t/t_1, R_O/H)$ is given in Figure B.2 for four bore holes in a rectangle. Other bore hole placements are given in the above reference.

The annual net heat load for the ground heat exchanger load can be determined using the procedure detailed in Table 3.4 or as follows:

Location: DOE Region IV (NRECA/EPRI *Heat Pump Manual*)

- Heating hours (HHR): 2,250
- Cooling hours (CHR): 800
- Design Heating Load: 50,000 Btu/hr
- Design Cooling Load: 30,000 Btu/hr
- Heat pump annual average COP_H = 2.9
- Heat pump annual average COP_C (EER/3.412) = 2.93

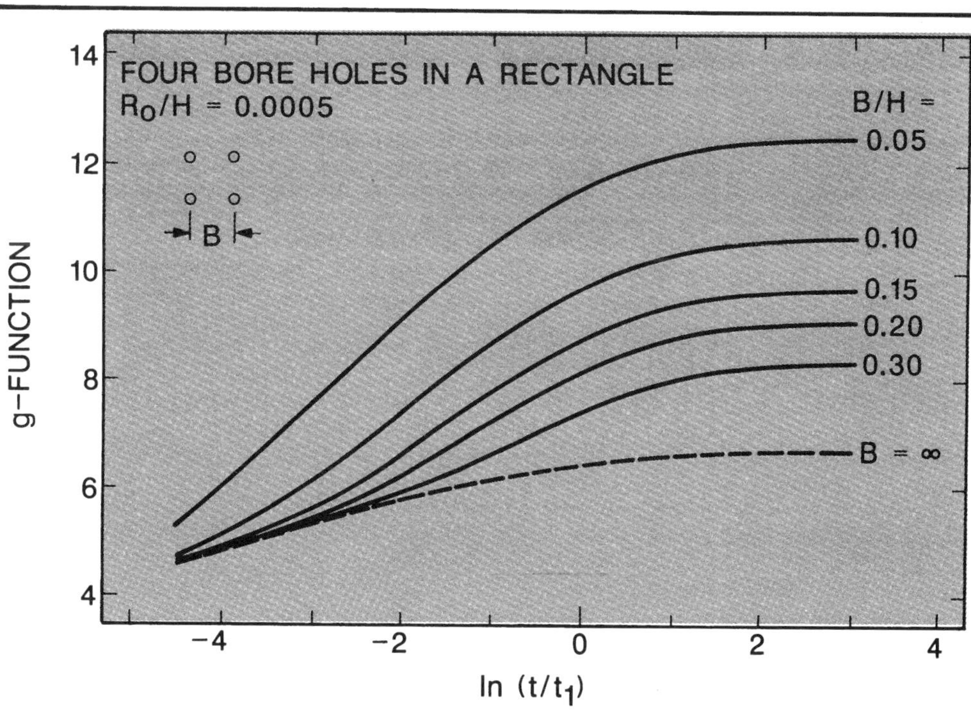

FIGURE B.2: Multiple Bore Holes

B

Bore Hole depth (H) = 150 feet

Loop: 1 in. SDR 13.5 cts, D_{OE} = 1.69 inches (Table 4.1)

Number of bore holes (NB) = 4

Soil thermal conductivity: K_S = 1.2 Btu/hr–ft–F

Soil thermal diffusivity: A = 0.6 ft^2/day

Annual net heat load (ANL):

ANL(Btu/yr) = DCL x [COP_C + 1 /COP_C] x CHR – DHL x [{COP_H–1}/COP_H] x HHR

ANL(Btu/hr) = 30,000 x (3.93/2.93) x 800 – 50,000 x (1.9/2.9) x 2,250

= – 40,000,000 Btu/yr

and

Q = [– 40,000,000 Btu/yr] / (365 x 24)

= – 4,566 Btu/hr

Characteristic time t_1 is:

$t_1 = H^2/9A = 150^2/(9 \times 219) = 11.4$ years

Where: A = 0.6 ft^2/day = 219 ft^2/yr

Thus:

DT = –4,566/(150 x 4 x 6.28 x 1.2) x $g(t/t_1, R_O/H)$

= –1.009 x $g(t/t_1, R/H)$

R_O(ft) = D_{OE}(inches)/12

R_O/H = 1.69/12/150 = 0.00044

> NOTE: Design Figure B.2 is for R_O/H = 0.0005. For other bore radii, the function g should be modified to:
>
> $g(t/t_1, R_O'/H) = g(t/t_1, R_O'/H) - L_N(R_O/R_O')$
>
> Where: R_O' is the new bore radius.
>
> L_N = Natural logarithm to the base e.
>
> Example: For a 150-foot bore hole, R_O = 0.0005 x 150 or 0.075 ft or 0.9 inches radius.
>
> For a PB loop of (D_{OE} = 1.69) inches dia or 0.845/12 = 0.0704 feet, the following adjustment is required:
>
> $g(t/t_1, R_O/H) = g(t/t_1, R_O'/H) - L_N(R_O/R_O')$
>
> $= g(t/t_1, R_O'/H) - L_N(0.0704/0.075)$
>
> $= g(t/t_1, R_O'/H) + 0.063$

For four bore holes in a rectangular configuration (Figure B.2) and a time (t) of 10 years:

$L_N(t/t_1) = L_N(10/11.4) = L_N(0.8772) = -0.13$

B = 15 feet

H = 150 feet

B/H = 15/150 = 0.10

And from Figure B.2:

$g(t/t_1, R_O/H) = 9.5 + 0.063 = 9.65$

Thus DT = –1.009 x 9.56 = –9.65°F

The –9.65°F represents an estimated decrease in the average earth's temperature after 10 years of operation. If this temperature decrease is not acceptable, then an increase in bore length or spacing is required.

APPENDIX C

C Specification of Polyethylene and Polybutylene Pipe

In This Appendix: Specification of polyethylene pipe; specification of polybutylene pipe

Specification of Polyethylene Pipe

Polyethylene pipe, because of its versatility and the large numbers of producers, comes in an almost limitless number of grades and strengths. Because of this wide range, the user of polyethylene pipe must consider a number of factors in selecting the material.

Manufacturer: While pipe material is only one aspect of a successful project, the "pipe system" offered by the manufacturer must include:

1. a recommended method of joining
2. specialized fittings such as tees, elbows, service valves, bell reducers, transition fittings, etc.
3. warranty information
4. engineering test data of pipe physical properties based on standards of the ASTM, Plastic Pipe Institute, etc.

Resins: Polyethylene resins are formed from ethylene (a natural gas product) by means of a polymerization process. Polymerization uses a combination of heat, pressure, and catalysts to combine single molecules into chain-like molecular structures known as polymers. These resins are the materials from which polyethylene pipe is made. Each potential polyethylene resin offers specific performance characteristics. Since the number of different polymerization processes is almost limitless, each manufacturer of polyethylene resins can choose from the available characteristics to produce resins designed for specific applications.

Applications of polyethylene materials include injection molding, blow molding, rotational molding, film extrusions, and profile extrusions. For this reason, it is important the supplier is aware of the application before a particular resin can be chosen. The resin selected for the buried pipe system must produce a pipe and piping system that has:

1. long life
2. high abrasion resistance
3. high environmental stress crack resistance

Fortunately, the plastics industry has addressed this issue and has established a uniform method of classifying polyethylene resins.

CLASSIFYING POLYETHYLENE RESINS

ASTM Standard D-1248, entitled "Polyethylene Plastics Molding and Extrusions Materials," details the classifi-

C

cation procedure. As presently written, *ASTM D-1248* defines resins as to their type, class, category, and grade:

1. **Type** - Identification by nominal density (gm/cm^3)
   Type I (0.910 to 0.925) - low density
   Type II (0.926 to 0.940) - medium density
   Type III (0.941 to 0.959) - high density
   Type IV (0.960 & higher) - high density
   NOTE: The density shown is for uncolored resins. Please note that Type III and Type IV are both described as high density.
2. **Class** - Identification by composition
   Class A - natural color only
   Class B - colors, including white and black
   Class C - black, weather resistant
   NOTE: Class C contains 2% or more carbon black.
3. **Categories** - Identification by nominal flow
   Category 1 - greater than 25
   Category 2 - greater than 10 to 25
   Category 3 - greater than 1 to 10
   Category 4 - greater than 0.4 to 1.0
   Category 5 - 0.4 maximum
   NOTE: Flow is defined as the amount of polyethylene material through a specified orifice: g/10–min at 190°C and a 2,160–g load.
4. **Grades** - Identification by other key characteristics: tensile strength, elongation, brittleness, environmental stress crack resistance.

Table C.1 defines two grades of pipe, P23 and P34. These two pipes, medium and high density, are the two most often used piping materials extruded. The test values shown in Table C.1 give physical properties of the two materials.

In an effort to further define the properties of P34 resins, *ASTM D-3350*, entitled "Polyethylene Plastics Pipe and Fittings Materials," was developed. The potential of up to six ranges to define a single parameter plus designation of five possible combinations of color and UV stabilizers allow a much more precise definition of a resin than previously possible.

The physical properties of a resin and the standards used to evaluate them using *ASTM D-3350* are:

1. Density (*ASTM D-1505*)
2. Melt Index (*ASTM D-1238*)
3. Flexural Modulus (*ASTM D-790*)
4. Tensile Strength at Yield (*ASTM D-638*)
5. Environmental Stress Crack Resistance (*ASTM D-1693*)
6. Hydrostatic Design Basis (*ASTM D-2837*)
7. a letter designating the method of UV stabilization

TABLE C.1: Polyethylene Pipe Grades

Property	Grade P23	Grade P34
Tensile Strength, psi (*ASTM D-638*)	1800	3200
Elongation, % (*ASTM D-638*)	400	500
Brittleness Temperature °C (*ASTM D-746*)	-60	-70
Environmental Stress Crack Resistance, hrs. (*ASTM D-1693*, Condition C)	none	192

C

Ranges of physical values of the above parameters are divided into cells. The combination of the cell values resulting from each test procedure with the color and UV stabilization designation form the cell classification number for a specific resin. Table C.2 gives the cell classification limits as defined in *ASTM Standard D-3350*.

To obtain the maximum benefits when utilizing ASTM D-3350 to specify a polyethylene resin, the effect of changes in specific physical characteristics must be understood. Unfortunately, most desirable performance traits of a resin are not a function of a single physical parameter. The "real world" performance expected from a piping system requires careful evaluation of the effect of each parameter. To complicate the specification problem even further, increases in the value of a physical parameter do not always indicate that the performance of the resin is affected in a desirable manner. For example, increases in polyethylene resin density have several recognized effects:

1. pipe stiffness increases
2. tensile strength at yield increases
3. surface hardness increases
4. burst strength increases
5. permeability decreases

Each of the above is a desirable physical characteristic for a pipe resin. Depending on the needs for a specific application, each takes on increasing or decreasing importance. However, when the density of an uncompounded polyethylene resin exceeds approximately 0.96 g/cc, it becomes unsuitable for production of pipe although it makes excellent sheet for abrasive applications. Resins above this value become brittle and no longer exhibit the balance of flexibility and long-term strength required for pipe.

TABLE C.2: Primary Properties — Cell Classification Limits (Taken from ASTM D-3350)

Property	Test Method	0	1	2	3	4	5	6
1. Density g/cm^3	D-1505		0.910-0.925	0.926-0.940	0.941-0.955	>0.955		
2. Melt Index	D-1238		>1.0	1.0 to 0.4	<0.4 to 0.15	<0.15	Note[1]	Note[2]
3. Flexural Modulus MPa (psi)	D-790		<138 (<20,000)	138 - <276 (20,000 to <40,000)	276 - <552 (40,000 to <80,000)	552 - <758 (80,000 to <110,000)	758 - <1103 (110,000 to <160,000)	>1103 >160,000
4. Tensile Strength @ Yield MPa (psi)	D-638		<15 (<2200)	15 - <18 (2200- <2600)	18 - <21 (2600-<3000)	21 - <24 (3000-<3500)	24 - <28 (3500-<4000)	>28 (>4000)
5. Environmental Stress Crack Resistance a. Test Condition b. Test Duration, h c. Failure, max. %	D-1693		A 48 50	B 24 50	C 192 20			
Hydrostatic Design Basis, MPa (psi), (23°C)	D-2837	NPR[3]	5.52 (800)	6.89 (1100)	8.62 (1250)	11.03 (1600)		

[1]Refer to 10.1.4 ASTM D-1693
[2]Refer to 10.4.2 ASTM D-1693
[3]NPR = Not Pressure Rated

C

Maximum benefits from use of the cell classification method of specification result when the user clearly defines the performance characteristics needed for a piping system in a given application. Systems subject to fluctuating temperatures (such as ground heat exchangers) encounter stress patterns not present in piping operating at a constant temperature. In each application, a polyethylene resin providing the performance needed must be used for a successful application.

A user of polyethylene pipe must determine the performance needed in a specific application. The performance of the polyethylene pipe available must then be compared to the application requirements. Each pipe manufacturer chooses from the resins available to obtain the properties desired for its products. The physical characteristics of the resin used to manufacture a specific product may affect the performance in application.

For example, an increase in density increases the hardness and tensile strength of a polyethylene resin. This could be a major consideration in pipes used in abrasive soils. Also, permeability is decreased which could reduce the amount of oxygen passing through the pipe wall, thus preventing unnecessary corrosion to the heat pump and pump circulation system's metal parts. Pipes subject to burial stresses, temperature fluctuations, gouging, or abrasion during installation, chemical solutions, etc., should be carefully evaluated for environmental stress crack resistance. Almost all pipe applications subject pipes to one or more of these conditions.

Obviously, knowledge of the physical characteristics of polyethylene, their effect on pipe performance, and the test values indicative of specified performance characteristics, will make selection of appropriate cell classification values easier. Unfortunately, many users do not have sufficient knowledge to develop a cell class specification independently. Also, if such a specification were developed, such a resin may not be commercially available.

A more realistic approach to pipe specifications and a more practical method for the typical user involves a review of the available products and their performance under ASTM test procedures. Major resin and pipe manufacturers provide data from product testing. This data normally will include a cell classification number. A comparison of the physical performance of the resins to system requirements allows the user to determine the minimum acceptable resin commercially available in pipe. The cell classification of this resin then becomes the minimum material specification. The use of the cell classification values to evaluate various products becomes a matter of comparing standardized test results. Resins having values that equal or exceed the values of the specified cell class number would be expected to equal or exceed the design performance requirements.

ENVIRONMENTAL STRESS CRACK RESISTANCE (ESCR)

ASTM Standard D-1693 states, "Environmental stress-cracking is a property that is highly dependent upon the nature and level of the stresses applied and on the thermal history of the specimen." All P34 resins are required to test under *ASTM D-1693*, Condition C. Condition C incorporates bending, notching, a surfactant, and temperature to simulate physical stresses that a material may encounter in actual applications. This is the most severe ESCR test currently approved by ASTM. Resins are allowed a 20% (F_{20}) failure rate after 192 hours of testing. Grade P34 resins encountered in the pipe market generally exceed this minimum by a large margin. Three of the most frequently extruded P34 resins show test values of zero failures (F_0) at a time greater than 1,000 hours, 50% (F_{50}) failures at a period greater than 1,500 hours, and zero failures (F_0) at greater than 5,000 hours of testing. If a user desires a material offering greater

C

stress resistance than the maximum ASTM test requirement, the allowable failure rate and minimum time of testing should be defined in the written specification.

POLYETHYLENE SPECIFICATION

It may seem that the specification of polyethylene pipe using physical performance parameters is very complicated. This is not the case. Using the cell classification system, a very brief but very descriptive specification is easily written. In addition, since it will be based on ASTM standards, this specification is easily interpreted by the manufacturer and the pipe purchaser.

The sample specification below contains the standard language for specifying polyethylene pipe. In it, the symbol "#" is used for cell number values, and "Z" is used for designation of the method of ultraviolet protection used in the specified pipe. Blanks appear where other numerical values must be supplied for the percentage of allowable ESCR and the minimum time period for the allowable failure rate.

> "Polyethylene pipe furnished under this specification shall be manufactured in accordance with dimensional specifications of *ASTM D-2513* or *ASTM F-714*, as appropriate, using a resin having a cell classification of ######Z as defined by *ASTM D-3350*. This resin shall exhibit not more than ___% failures (F___) when tested for ___ or more hours under *ASTM D-1693*, Condition C. Acceptable products are ABC Pipe as manufactured by ABC Pipe Company, Somewhere, USA or equal."

GUIDING PRINCIPLES

Before selecting any pipe material, the following should be considered:

1. Does the manufacturer have a pipe material specifically intended for ground-coupled operation? Does the manufacturer understand the range of operating temperatures, the kinds of fluids that will be circulated, the types of soils that may be encountered, and the expected life of the piping system?
2. Does the manufacturer provide warranties of his material and appropriate test data as specified by various testing standards?
3. Have appropriate minimum standards been adopted for pipe selection? Has a cell classification system been adopted to allow product suppliers to compete fairly?

NOTE: The manner in which materials are identified in a cell classification is as follows (refer to Table C.2):

PE355434C

- PE -- Polyethylene
- 3 -- Density (0.941 - 0.955 g/cm3)
- 5 -- Melt Index (ASTM D-1693, 10.1.4)
- 5 -- Flexural Modulus (758–<1103 MPa) (110,000 to 160,000 psi)
- 4 -- Tensile Strength at Yield (21–<24 MPa) (3,000–<3500 psi)
- 3 -- Environmental Stress Crack Resistance (Condition C, 192 hrs, 20% max. failure)
- 4 -- Hydrostatic Design Basis at 23°C (11.03 MPa (1600 psi))
- C -- Color and UV Stabilizer (colored, C-Carbon blk)

C

4. Has the manufacturer identified his product by its cell classification number?

   Since grade P23 does not have a requirement to be tested under *ASTM Standard D-1693* Condition C, specifying grade P23 leaves some uncertainty as to the design life of the material. This is not to say that some grades of P23 are not acceptable, only that specific design values based on *ASTM Standard D-1693* cannot be assumed.

   Grade P34 allows the designer to set specific guidelines based on values of Environmental Stress Crack Resistance. Based on values of commercially extruded pipe, a guideline or standard can be set.

The International Ground Source Heat Pump Association (IGSHPA) recommends the minimum cell classification number acceptable for polyethylene pipe is PE355434C or PE345434C when tested under ASTM 3350.

FUSION JOINING OF POLYETHYLENE PIPE

While most polyethylene resins are capable of fusion to resins of other manufacturers, such joints frequently require fusion procedures to make them compatible that differ from the recommended fusion procedures of either manufacturer. In addition, the method of joining should be considered. All medium-density and high-density resins with lower molecular weight are suitable for joining using butt- or socket-fusion techniques. Some higher molecular weight resins can only be fused using the butt-fusion technique.

Specification of Polybutylene Pipe

Polybutylene pipe in the United States has a single manufacturer. Therefore, the number of resins for extruding is limited. This also allows for closer control of the raw resin material.

The International Ground Source Heat Pump Association (IGSHPA) recommends a minimum specification for polybutylene pipe:

"Polybutylene pipe furnished under this specification shall be manufactured in accordance with ASTM Standard D-2581. The material shall be:

1. Either Class B (general purpose and dielectric, in colors) or Class C (weather resistant, black in color containing not less than 2% carbon black)
2. Type II (density, 0.91 to 0.92 g/cm^3)
3. Grade 1 (flow rate 0.25 to 0.75 g/10 min)"

FUSION JOINING OF POLYBUTYLENE PIPE

Polybutylene pipe in all buried applications should be heat fused. In the sizes used in ground heat exchangers, the socket fusion procedure is used. See Section 5 for socket fusion procedures for polybutylene pipe.

D Standardized Parallel System Header Design

In This Appendix: Standardized parallel system header design; flushing flow rates and pressure head of standardized header; fabricating the ground heat exchanger parallel header

Standardized Parallel System Header Design

The advantages of a standardized parallel header design for ground heat exchangers from one ton of heat pump capacity and up are:

1. All parallel path connections (saddle fusions) are premanufactured and drilled out prior to arrival at the field site. Therefore, quality control is improved.
2. Only a minimum number of field-fabricated butt joints are required.
3. The header design minimizes the water flow rate required for air purging.

The design recommendations are as follows:

1. The design assumes that one loop will be used for each ton of heat pump capacity and that the nominal heat pump operating water flow rate will be 3 GPM/ton.
2. The maximum recommended bore hole depths or length of horizontal pipe loops are:
   a. 250 feet using ¾-inch SDR-11 polyethylene pipe or 1-inch SDR-13.5 cts polybutylene pipe
   b. 400 feet using 1-inch SDR-11 polyethylene pipe or 1 ¼-inch SDR-13.5 cts polybutylene pipe

   In all cases, the specific fluid used and the recommended lower operating temperature will determine the pipe diameter and length. For bore holes greater than this depth, larger diameter pipe is required and a detailed analysis must be made.
3. One-ton ground heat exchangers will be of the series type. They will be fabricated from either ¾- or 1-inch SDR-11 polyethylene pipe, or 1- or 1 ¼-inch SDR-13.5 cts polybutylene pipe.
4. Two-ton ground heat exchangers are fabricated from 1-inch SCH 40 polyethylene header pipes and ¾- or 1-inch SDR-11 pipe loops, or 1¼-inch SDR-13.5 cts polybutylene pipe header pipes and 1-inch SDR-13.5 polybutylene pipe.
5. The basic design module for ground heat exchangers for three tons and higher is the three-ton preassembled unit shown in Figure D.1. The unit is shown butt fused for polyethylene pipe, but can also be socket fused with appropriate socket fittings using polybutylene pipe.

This unit is premanufactured and preassembled. It consists of one 1 ½- x ¾-inch bell reducer, two ¾-inch tees, and one 90° elbow. If the loop

D

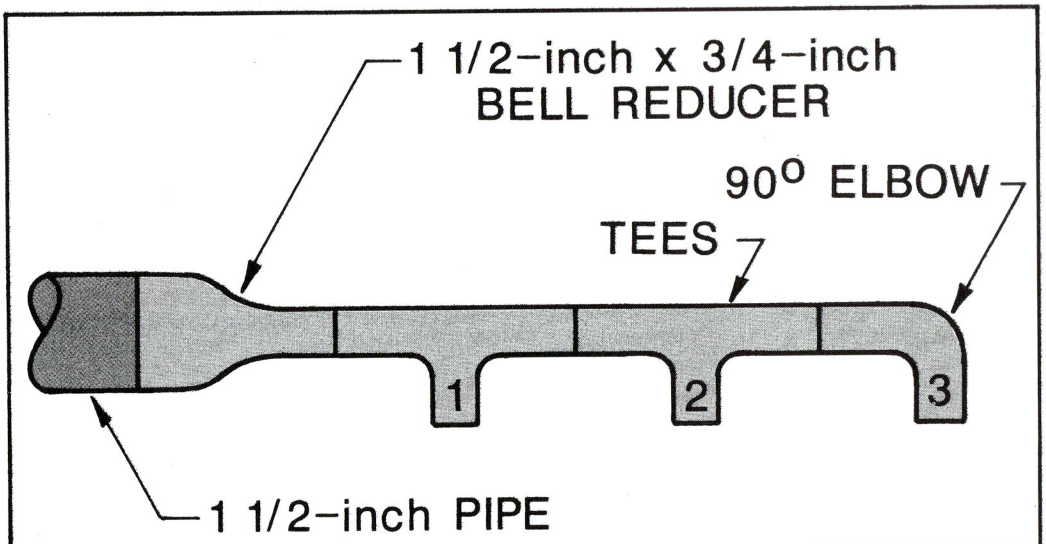

FIGURE D.1: Three-Ton Header Design for PE Pipe

water pressure drop becomes excessive, the ¾-inch pipe size can be replaced with 1-inch pipe for bore holes greater than 200 feet in depth.

Since ¾-inch SDR-11 (1.077 inch ID) polyethylene pipe and 1-inch SDR-13.5 cts (0.957 inch ID) polybutylene pipe have approximately the same internal diameter, it is necessary to use the larger sized polybutylene pipe to have the same fluid pressure drop.

The bell reducer shown in Figure D.1 is used to prevent any trapped air at the end of the header pipe. By using two tees and the 90° elbow at the bell reducer end, the fluid flow required for air purging in the 1 ½-inch header section is 12.7 GPM.

6. Heat exchangers up to seven tons can be constructed by adding the required number of parallel adapters (see Figure D.2) or reducing tees (polybutylene pipe) to the basic three-ton heat exchanger unit (see Figure D.1). Table D.1 summarizes the required components necessary for heat exchangers in the range of one to ten tons.
7. For heat exchanger sizes above seven tons, additional parallel adapters or reducing tees fabricated from 2-inch header pipe and ¾- or 1-inch service saddles are added as shown in Figure D.3. Heat exchangers from seven to ten tons should use 2-inch main header pipe and 1 ½-inch subheader pipe.

Above ten tons, the ground heat exchangers described above should be installed in parallel to give the required capacity.

For ground heat exchangers of less than ten tons, remove the service saddles from left to right as shown in Figure D.3. For example, a seven-ton heat exchanger would consist of the basic three-ton module plus the four service saddles on the 1 ½-inch header pipe. The spacing between the service saddles on the 2- and 1 ½-inch headers is optional. That is, the flushing flow rate or the operating pressure heads are not affected greatly.

Do not space the ¾-inch tees and 90° elbow. Spacing these components can result in excessive pres-

D

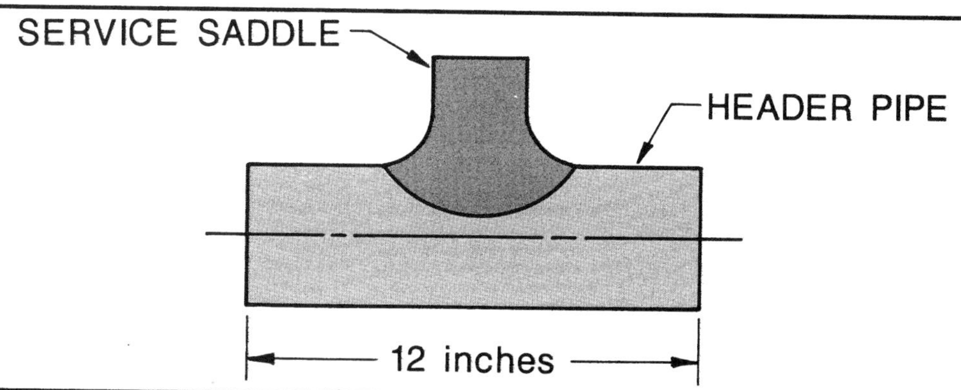

FIGURE D.2: Parallel Adapter

TABLE D.1: Required Heat Exchanger Components

Hex Size (Tons)	Parallel Adapters or Tees	Tees (Loop Size)	Bell Reducers 1½ x ¾	Elbows (Optional) ¾-inch
1	Series system recommended, ¾- or 1-inch pipe			
2	Parallel system, 1-inch headers, ¾-inch loops a. Two 1 x ¾-inch bell reducers b. Two ¾-inch tees c. Two 90° elbows			
3	—	4	2	2
4	2	4	2	2
5	4	4	2	2
6	6	4	2	2
7(1)	8	4	4(2)	2
8(3)	10(4)	4	4(2)	2
9	12(4)	4	4(2)	2
10	14(4)	4	4(2)	2

(1) If the supply and return headers exceed 200 feet, a two-inch header should be used.

(2) Use 2 each 2 inches by 1½ inches;
2 each 1½ inches by ¾ inch.

(3) Use 2-inch supply and return headers up to the subheaders for all ground heat exchangers from 7 tons to 10 tons.

(4) Use 10 each 1½-inch by ¾-inch parallel adapters. The remaining parallel adapters or reducing tees will be 2 inches by ¾ inch.

D

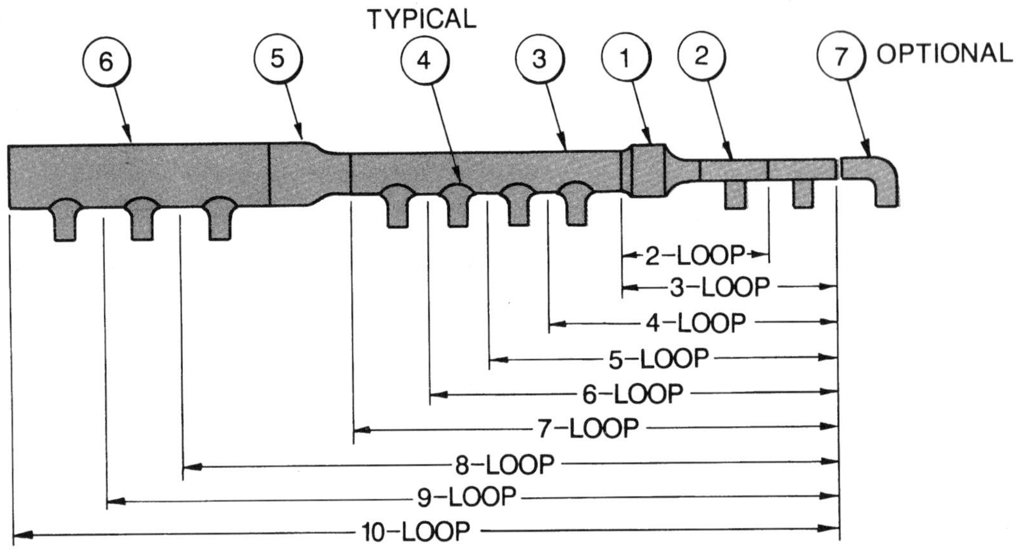

ITEM	SIZE AND DESCRIPTION
1	REDUCER, 1 1/2- OR 1 1/4-INCH X 1- OR 3/4-INCH
2	TEES, 1- OR 3/4-INCH
3	PIPE, 1 1/2- OR 1 1/4-INCH
4	STRAIGHT OUTLET SERVICE SADDLES OUTLETS, 1-INCH OR 3/4-INCH
5	REDUCER, 2-INCH X 1 1/2-INCH
6	PIPE, 1-INCH
7	ELBOW, 1- OR 3/4-INCH ON ANY OUTLETS (OPTIONAL)

FIGURE D.3: 10-Ton Ground Heat Exchanger

Source: *McElroy No. 2 CU Fusion Manual*[17]

sure loss and unequal flow in the loops connected to the three-ton module.

If it is necessary to space the tees and elbow, refer to Figure 7.2 in Section 7 for recommended methods.

Flushing Flow Rates and Pressure Head of Standardized Header

Table D.2 gives the pressure head for ¾- and 1-inch SDR-11 polyethylene pipe necessary to purge any air from the parallel loops of the ground heat exchanger. The required flushing flow rate using this header design (Figure D.3) is 4.25 GPM times the number of loops in parallel.

Table D.3 gives the pressure head for the header pipes in terms of ground heat exchanger capacity and header pipe length for two pipe sizes.

Given these two pressure losses for the loops and header pipes, the following requirements result:

1. The required flushing flow rate for this header design and loop system as shown in Figure D.3 equals:

   4.25 GPM x number of tons
2. The flushing head required equals:

   Header pressure loss + loop pressure loss

D

TABLE D.2: Pressure Head for Parallel Loops

Pipe Size	Ft Head/100 Ft @ 4.25 GPM	Ft of Head 200-Ft Bore Hole
¾-inch SDR-11	2.8	11.2
1-inch SDR-11	0.93	3.72

TABLE D.3: Pressure Head for Header Pipes

Pipe Size	Flow/Cap GPM/Tons	Ft Head/100 ft
2-inch	42.5/10	3.3
	38.3/9	2.7
	34.0/8	2.2
	29.8/7	1.7
	25.5/6	1.3
	21.3/5	0.9
	17.0/4	0.6
	12.8/3	0.4
	8.5/2	0.1
	4.3/1	—
1½-inch	29.8/7	5.8
	25.5/6	4.4
	21.3/5	3.2
	17.0/4	2.1
	12.8/3	1.3
	8.5/2	0.6
	4.3/1	0.1

Example: 10-ton heat exchanger with 400 feet of 2-inch headers and 200-foot bore holes (400 feet of pipe)

A. The flushing flow rate is 10 x 4.25 or 42.5 GPM

B. The pressure (ft of head) would be the sum of header pressure loss and the loop pressure loss or:

400 ft (3.3 psi/100 ft) + 11.2 = 24.4 ft

Therefore a pump with a flow of 42.5 GPM and at a head of 24.4 feet is required to remove all entrapped air. Table D.4 gives flushing requirements for a 10-ton ground heat exchanger for ¾- and 1-inch pipe with varying header lengths.

Tables D.5 and D.6 give pressure drop values for commonly used polyethylene pipe. Values are given in feet of head / 100 feet and psi/100 feet. These tables and Table D.3 can be used to quickly estimate pressure and flow requirements for other heat exchanger sizes.

D

TABLE D.4: Flushing Requirements for 10-Ton Ground Heat Exchangers with 200-Foot Bore Holes or 400 Feet of Pipe			
Loop Pipe Size (SDR-11)	Total Header Length (ft) (2-inch pipe SCH 40)	Flow (GPM)	Pressure (Feet)
¾" (0.86 in.)	200 400 800	42.5	17.8 24.4 37.6
1" (1.077 in.)	200 400 800	42.5	10.3 16.9 30.1

Fabricating the Ground Heat Exchanger Parallel Header

The following procedure is recommended for fabricating the parallel ground heat exchanger header (Figure D.2):

1. Fabricate the necessary number of parallel adapters using the following procedure:
   a. Fuse 1 ½- x 3¾-inch service saddle to a 12-inch section of header pipe.
   b. **Drill out service saddle hole** to provide clear flow path to header. Be careful not to drill through the other pipe wall.
   c. Remove all drilled-out shavings and foreign materials.
2. Add loop size elbows to parallel adapters (see Figure 6.7). This allows for a minimum bell hole width for connecting the loops to the header and greatly reduces the possibility of pipe crimping due to short radius bending or improper backfilling.

NOTE: Minimum bend radius for polyethylene butt-fused pipe is 25 pipe diameters.

3. Again check for a clear path through the service saddle.
4. Add "pigtails" to elbows. This allows more freedom of movement in the bell hole or pit. With this procedure, all the fusion joining can be at ground level.
5. Complete all loop-size butt fusion connections. Following this procedure reduces the number of times the fusing machine must be set up, *i.e.*, changing pipe inserts.
6. Fuse all 1 ½-inch butt fusion connections.

This recommended procedure is important in reducing the time for connecting the loops of the parallel system to the header. As a task that can be done at the shop and not in the field, it is generally faster and more reliable.

D

TABLE D.5: Feet of Head/100 Feet — Hazen-Williams Equation

Flow (GPM)	¾ Inch (0.86)	1 Inch (1.077)	1½ Inch (1.61)	2 Inch (2.067)
1.00	0.19	0.06	0.01	0.00
2.00	0.69	0.23	0.03	0.01
3.00	1.47	0.49	0.07	0.02
4.00	2.50	0.84	0.12	0.04
5.00	3.78	1.27	0.18	0.05
6.00	5.30	1.78	0.25	0.07
7.00	7.05	2.36	0.33	0.10
8.00	9.03	3.02	0.43	0.13
9.00	11.23	3.76	0.53	0.16
10.00	13.64	4.57	0.65	0.19
11.00	16.27	5.45	0.77	0.23
12.00	19.11	6.40	0.91	0.27
13.00	22.17	7.43	1.05	0.31
14.00	25.42	8.52	1.21	0.36
15.00	28.88	9.68	1.37	0.41
16.00	32.55	10.90	1.55	0.46
17.00	36.41	12.20	1.73	0.51
18.00	40.47	13.56	1.92	0.57
19.00	44.73	14.99	2.12	0.63
20.00	49.18	16.48	2.33	0.69
21.00	53.82	18.03	2.56	0.76
22.00	58.66	19.65	2.79	0.83
23.00	63.69	21.34	3.02	0.90
24.00	68.91	23.09	3.27	0.97
25.00	74.31	24.90	3.53	1.05
26.00	79.91	26.77	3.79	1.13
27.00	85.68	28.71	4.07	1.21
28.00	91.65	30.71	4.35	1.29
29.00	97.79	32.76	4.64	1.38
30.00	104.12	34.89	4.94	1.47
31.00	110.64	37.07	5.25	1.56
32.00	117.33	39.31	5.57	1.65
33.00	124.20	41.61	5.90	1.75
34.00	131.25	43.97	6.23	1.85
35.00	138.48	46.40	6.57	1.95
36.00	145.89	48.88	6.93	2.06
37.00	153.48	51.42	7.29	2.16
38.00	161.24	54.02	7.66	2.27
39.00	169.18	56.68	8.03	2.38
40.00	177.29	59.40	8.42	2.50
41.00	185.58	62.18	8.81	2.62
42.00	194.04	65.01	9.21	2.74
43.00	202.67	67.90	9.62	2.86
44.00	211.48	70.85	10.04	2.98
45.00	220.45	73.86	10.47	3.11
46.00	229.60	76.93	10.90	3.24
47.00	238.92	80.05	11.34	3.37
48.00	248.41	83.23	11.79	3.50
49.00	258.07	86.46	12.25	3.64
50.00	267.90	89.76	12.72	3.78
51.00	277.89	93.11	13.19	3.92
52.00	288.06	96.51	13.68	4.06
53.00	298.39	99.97	14.17	4.21
54.00	308.89	103.49	14.67	4.35
55.00	319.56	107.06	15.17	4.50
56.00	330.39	110.69	15.69	4.66
57.00	341.38	114.38	16.21	4.81
58.00	352.55	118.12	16.74	4.97
59.00	363.87	121.91	17.28	5.13
60.00	375.37	125.76	17.82	5.29

D

TABLE D.6: PSI/100 Feet — Hazen-Williams Equation

Flow (GPM)	¾ Inch (0.86)	1 Inch (1.077)	1½ Inch (1.61)	2 Inch (2.067)
1.00	0.08	0.03	0.00	0.00
2.00	0.30	0.10	0.01	0.00
3.00	0.64	0.21	0.03	0.01
4.00	1.08	0.36	0.05	0.02
5.00	1.64	0.55	0.08	0.02
6.00	2.30	0.77	0.11	0.03
7.00	3.05	1.02	0.14	0.04
8.00	3.91	1.31	0.19	0.06
9.00	4.86	1.63	0.23	0.07
10.00	5.91	1.98	0.28	0.08
11.00	7.05	2.36	0.33	0.10
12.00	8.28	2.77	0.39	0.12
13.00	9.60	3.22	0.46	0.14
14.00	11.01	3.69	0.52	0.16
15.00	12.51	4.19	0.59	0.18
16.00	14.09	4.72	0.67	0.20
17.00	15.76	5.28	0.75	0.22
18.00	17.52	5.87	0.83	0.25
19.00	19.37	6.49	0.92	0.27
20.00	21.29	7.13	1.01	0.30
21.00	23.31	7.81	1.11	0.33
22.00	25.40	8.51	1.21	0.36
23.00	27.58	9.24	1.31	0.39
24.00	29.84	10.00	1.42	0.42
25.00	32.18	10.78	1.53	0.45
26.00	34.60	11.59	1.64	0.49
27.00	37.10	12.43	1.76	0.52
28.00	39.68	13.30	1.88	0.56
29.00	42.34	14.19	2.01	0.60
30.00	45.09	15.11	2.14	0.64
31.00	47.91	16.05	2.27	0.68
32.00	50.80	17.02	2.41	0.72
33.00	53.78	18.02	2.55	0.76
34.00	56.83	19.04	2.70	0.80
35.00	59.96	20.09	2.85	0.85
36.00	63.17	21.16	3.00	0.89
37.00	66.46	22.27	3.16	0.94
38.00	69.82	23.39	3.31	0.98
39.00	73.25	24.54	3.48	1.03
40.00	76.77	25.72	3.64	1.08
41.00	80.35	26.92	3.82	1.13
42.00	84.02	28.15	3.99	1.18
43.00	87.76	29.40	4.17	1.24
44.00	91.57	30.68	4.35	1.29
45.00	95.46	31.98	4.53	1.35
46.00	99.42	33.31	4.72	1.40
47.00	103.45	34.66	4.91	1.46
48.00	107.56	36.04	5.11	1.52
49.00	111.74	37.44	5.31	1.58
50.00	116.00	38.86	5.51	1.64
51.00	120.33	40.31	5.71	1.70
52.00	124.73	41.79	5.92	1.76
53.00	129.20	43.29	6.13	1.82
54.00	133.75	44.81	6.35	1.89
55.00	138.37	46.36	6.57	1.95
56.00	143.06	47.93	6.79	2.02
57.00	147.82	49.53	7.02	2.08
58.00	152.65	51.14	7.25	2.15
59.00	157.56	52.79	7.48	2.22
60.00	162.53	54.45	7.72	2.29

APPENDIX E

E DESUPERHEATERS

In This Appendix: Desuperheaters for domestic hot water heating; desuperheater design and selection; system layout and installation

Desuperheaters for Domestic Hot Water Heating

Domestic hot water heating using a heat pump can significantly decrease energy costs during both heating and cooling seasons. A desuperheater is a heat exchanger that removes the high-grade (high-temperature) superheat available in the refrigerant gases exiting the heat pump compressor. Depending on the heat pump design and operating conditions, superheat temperatures of 200°F and greater may be reached. By removing the superheat, desuperheaters can provide a source of hot water typically at a temperature of 160°F.

For many years, desuperheating systems (commonly known as "heat reclaimers") have been successfully retrofittted on central air conditioning systems. Industry members reported some 50,000 units installed during 1985 on central air conditioning systems to provide hot water. Frequently, such systems were reported to pay for themselves in only one year.

Desuperheaters' function with ground-source heat pumps is very similar to their function with central air conditioning. During cooling, heat pumps operating to provide air conditioning generate unused heat. This heat can be reclaimed and can therefore be considered a "free" source of energy for generating domestic hot water.

During the heating mode, the functions of the condenser and evaporator are reversed by a four-way reversing valve. This requires the desuperheater heat exchanger to be installed between the compressor and the reversing valve. Heating of hot water is not "free" but is provided at a reduced cost based on the heating COP of the heat pump.

On ground-coupled heat pumps, desuperheaters may provide most of the domestic hot water required for a typical residence. During summer cooling cycles, it is not uncommon for the homeowner to turn off the power serving the hot water heater's electric heating elements. During summer cooling cycles, the entering water temperatures are higher, resulting in higher superheat temperatures. In winter, domestic hot water production will be reduced because of the lower entering water temperatures from both the domestic water supply system and the ground heat exchanger.

E

Desuperheater Design and Selection

There are many factors that influence the proper design and selection of desuperheaters. The most important are compactness, cost, and an acceptable level of heat transfer. Other factors include:

Vented Double Wall: Many local and/or state codes require that the desuperheater be constructed with a "vented double wall." In the event of failure in one wall, either refrigerant or water is vented to the atmosphere from the heat exchanger. This allows a leak to be detected prior to any mixing of domestic water with refrigerant and/or oil.

Field or Factory Installed: Given a choice, a factory-installed unit on a heat pump is preferred. Factory installation is normally under more controlled conditions, and the need for "blowing the charge and recharging" in the field is not necessary. For field-installed desuperheater units, careful cutting of the lines using tubing cutters, brazing in a nitrogen atmosphere, evacuation, and a weighed freon charge will be required to provide the same heat pump unit reliability as a factory-installed unit. It is not impossible, but extreme care and good training are required.

Ratings and Sizing: Desuperheater ratings are based on the nominal cooling capacity of the system. For example, a five-ton desuperheater is usually appropriate for application on any five-ton cooling system. However, a five-ton desuperheater can also be installed on a smaller system, but the reverse is not true.

Desuperheaters should not be undersized. This can result in an excessively high pressure drop in the refrigerant circuit which in turn results in poor performance of the heat pump circuit.

Retrofit Applications: In retrofit applications, the desuperheater can be purchased as part of a total heat reclaiming unit which includes a cabinet, water circulating pump, and thermostat. If a pump is purchased separately, the manufacturer's recommendations should be followed for GPM and head requirements. A nominal flow rate of about 0.75 GPM per ton of cooling is recommended by manufacturers of desuperheating equipment.

Ratings: A standard rating system for desuperheaters has yet to be established by the industry. Therefore, it is often difficult to accurately compare the performance of desuperheaters. A common way of stating performance is by showing Btus transferred as a function of the water flow rate under a given set of conditions.

Figure E.1 presents the performance of a typical commercial desuperheater. As a rule of thumb, a desuperheater will transfer approximately 3,500 Btus per hour (or 6 GPH) per ton of cooling at an entering water temperature of 70°F and an exit water temperature of 140°F for R-22 systems. This design value assumes specific compressor discharge temperatures and will vary for heating and cooling cycles.

$$\text{GPH} = \frac{(3500 \text{ Btu/hr})}{(8.34 \text{ lb/gal})\,(1.0 \text{ Btu/lb–°F})\,(140 - 70°\text{F})} = 6.0$$

Manufacturer's Performance Data: Some manufacturers provide desuperheater capacity as a function of heat pump entering water (or liquid) temperature. Table E.1 shows measured performance data using a factory-installed desuperheater unit. Heat pump size and performance are as follows:

Cooling Capacity: 50,500 Btu/hr @ 70° EWT
Nominal Capacity (Tons): 50,500/12,000 = 4.2
Heat Pump Flow Rate: 9 GPM
Desupheater Flow Rate: 1.68 GPM or 0.4 GPM/Ton

Based on a nominal capacity of 4.2 tons, desuperheater capacity ranges from 1,095 to 2,881 Btu/hr/ton for the ranges of entering water temperatures for the heating and cooling seasons. These capacities are based on the manufacturer's desuperheater flow rate of 0.4 GPM/ton. The manufacturer has the option to increase desuper-

E

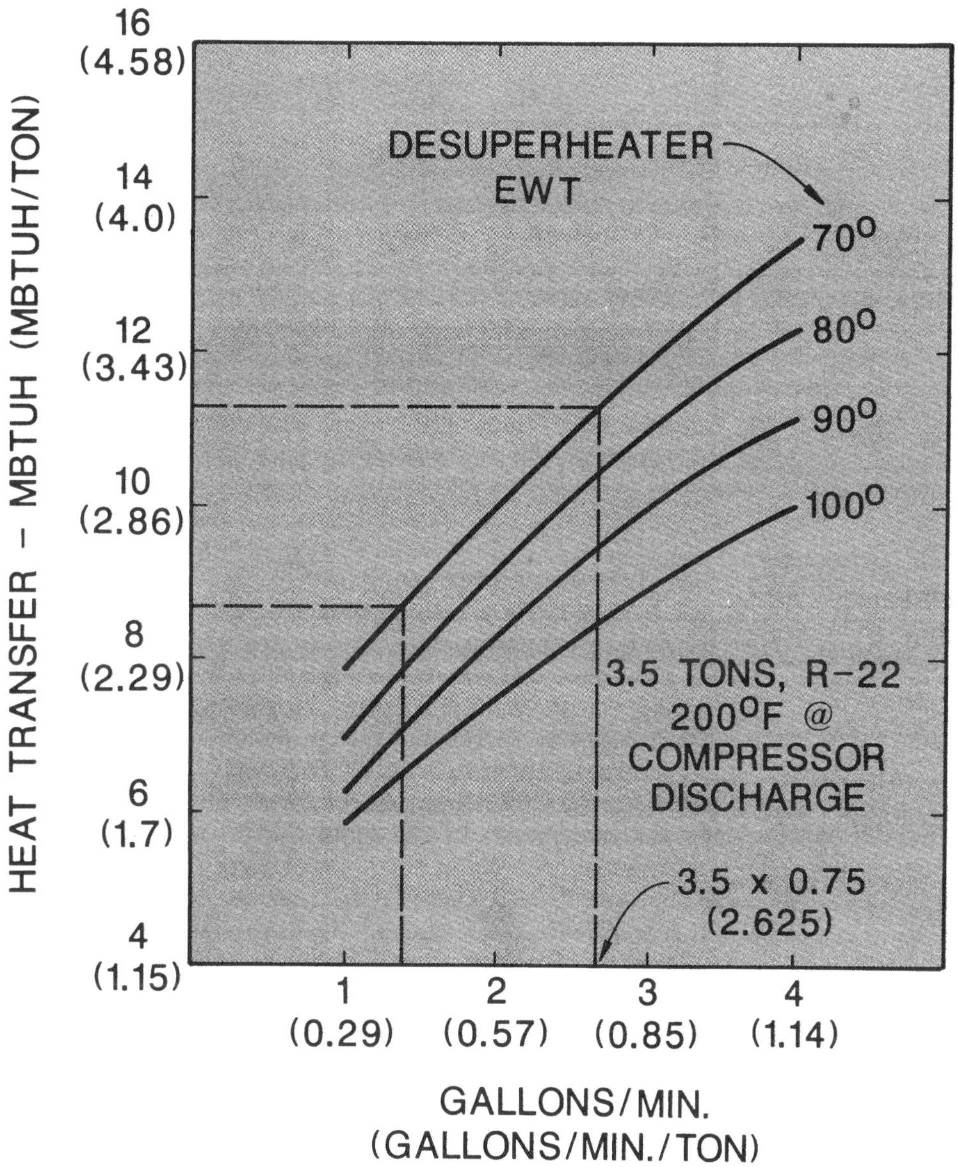

FIGURE E.1: Desuperheater Performance

heater capacity with higher water flow rates. This increased water flow through the desuperheater would require a larger capacity circulating pump and an increase in pump power. The economics of this decision will depend on the amount of circulating power required to raise the desuperheater flow. The manufacturer should always be consulted before any adjustments are made to the desuperheater flow rate.

Control: The desuperheater circulating pump is cycled with the compressor. In some instances, installers have discon-

E

TABLE E.1: Measured Desuperheater Performance		
Entering Liquid Temperature °F	Heat Pump Capacity (Btu/hr)	Desuperheater Capacity (Btu/hr) – (Btu/hr/ton)
Heating: Entering Air Temp 70°F		
30	34,100	4,600 — 1,095
50	45,800	6,900 — 1,643
70	54,700	8,500 — 2,023
Cooling: Entering Air Temp 80°F DB, 67°F WB		
70	52,200	7,500 — 1,786
90	46,900	9,800 — 2,333
110	41,600	12,100 — 2,881

nected the unit during heavy heating demands to achieve higher and more comfortable air temperatures. Since the capacity of the desuperheater is already reduced during periods of high heating demand, the increase in air temperature probably is not large. From an efficiency standpoint, it may be more practical and comfortable for the homeowner to use supplemental heat during periods of insufficient heat pump capacity and keep the desuperheater running with the compressor.

From an efficiency standpoint, the use of a desuperheater increases the heat pump cycle time, which increases the COP. Summer heat rejection loads to the ground heat exchanger are reduced by using a desuperheater, which makes its use and cost easily justified.

Selection: The following checklist may be helpful in selecting a desuperheating system:

1. Be sure the rated capacity of the desuperheater exceeds the heat pump's rated cooling capacity.
2. Check the published performance. Refrigerant and water-side pressure drop should not exceed 6 psi as a general guideline.
3. The heat exchanger should be of vented, double-wall construction, and UL-, ETL-, ARI- or CSA-listed for potable water use.
4. Check materials of construction. All copper desuperheaters should provide many years of trouble-free operation. The desuperheater container and its heat exchanger components should be well insulated to prevent heat loss.
5. Check the warranty. An excellent warranty covers defects in material and workmanship for up to five years. Check with the heat pump manufacturer on factory-installed units. Also, be sure that installing a desuperheater does not void the heat pump equipment warranty.

E

System Layout and Installation

Always install the desuperheater as close to the compressor discharge as possible and upstream of the reversing valve (Figure E.2). A word of caution: Verify that compressor discharge temperatures are at least 20°F higher than the desired end use water temperature. For example, a compressor discharge of 160°F or greater will be required to produce 140°F hot water.

In residential applications, the desuperheater should generate enough hot water to satisfy nearly all domestic needs while the heat pump is operating. It is acceptable to circulate water directly from the desuperheater to the existing domestic hot water tank (Figure E.3). The aquastat on the lower hot water tank heating element should be lowered 10° to 20°F below the normal setting. Auxiliary heating in the tank will activate only when water usage is high and the temperature falls below the set point on the aquastat.

A high-water-temperature thermostat should be used to prevent overheating of the hot water. This thermostat is available on package heat reclaimers and is usually factory preset for a high cut-off temperature of about 160°F. If the desuperheater is unable to meet the hot water demand, it is suggested an intermediate tank (Figure E.4) be used. Preheated water stored in the intermediate tank is then fed directly to the existing hot water heater. A good rule of thumb for water storage is 10

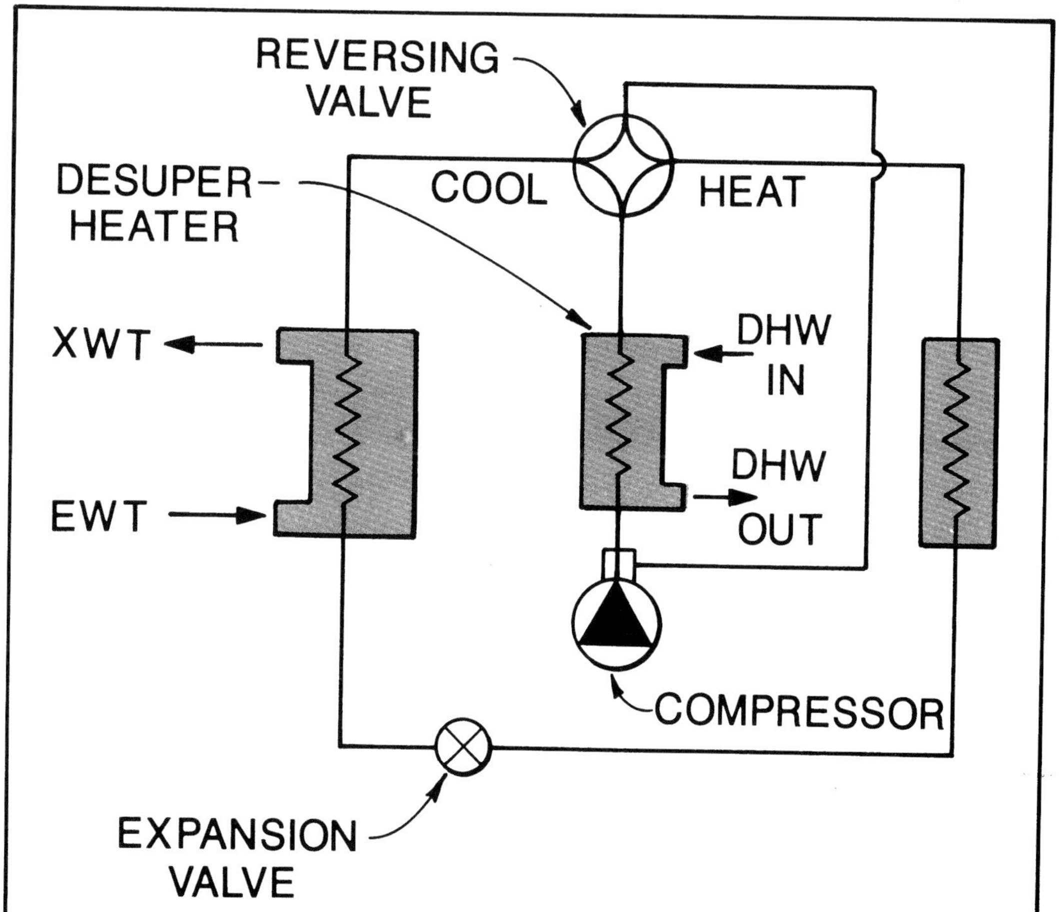

FIGURE E.2: Desuperheater Placement

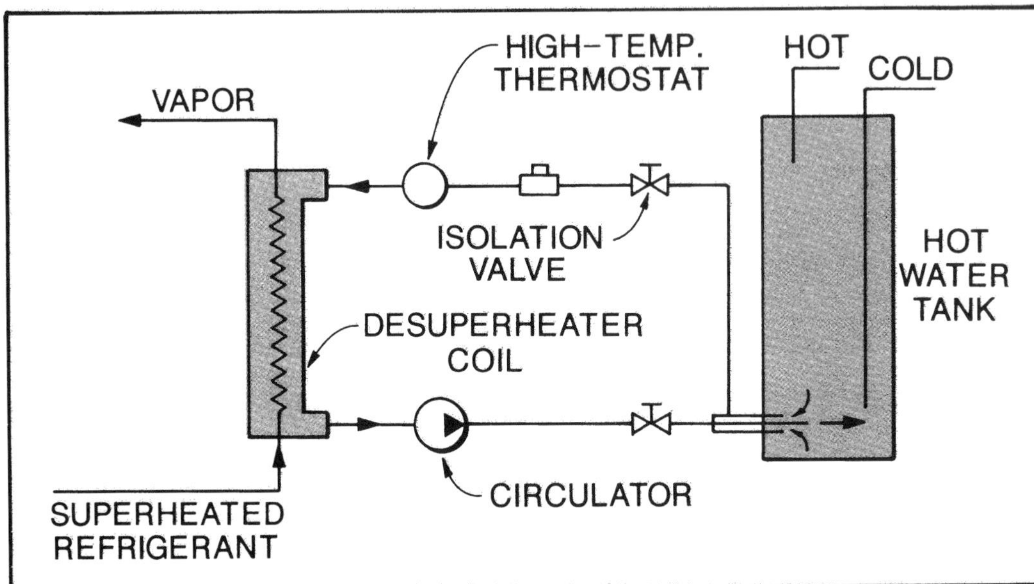

FIGURE E.3: Desuperheater System Configuration

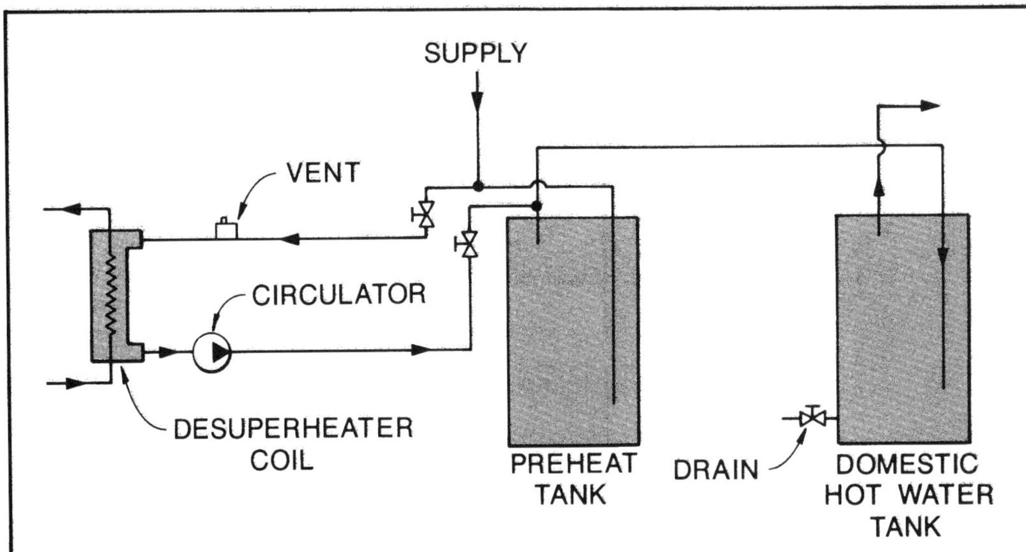

FIGURE E.4: Desuperheater Two-Tank System

gal/ton for R-22 systems and 7 gal/ton for R-12.

Figure E.5 details the specifics of desuperheater system installation. Water lines are short and insulated. Long water lines or separation distances result in additional circulating pump power and distribution heat loss.

Figure E.6 details a water heater hookup kit that simplifies the connection

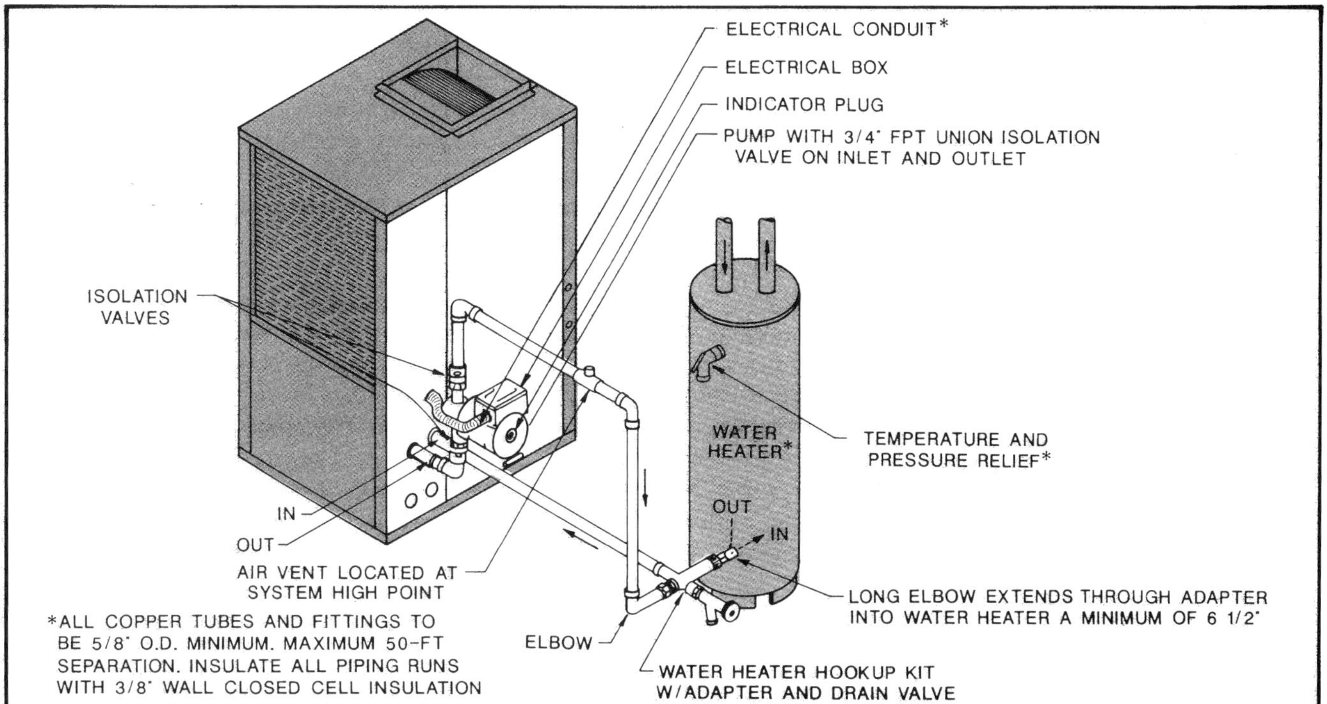

FIGURE E.5: Domestic Hot Water Hookup
Source: Closed Loop Earth Coupled Applications, Specifications/Installation Instructions[20]

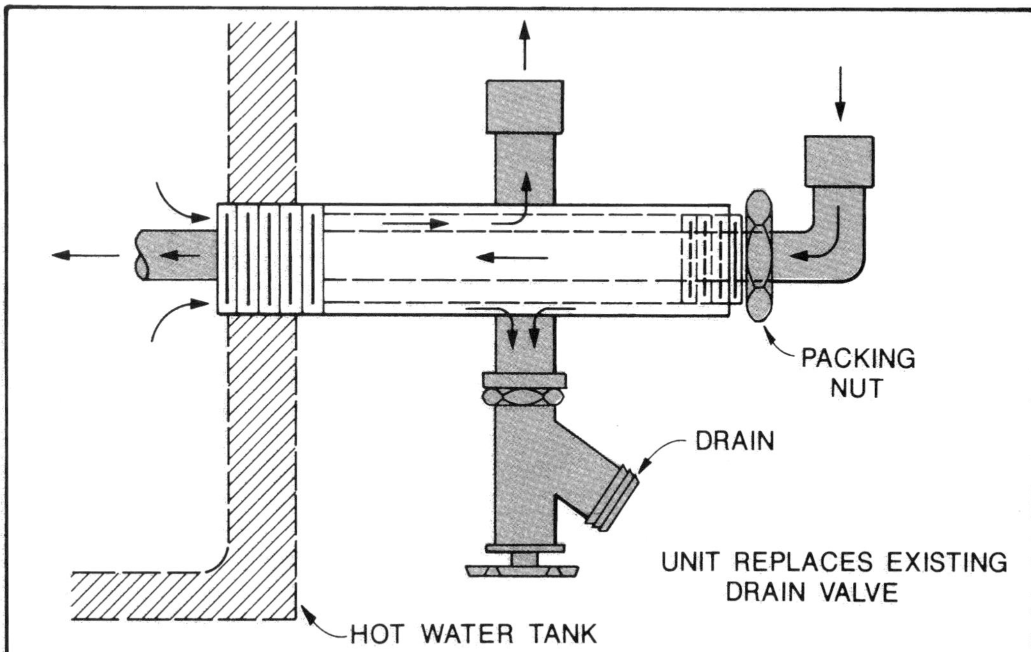

FIGURE E.6: Water Heater Hookup Kit

E

to the hot water tank. The kit replaces the existing drain valve on the hot water tank. The hot water tank drain valve is then installed on the kit.

Figure E.7 describes the connection between the desuperheater coil, the circulating pump, and the water heater hookup kit installed on the domestic hot water tank. By using this hookup fitting, the following advantages exist:

1. No water lines need to be cut if the unit is being installed as a retrofit.
2. Hot water from the desuperheater is introduced into the bottom of the hot water tank. This reduces the danger of introducing extremely hot water into a water line running directly into the household. **CAUTION: Never connect the outlet of the desuperheater into the hot water line at the top of the hot water heater unless a special fitting is provided that discharges the water into the tank before it can reach the hot water outlet.** Figure E.8 is one method used when connecting to the hot water outlet at the top of the tank. Water in the domestic hot water preheater can reach temperatures greater than 200°F when summer hot water demands have been met and the domestic hot water circulating pump has been turned off after the high-temperature thermostat limit has been met.

An air vent should be installed at the system's high point to allow air to be removed should any interruption in water service occur in the domestic water supply.

Water Tank Preparation: Before connecting the desuperheater unit to the domestic hot water tank, the following procedure is recommended:

1. Turn off the electrical or fuel supply to the heater.
2. Attach a garden hose to the water tank drain connection and run the other end of the hose to an open drain or outside.
3. Close the cold water inlet valve to the water heater tank or turn off the building's water supply.
4. Drain the tank by opening the drain valve on the bottom of the tank, and

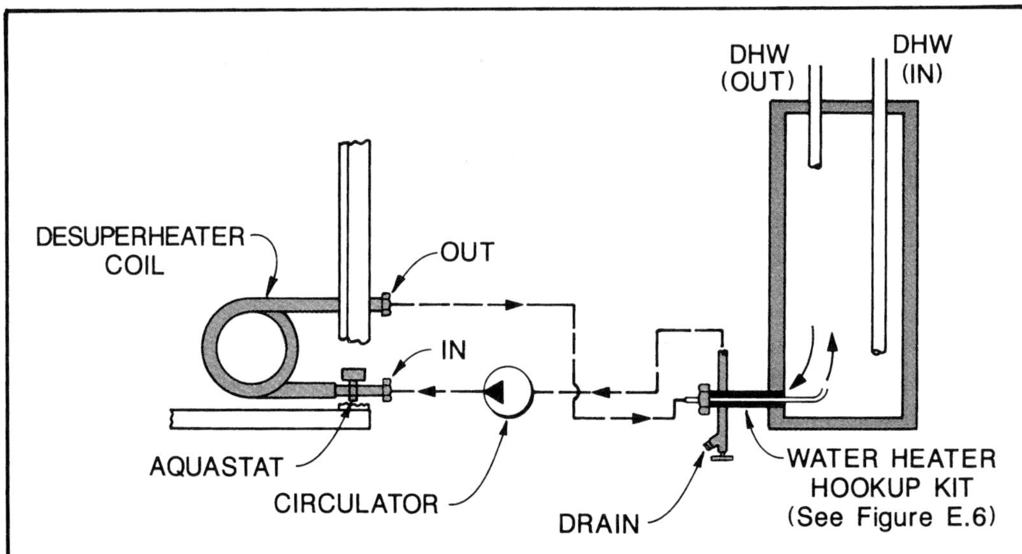

FIGURE E.7: DHW Tank, Circulator, and Coil Arrangement

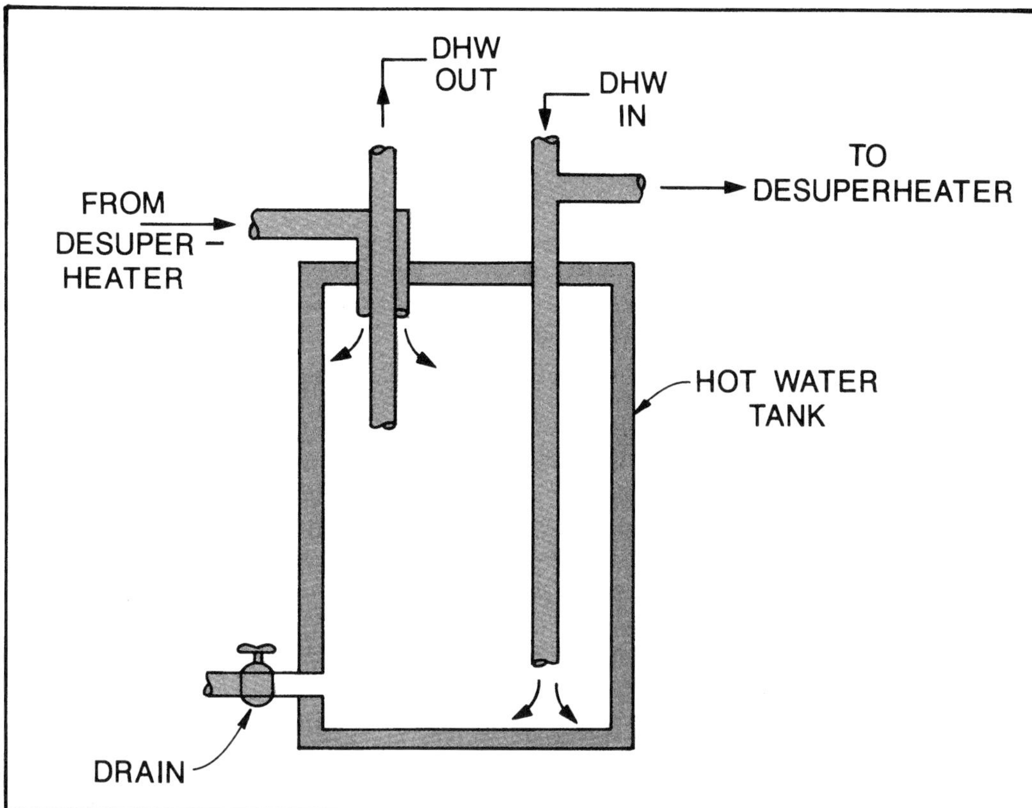

FIGURE E.8: Recommended Desuperheater Connection

then opening the pressure relief valve or a hot water faucet.

5. If the tank is not new, it should be cleaned, once drained, by flushing with water from the cold water inlet until the water draining from the tank is clear and free of sediment.
6. After the tank is flushed and drained, disconnect the garden hose.
7. Install the water hookup kit:
   a. Remove the drain valve at the bottom of the water tank.
   b. Install the fitting as shown in Figures E.5 and E.6.
   c. Be sure when connecting the desuperheater that the water returning is discharged as shown in Figure E.7.

Installing the Domestic Hot Water Circulating Pump: The following procedure should be followed:

1. Check with the manufacturer for any specific instructions on mounting the pump. In some cases, the pump shaft must always be in a horizontal position.
2. Route and install the interconnecting plumbing as shown in Figure E.5. All plumbing must be in accordance with local codes.
3. For the best service and the most economical operation, the water lines should be insulated.
4. The length of run should be as short as possible. Some manufacturers recommend that runs should be less than 50 feet.
5. Make sure both isolation valves are open.

E

Water Tank Refill:

1. Close the tank drain valve.
2. Open the cold water supply to the tank.
3. Open a cold and hot water faucet in the house to vent any air in the system.
4. Some pumps need to be air bled. If provisions are on the pump to vent any air in the bearings or rotor housing on enclosed pumps, follow the manufacturer's recommendations.
5. Before restoring the electrical or fuel supply to the water heater, adjust the temperature setting on the tank thermostat to ensure maximum utilization of heat available from the refrigeration system. On tanks with both upper and lower elements and thermostats, turn the lower element down to 100°F and adjust the upper element to 120°F. On tanks with a single element, lower the thermostat setting to 120°F or to the "LOW" position.
6. After thermostat adjustments are completed, replace the access cover and restore the electrical or fuel supply to the water heater.

Electrical Connection: The circulating pump normally is connected to the compressor contactor which limits the desuperheater unit to operation during heat pump operation.

The circulator should be properly fused, and the pump should be grounded.

Initial Start-up of the Desuperheater System: The following procedure is recommended:

1. Make sure all valves in the desuperheater water circulation system are open. On some pumps it is critical that they not be run dry.
2. Turn on the heat pump using the thermostat.
3. The circulating pump should be running if all the conditions of the desuperheater electrical system have been met. In some cases, the pump is not started until a minimum water temperature in the desuperheater is met. Be sure that the system's operation is well understood before it is assumed a problem exists.
4. Check the temperature difference between inlet and outlet of the desuperheater unit. The temperature will depend on the flow rate and capacity of the desuperheater. Typical values are between 5° and 15°F.
5. The major cause of system malfunction will be air locking in the circulating pump. If this happens, remove any air by following the manufacturer's instructions.

Economics: Typically, a desuperheater will supply 100% of the domestic hot water requirements during the summer cooling operation in southern climates. It is common to have an excess amount of hot water available during the peak summer months.

In northern climates, the amount of domestic hot water will depend on unit sizing, entering heat pump liquid temperature, entering water temperature to the domestic hot water tank, and the quantity of domestic hot water used. Based on the reduced desuperheater capacity at lower heat pump entering liquid temperatures, the amount of domestic hot water generated during heating seasons will not be as large as in southern climates.

REFERENCES

1. Graham, Rickey. *Waste Heat Recovery Research Project.* Gulfport, Miss.: Mississippi Power Corporation, 1985.
2. American Society of Heating, Refrigerating, and Air-Conditioning Engineers (ASHRAE). *ASHRAE Handbook of Fundamentals, 1985.* Atlanta: ASHRAE, 1985.
3. Air Conditioning Contractors of America. *Manual J, 1986.* Washington: ACCA, 1986.
4. American Society of Heating, Refrigerating, and Air-Conditioning Engineers (ASHRAE). *Cooling and Heating Load Calculation Manual, GRP 158.* Atlanta: ASHRAE, 1979.
5. McQuiston, F.C. *A Study and Review of Existing Data to Develop a Standard Methodology for Residential Heating and Cooling Load Calculations.* ASHRAE Transactions, Vol. 90, Part 2. Atlanta: American Society of Heating, Refrigerating, and Air-Conditioning Engineers, 1984.
6. McQuiston, *F.C. LDCAL5 Documentation:* Cooling and Heating Load Calculations (LDCAL5). Stillwater, Okla.: Oklahoma State University.
7. Departments of the Air Force, Army, and Navy. *Engineering Weather Data.* Air Force Manual 88-29, #008-070-00420-8. Washington: U.S. Government Printing Office, 1978.
8. WaterFurnace International. *WaterFurnace Liquid Source Heat Pumps.* Fort Wayne, Ind.: WaterFurnace International, 1987.
9. Collins, W.D. *Temperature of Water Available for Industrial Use in the United States.* U.S. Geological Survey Water Supply Paper 520-F. Washington: U.S. Geological Survey, 1925.
10. Kusuda, T. and P.R. Achenback. *Earth Temperatures and Thermal Diffusivity at Selected Stations in the United States.* ASHRAE Transactions, Article No. 1914. Atlanta: American Society of Heating, Refrigerating, and Air-Conditioning Engineers, 1965.
11. Labs, K. "Underground Building Climate." *Solar Age* 4(10):44–50 (October 1979).
12. Oklahoma State University. *Earth Coil Design Manual.* Stillwater, Okla.: International Ground Source Heat Pump Association, 1983.
13. Salomone, L.A., W.D. Kovacs, and Tamami Kusuda. "Thermal Performance of Fine Grained Soils." *Journal of Geotechnical Engineering* 110(10):359–74 (March 1984).
14. American Institute of Electrical Engineers (AIEE). *Soil Thermal Characteristics in Relation to Underground Power Cables.* AIEE Transactions. New York: AIEE, 1960.
15. Grundfos Pumps Corporation. Clovis, Cal.: 1987.
16. Bard Manufacturing Company. *Earth Coupled Loop System Design.* Bryan, Ohio: Bard Manufacturing Company, 1987.
17. McElroy Manufacturing, Inc. *McElroy No. 2 CU Fusion Manual.* Tulsa, Okla.: McElroy Manufacturing, Inc., 1986.
18. Vanguard Plastics Inc. *Socket Fusion Techniques: A Guide for Instruction and Certification in Vanguard Polybutylene Pipe Socket Fusion.* McPherson, Kans.: Vanguard Plastics Inc., 1986.
19. Niagara Mohawk Power Corporation, New York State Energy Research and Development Authority, Rochester Gas and Electric Corporation. *Manual of Acceptable Practices for Installation of Residential Earth-Coupled Heat Pump Systems.* Prepared by W.S. Flemming and Associates, Inc. Syracuse: [n.p.], 1986.

REFERENCES

20. Command-Aire Corporation. *Closed Loop Earth Coupled Applications, Specifications/Installation Instructions*. Form 5M0486B. Waco, Tex.: Command-Aire Corporation, 1986.
21. Perry, Robert H. and Cecil H. Chilton. *Chemical Engineers' Handbook*. 5th ed. New York: McGraw-Hill Book Company, 1973.
22. Eskilson, Per. *Thermal Analysis of Heat Extraction Boreholes*. Lund, Sweden: Department of Mathematical Physics, University of Lund, 1987.

General Reading

In addition to the references cited in specific sections of this manual, the following works are suggested for further reading:

23. Bierbaum, A. *Groundwater Heat Pump Selection and Application Handbook*. Urbandale, Iowa: GWHP Publishing Co., 1984.
24. Dexheimer, R. Donald. *Water Source Heat Pump Handbook*. Worthington, Ohio: National Water Well Association, 1985.
25. Hannifan, J. Mark and G.C. Thomann. *Kansas Ground Source Heat Pump Manual*. Topeka, Kans.: Kansas Electric Utilities Research Program, 1986.
26. Lear Siegler Mammoth Heat Pumps Division. *Ground Water Applications Manual* (A-3-SAT-1184). Minneapolis: Lear Siegler Mammoth Heat Pumps Division, 1986.

GLOSSARY OF TERMS

Accumulator: A storage vessel that holds excess liquid refrigerant to help prevent the liquid from flooding back into the compressor.

Add-On Heat Pump: A heat pump used in combination with a furnace or boiler. The heat pump operates above the balance point temperature and the furnace or boiler also operates when the outside temperature drops below the balance point.

Ambient Air: The surrounding air (usually outdoor air or the air in a specific location).

Air-Source Heat Pump: A heat pump that uses an air-to-refrigerant heat exchanger to extract and reject heat to the outside air.

ARI: Air-Conditioning and Refrigeration Institute.

ASHRAE: American Society of Heating, Refrigerating, and Air-Conditioning Engineers.

ASTM: American Society for Testing and Materials.

Balance Point: The temperature above which the heat pump can provide enough heat for the home without the use of supplemental heating.

Bend: A fitting either molded separately or formed from pipe for the purpose of accommodating a directional change.

Bin: In the bin method, a temperature increment, usually 5°F, into which the range of temperatures for an area are divided. Bins are used to produce a frequency distribution of hourly, monthly, or annual outdoor temperature occurrences for a specified location.

Blowers: Fans used to force air across the heat exchanger. With a ground-source heat pump, the only blower used is to force air through the central heating system.

British Thermal Unit (Btu): The quantity of heat required to raise the temperature of one pound of water one degree Fahrenheit at a specified temperature.

Burst Strength: The internal pressure required to cause a pipe or fitting to fail.

Capillary Tube: A small bore tube of varying length, generally 0.5 to 2 millimeters in diameter, used as an expansion device to regulate the flow of liquid refrigerant to the evaporator.

Cavitation: The formation of bubbles due to partial vacuums in a flowing liquid as a result of separation of fluid particles.

Circulating Pump(s): The pump(s) that circulates the fluid in the closed-loop system during normal operation.

Closed-Loop System: In a ground-coupled system, a pressurized heat exchanger consisting of the ground heat exchanger, the circulating pump, and the water-source heat pump in which the heat transfer fluid is not exposed to the atmosphere.

Coefficient of Performance (COP): The ratio of heating (or cooling) provided by a heat pump (or other refrigeration device, *e.g.*, air conditioner or refrigerator) to the energy consumed by the system under designated operating conditions.

Coil: A heat exchanger used to transfer energy from one source to another. In ground-source heat pumps, water-to-refrigerant and refrigerant-to-air coils are used.

Compressor: The central component of a heat pump system. The compressor increases the pressure of a refrigerant fluid, and simultaneously reduces its volume, while causing the fluid to move through the system.

GLOSSARY OF TERMS

Compressor, Reciprocating: A positive displacement compressor in which the change in internal volume of the compression chamber(s) is accomplished by the reciprocating motion of one or more pistons.

Compressor, Rotary: A positive displacement compressor in which the change in internal volume of the compression chamber(s) is accomplished by the rotation of a rolling piston or rotating sliding vanes.

Condenser: A heat exchanger in which hot, pressurized (gaseous) refrigerant is condensed by transferring heat to cooler surrounding air, water, or earth.

Crack: Any narrow opening or fissure in the surface of a pipe that is visible to the naked eye.

Cycling Efficiency: Ratio of actual efficiency to steady-state efficiency. The actual efficiency of a heating or cooling system is somewhat lower due to start-up and shut-down losses.

Damper: A device used to vary the airflow rate through an air outlet, inlet, or duct.

Defrost Cycle: The control-activated process of removing accumulated frost and ice from the outdoor heat exchanger of air-source heat pumps.

Degree Day: A measure of the severity and duration of an outdoor temperature deviation above or below a fixed temperature (65°F), used in estimating the heating or cooling requirement and fuel consumption of a building for either summer or winter conditions.

Design Temperature, Summer: A specific temperature used in calculating the cooling load of a building. The summer design temperature is typically the outdoor air temperature that is exceeded 2.5% or 5% of the time.

Design Temperature, Winter: A specific temperature used in calculating the heating load of a building. The winter design temperature is typically the outdoor temperature that is exceeded 97.5% or 95% of the time.

Desuperheater: A device for recovering superheat from the compressor discharge gas of a heat pump or central air conditioner for use in heating or preheating water. Also known as a heat recovery water heater.

DOE: U.S. Department of Energy.

Diffuser (Air): A supply air outlet composed of deflecting elements discharging air in various directions and patterns to accomplish mixing of supply and room air.

Direct Expansion Earth-Coupled Heat Pump: A heat pump system in which the refrigerant is circulated in pipes buried underground.

Effective Length: The design parameter for calculating the friction loss for a run. Effective length, expressed in feet, is the sum of the actual length and the equivalent length.

Energy Efficiency Ratio (EER): A dimensional quantity usually used in specifying cooling performance. The EER is the ratio of cooling provided by the system (in Btu) to the energy consumed by the system (in watt-hours) under designated operating conditions.

EPA: Environmental Protection Agency.

Equivalent Length: Used in duct system calculation procedures to express the same pressure loss for pipes of various diameters.

Environmental Stress Cracking: The development of cracks in a material that is subjected to stress or strain in the presence of specific chemicals.

Evaporator: A heat exchanger in which cold, low-pressure (liquid) refrigerant is vaporized to

GLOSSARY OF TERMS

absorb heat from the warmer surrounding air, earth, or water.

Expansion Device: A device that reduces the pressure of liquid refrigerant entering the evaporator and meters and regulates the flow of refrigerant so that it can properly absorb heat.

Fabrication: The manufacture of products; refers herein to the manufacture of plastic products from molded parts, rods, tubes, sheeting, extrusions, or other forms by appropriate operations such as punching, cutting, drilling, and taping including fastening plastic parts together or to other ports by mechanical devices, adhesives, heat sealing, or other means.

Fitting: A piping component used to join or terminate sections of pipe or to provide changes of direction or branching in a pipe system.

Fuse: To make a plastic pipe joint by heat and pressure.

Grille: A louvered or perforated covering of an inlet or outlet opening through which air flows.

Ground-Coupled Heat Pump: A heat pump that uses the earth itself as a heat source and heat sink. It is coupled to the ground by means of a closed-loop heat exchanger (ground coil) installed horizontally or vertically underground.

Heat Exchanger: A device, often a coil, specifically designed to transfer heat between two physically separated fluids of different temperatures.

Heat Joining: Making a joint by heating the mating surfaces of the pipe components to be joined and pressing them together so that they fuse and become essentially one piece.

Heat Pump: A mechanical device used for heating and cooling which operates by pumping heat from a cooler to a warmer location. Heat pumps can draw heat from a number of sources, *e.g.*, air, water, or earth, and are most often either air-source or water-source.

Heat Pump Water Heater, Integral: A heat pump built into a water heater.

Heat Pump Water Heater, Remote, Integral: A heat pump contained in a separate cabinet and connected to an existing water heater by pipes or hoses.

Heat Sink: The medium — air, water, earth, etc. — which receives heat from a heat pump.

Heat Source: The medium — air, water, earth, etc. — from which heat is extracted by a heat pump.

Heating Seasonal Performance Factor (HSPF): The ratio of total heating provided by a heat pump during the heating season (in Btu) to the total energy consumed by the system (in Watt-hours), taking into account regional weather conditions and supplemental resistance heat.

Hoop Stress: The tensile stress in the wall of the pipe in the circumferential orientation due to internal hydrostatic pressure.

HVAC: Heating, ventilating, and air conditioning.

Hydronic: A heating or cooling distribution system using liquid piped throughout the house to radiators or convectors.

Isolation Hangers: Insulated, tubular holders for refrigerant piping used to prevent transmission of vibration noise from pipes to the structure.

Joint: The location at which two pieces of pipe or a pipe and a fitting are connected.

Joint, Butt-Fused: A joint in which the prepared ends of the joint components are heated and then placed in contact to form the joint.

GLOSSARY OF TERMS

Joint, Clamped Insert-Fitting: A mechanical joint using external metal clamps or other mechanical devices to form a pressure seal between an insert fitting and the outside surface of the pipe.

Joint, Heat-Fused: A joint made using heat and pressure only. There are three basic types of heat-fused joints: butt-fused, socket- or insert-fused, and saddle-fused.

Joint, Saddle-Fused: A joint in which the curved base of the saddle fitting and a corresponding area of the pipe surface are heated and then placed together to form the joint.

Joint, Socket-Fused: A joint in which the two pieces to be heat fused are connected using a third fitting or coupling with a female end.

Life-Cycle Costing: A method of analyzing the cost of HVAC systems that considers all the significant costs of ownership, including the time value of money, initial capital investment, energy costs, and maintenance costs over the service life of each system under consideration.

Multizone Heat Pump: A central, split-system heat pump consisting of an outdoor heat exchanger, compressor unit, and multiple (three to five) indoor air handling units, each of which can be independently controlled and operated for space conditioning.

Open-Fluid System: In a ground-coupled heat pump, a heat exchanger system in which the transfer fluid is exposed to the atmosphere.

P/T Ports: Pressure/temperature ports used to monitor the performance of a heat pump system. A small-diameter opening that allows a probe to be inserted during system operation.

Packaged Heat Pump: A self-contained heat pump unit available as either a free-delivery, ductless system for single rooms or as a larger, central ducted unit that can heat or cool an entire home.

Packaged Terminal Heat Pump: A window or through-the-wall-mounted air-to-air heat pump unit designed to heat or cool a single room or zone.

Parallel System: A flow condition where two or more fluid paths are possible in the closed-loop circuit.

Performance Factor: The ratio of useful output capacity of a system to the input required to obtain it. Units of capacity and input need not be consistent.

Pipe Heat Transfer Resistance: The resistance to heat flow resulting from pipe wall thermal properties and dimensions.

Plastic: A material that contains as an essential ingredient one or more organic polymeric substances of large molecular weight, is solid in its finished state, and at some stage in its manufacture or processing into finished articles can be shaped by flow.

Plastic Pipe: A hollow cylinder of a plastic material in which the wall thicknesses are usually small when compared to the diameter and in which the inside and outside walls are essentially concentric.

Plastic Tubing: A particular size of smooth wall plastic pipe in which the outside diameter is essentially the same as the corresponding size of copper tubing.

Power Flushing: A higher than normal water flow and pressure in a ground heat exchanger used to flush air and debris from the closed-loop piping system.

Polymer: A substance consisting of molecules characterized by the repetition of one or more types of monomeric units.

GLOSSARY OF TERMS

Polyolefin: A polymer prepared by the polymerization of an olefin(s) as the sole monomer(s).

Pressure: When expressed with reference to pipe, the force per unit area exerted by the medium in the pipe.

Pressure Drop: The decrease in pressure down the length of a pipe resulting from fluid flow.

Pressure Rating: The estimated maximum pressure that the medium in the pipe can exert continuously with a high degree of certainty that failure of the pipe will not occur.

Pump Kit: A prefabricated pumping and service unit for the closed-loop circuit which provides standardized piping, inlets, and outlets for flushing, air venting, etc.

Purge Pump: A high-pressure and high-flow-rate pump used to flush air and debris from the closed-loop circuit of a closed-loop/ground-source (cl/gs) heat pump system.

Refrigerant: A fluid of extremely low boiling point used to transfer heat between the heat source and heat sink. It absorbs heat at low temperature and low pressure and rejects heat at a higher temperature and higher pressure, usually involving changes of state in the fluid (*i.e.*, from liquid to vapor and back).

Register: A combination grille and damper device covering an air inlet or outlet opening.

Return (Air): Air returned to the space conditioning unit from the conditioned space.

Schedule: A pipe size and wall thickness classification system (outside diameters and wall thicknesses) originated by the iron pipe industry.

Seasonal Energy Efficiency Ratio (SEER): A measure of seasonal cooling efficiency under a range of weather conditions assumed to be typical of a location, of performance losses typical of a location, and of performance losses due to cycling under part-load operation.

Series System: A system in which the circulating fluid from the heat pump(s) has a single flow path through the ground heat exchanger.

Simple Payback Method: A method for analyzing the cost of HVAC systems which considers only the time it takes for annual energy and maintenance cost savings to offset an initial difference in cost between two systems.

Soil/Field Resistance: The resistance to heat flow resulting from soil thermal properties and underground pipe placement.

Standard Dimension Ratio (SDR): A specific ratio of the average specified outside diameter to the minimum specified wall thickness (OD/t) for outside-diameter controlled plastic pipe.

Suction Line: The tube or pipe that carries the refrigerant vapor from the evaporator to the compressor inlet.

Supplemental Heating: A heating system component used when a heat pump is operating below the balance point, during defrost, or as an emergency backup when the main system is inoperable. Usually electric resistance heat, but natural gas, LPG, or oil heating systems are also used.

Therm: A quantity of heat equivalent to 100,000 Btu.

Throw: The distance an airstream travels after leaving a supply outlet before the velocity is reduced to a specific terminal velocity (usually 50 ft/min).

Thermostat: An instrument that responds to changes in temperature and is used to directly or indirectly control indoor temperature by operating a space conditioning system.

GLOSSARY OF TERMS

Ton of Refrigeration: A measure of cooling delivered by a heat pump (or other air conditioning system) equal to 12,000 Btu per hour.

U-Bend: A prefabricated closed-return pipe assembly used in vertical heat exchangers to connect the two pipes at the bottom of the bore hole.

Unitary Heat Pump: A complete factory-assembled heat pump.

Valve, Expansion: A device for regulating the flow of liquid refrigerant to the evaporator. Two types of valves are commonly used: an electronic valve that responds to variation in electric resistance reflecting changes in refrigerant temperature, and a thermostatic valve that uses a refrigerant-filled bulb to sense changes in refrigerant temperature.

Valve, Reversing: An electrically operated valve that allows the heat pump to switch from heating to cooling, or vice versa, by changing the refrigerant's direction of flow.

Water-Source Heat Pump: A heat pump that uses a water-to-refrigerant heat exchanger to extract heat from the heat source.

Water-Source Heat Pump, Closed Loop: Closed-loop systems circulate a heat transfer fluid (such as water or a water-antifreeze mixture) continuously to extract or reject heat from a ground or water heat source or sink.

Water-Source Heat Pump, Open Loop: Open-loop systems pump groundwater or surface water from a well, river, or lake through a water-to-refrigerant heat exchanger and return the water to its source, a drainage basin, pond, or storm sewer.

INDEX

INDEX

INDEX

INDEX

INDEX

INDEX

INDEX

INDEX

INDEX

INDEX